The Facts On File

DICTIONARY OF GEOLOGY AND GEOPHYSICS

DOROTHY FARRIS LAPIDUS

Donald R. Coates, Ph.D.
Scientific Adviser

Edmund H. Immergut, Ph.D.
Series Editor

Facts On File Publications
New York, New York ● Oxford, England

This book is dedicated to Tessie.

The Facts On File Dictionary of Geology and Geophysics

Copyright © 1987 by Dorothy Farris Lapidus

All rights reserved. No part of this book may
be reproduced or utilized in any form or by any
means, electronic or mechanical, including
photocopying, recording or by any information
storage and retrieval systems, without permission
in writing from the Publisher.

Library of Congress Cataloging in Publication Data
Lapidus, Dorothy Farris
    The Facts on File dictionary of geology & geophysics.

        1. Geology--Dictionaries.   2. Geophysics--Dictionar-
ies.   I. Facts on File, Inc.   II. Title
QE5.l45   550'.3'21   82-7389

ISBN 0-87196-703-0 hc
ISBN 0-8160-1929-0 pb

Printed in the United States of America

10 9 8 7 6 5 4 3 2 1

PREFACE

During the past few decades, the results of particular studies have had a profound impact on developments in many areas of geology and geophysics. Certain concepts have had to be abandoned, and others modified or revised.

The concept of sea-floor spreading, the constant generation of new sea floor, offered a reason for why oceanic crust is younger than continental crust, and not older, as had been believed. The same concept was one of the ideas that was basic to an acceptance of the notion of continental drift. Both these hypotheses were later joined by the theory of plate tectonics, a singular contribution of which was the derivation of some sense of order from the distribution patterns of earthquakes and volcanoes.

Many geological terms have now been redefined in the context of the components of modern geology. Where the association is not direct, additional references are given.

I am deeply grateful to Prof. Donald Coates for all his help, particularly in the areas of geomorphology and glaciology. My thanks to Dr. Reed Craig for his generous assistance, and to Milton Kerr for all drawings and figures.

Dorothy Farris Lapidus

Table of Geological Eras

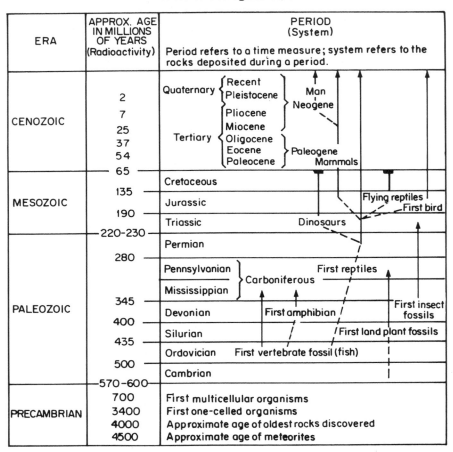

ERA	APPROX. AGE IN MILLIONS OF YEARS (Radioactivity)	PERIOD (System) Period refers to a time measure; system refers to the rocks deposited during a period.
CENOZOIC	2 7 25 37 54 65	Quaternary { Recent / Pleistocene } Man Neogene Tertiary { Pliocene / Miocene / Oligocene / Eocene / Paleocene } Paleogene Mammals
MESOZOIC	135 190 220–230	Cretaceous Jurassic Flying reptiles / First bird Triassic Dinosaurs
PALEOZOIC	280 345 400 435 500 570–600	Permian Pennsylvanian } Carboniferous — First reptiles Mississippian Devonian First amphibian / First insect fossils Silurian / First land plant fossils Ordovician First vertebrate fossil (fish) Cambrian
PRECAMBRIAN	700 3400 4000 4500	First multicellular organisms First one-celled organisms Approximate age of oldest rocks discovered Approximate age of meteorites

A

aa Hawaiian term for lava flows characterized by a rough, jagged, or spiny surface. Cf: *pahoehoe.* See *lava flow.*

a-axis 1. (Crystallog.) One of the referential axes used in the description of a crystal; it is directed horizontally, from front to back. 2. (Struct. petrol.) In deformed rocks, the direction of maximum displacement. Striations on a slickensided surface run parallel to a.

Abbé refractometer An instrument that determines the refractive index of liquids, minerals, and gemstones. A sample is enclosed between two similar glass prisms, and the total refraction at the interface is noted.

ablation 1. Reduction in the volume of snow, ice, or *neve* from glaciers, snow fields, or sea ice. It results from evaporation, sublimation, deflation, and melting caused by rain and warm air, and by *calving.* 2. Removal by vaporization of the molten surface layers of meteorites during passage through the atmosphere.

ablation cone Debris-covered cone of ice, snow, or *firn* formed by differential ablation.

ab-plane (Struct. petrol.) In deformed rocks, the surface along which differential movement occurs. a is the direction of maximum displacement (direction of tectonic transport); b is that direction in the plane of movement at right angles to the direction of displacement.

abrasion Mechanical wearing down or grinding away of rock surfaces by the friction of rock particles carried by wind, running water, ice, or waves. Abrasion polishes, smooths, scratches, or pits exposed rock faces. Syn: *corrosion.* The term also denotes the effect of abrading; e.g., abrasion left by glacial action.

abrasion pH Acidity resulting from the absorption of OH^- or H^+ ions at the surfaces of finely ground minerals suspended in water.

abrasion platform See *wave-cut platform.*

absarokite Variety of porphyritic alkalic basalt containing a small amount of orthoclase in the *groundmass.*

absolute age Geologic age of a rock, fossil, or formation expressed in units of time, usually years. It commonly refers to ages determined by radioactive dating, but tree rings may also be used. The term "absolute" is somewhat misleading, since it implies an exactness that is not always readily achieved. Cf: *relative age.*

absolute humidity The content of water in the atmosphere, expressed as the mass of water vapor per unit volume of air. Cf: *relative humidity.*

absolute permeability Ability of a rock to transmit a fluid when the saturation with that fluid is 100%. See also: *effective permeability; relative permeability.*

absolute temperature A temperature measured with respect to absolute zero ($-271.18°$ C), and expressed as "degrees absolute" (e.g., 225° A) or "degrees Kelvin" (e.g., 225° K).

absorption 1. Assimilation or incorporation by molecular or chemical action, as of liquids in solids or of gases in liquids. Cf: *adsorption.* 2. Reduction in light intensity during transmission through a medium. 3. The process by which one energy form may be converted into another; e.g., seismic wave energy to heat. 4. Penetration of surface water into the lithosphere.

absorption coefficient　The ratio of the amount of energy absorbed by a given material to the total amount incident on that material. Syn: *absorptance*.

abstraction　1. The simplest type of *stream capture*. 2. That part of precipitation that is stored, transpired, evaporated, or absorbed; i.e., does not become direct runoff.

abyssal　1. Pertaining to the region deep within the earth, said esp. of an igneous intrusion that occurs deep within the crust. 2. Pertaining to ocean depths of about 2000 to 6000 m and to the organisms of that environment. This includes all areas below the base of the continental slope, with the exception of oceanic trenches. The upper region of the abyssal zone, at about 2000 m, is referred to as the *bathyal*. The abyssal zone may also be defined as the zone, excluding trenches, where water temperature never exceeds 4° C. See *continental shelf*. See also: *hadal*.

abyssal cones　See *submarine fans*.

abyssal fans　See *submarine fans*.

abyssal hills　Submarine geomorphic features found only in the deep sea (3000 to 6000 m). They are typically dome shaped, rising as much as 1 km above the surrounding abyssal plain, and may be up to several kilometers wide at the base. Abyssal hills occur in all oceanic basins, but are most prevalent in the Pacific. They are common along the outer margins of the Mid-Atlantic Ridge. Their origin is uncertain. Marine geologists and geophysicists tend toward the theory of a volcanic origin for these structures, but it is also possible that the hills are lithified and compacted sedimentary material. See also: *abyssal plain*.

abyssal plain　A flat area of the ocean-basin floor with a slope of less than 1°. Topographic and sedimentary studies of abyssal plains indicate they were formed by *turbidity currents* that obscured the

pre-existing geomorphic features. Long sediment cores taken from these plains all contain silts and sands apparently issuing from shallow continental areas. Trench abyssal plains are those that lie on the bottom of deep-sea trenches.

Acadian　Obsolete name for the Middle Cambrian in North America. *Albertan* is the modern equivalent.

Acadian orogeny　A Middle Paleozoic disturbance that formed a mountain chain from northern Appalachia through New England and the Canadian Maritimes. The orogeny was named for Acadia, the area of the Canadian Maritime Provinces where it was first recognized. The *Taconian* plus Acadian orogenies of North America correspond to the *Caledonian orogeny* of Northwest Europe. Extensive igneous activity accompanied the Acadian orogeny at its climax, in the mid- to late Devonian. It is now ascribed to the collision of the northeastern portion of the North America plate with western Europe. See also: *Antler orogeny*.

acceleration　1. (Physics) Ratio of the increment in velocity to the increment in time; $a = \Delta v/\Delta t$. Velocity has both magnitude and direction, and is thus a vector quantity. 2. (Biol.) During the process of evolution, the increasingly early appearance of modifications in the life cycle of sequential generations. Extreme crowding of primitive phylogenetic characteristics is called *tachygenesis*. This phenomenon, carried to the point where certain stages of the sequence are eliminated completely, is called *brachygenesis*.

acceleration due to gravity　Acceleration due to earth's gravitational attraction of a freely falling body in a vacuum. The value adopted by the International Committee of Weights and Measures is 980.665 cm/sec^2; however, the actual value varies with altitude, latitude, and composition of the underlying rocks.

accelerometer A device that measures acceleration by determining the relative movement of a weight within a frame when the system containing the device experiences a change in velocity. Such a mechanism functions according to Newton's second law of motion.

accessory mineral Any mineral in a rock not essential to classification of the rock. When present in small amounts it is a *minor accessory*. If the amount is greater or has particular significance, the mineral is called a *varietal accessory* and may be added to the name of the rock (e.g., biotite in biotite granite). The typical formation of accessory minerals occurs during solidification of rocks from magma. Common minor accessory minerals include garnet, topaz, zircon, rutile, and tourmaline. Varietal accessories include biotite, amphibole, olivine, and pyroxene. In sedimentary rocks, accessory minerals are mostly *heavy minerals*. Cf: *secondary mineral*.

accidental inclusion A *xenolith*.

acclivity Ascending slope; opposite of *declivity*.

accordant Matching or in agreement. Describes strata faces of similar orientation, or two streams whose surfaces are at the same level at the junction location. Ant: *discordant*.

accordant summit level A hypothetical level that intersects the summits or hilltops of certain regions. In an area of high topographic relief it may imply that the summits are remnants of a plain formed during an earlier erosion cycle. See also: *summit concordance*.

accreting plate boundary A borderline between two crustal plates that are separating, and along whose seam new oceanic lithosphere (sea-floor plates) is being created. Syn: *divergent plate boundary*. See also: *mid-oceanic ridge*.

accretion 1. Gradual enlargement of a land area through the accumulation of sediment carried by a river or stream. 2. The increase in size of inorganic bodies by the addition of new material to the exterior. 3. The theory that the continents have increased their surface area during geologic history. The mechanism responsible for such growth is related to the formation and destruction of geosynclinal belts that form near the craton. Also called *continental accretion*. 4. Increase in mass of a celestial body by the incorporation of smaller bodies that collide with it. Gravitational attraction accelerates the accretion process when the body becomes sufficiently large. Accretion is now thought to have been a significant factor in the formation of the planets from dust grains. Cf: *Nebular Hypothesis; Tidal Wave Theory*.

accretionary hypothesis See *accretion*.

accretionary lapilli Also called *volcanic pisolites*, these are spheroidal, concentrically layered pellets composed mainly of vitric dust and ash, usually between 2 and 10 mm in diameter. They are formed primarily through accretion of ash and dust by condensed moisture in eruption clouds. Formless nuclei of coarse particles fall through the fine debris and acquire shells of progressively finer ash. These concentric shells indicate the increasing temperature and decreasing humidity of the cloud at lower levels. Cf: *pisolite*.

accretion vein A vein formed by the repeated filling of channel ways with mineral deposits, and by the reopening of these channels by fractures that develop within the zone of mineralization.

accumulation 1. The sum of all processes that contribute mass to a glacier or to floating ice or snow cover; these include snowfall, avalanching, and the transport of snow by wind. Cf: *ablation*. 2. The amount of precipitation or hoarfrost added to a glacier or snowfield.

accumulation area   The part of a glacier or snowfield in which, during a year's time, the mass balance is positive; i.e., accumulation exceeds ablation.

ACF diagram   Triangular diagram used to represent mineral assemblages (metamorphic facies) resulting from the metamorphism of chemically diverse rocks. Each diagram used indicates the minerals that will form as the result of subjecting a rock to a certain range of pressure and temperature. Most chemical components occurring in the majority of rocks can be represented by an equilateral triangle with corners A, C, and F. Corner A represents the percentage of Al_2O_3 plus Fe_2O_3 in the rock; C is the percentage of CaO, and F the percentage of FeO, MnO, and MgO. One requirement for using an ACF diagram is that the rocks contain silica in amounts sufficient to form not only all the silicate minerals present, but quartz as well. Cf: *AFM diagram; A'KF diagram.*

ac-fracture   (Struct. petrol.) In deformed rocks, a tension fracture that parallels the ac-plane and is normal to *b.*

Acheulian   Stratigraphic stage name for European Lower Pleistocene.

achondrites   Stony meteorites in which *chondrules* (inclusions) are absent. They are categorized on the basis of calcium content, and are similar to near-surface igneous rocks on earth and to moonrocks; evidence indicates they were formed at or near the surface of a differentiated planet.

acicular   Needle-shaped, as in certain crystals.

acid   *Acidic.*

acidic   It should be noted that the first three definitions of the term *acidic* have no reference to the hydrogen ion content, or pH, of a substance, as used in chemistry. 1. A term describing igneous rocks that contain more than 60% SiO_2,

as distinct from *intermediate* and *basic* rocks. Acidic is sometimes incorrectly used as equivalent to *felsic* and to *silica-oversaturated,* but these terms include some rock types not usually considered acidic. Cf: *silica-saturated.* See *silica concentration.* 2. Commonly but loosely applied to any igneous rock in which light-colored minerals predominate, and which has a relatively low specific gravity. Syn: *acid; silicic.* This usage arose because of the general correspondence of acidic to light color. 3. (Metall.) Said of a slag in which the proportion of silica exceeds the amount required to form a "neutral" slag with the earthy bases present. 4. Used in the chemical sense of a high hydrogen ion concentration (low pH), when referring to hydrothermal, pegmatitic, or other aqueous fluids.

acidization   The forcing of acid into limestone, dolomite, or sandstone in order to increase porosity and permeability by removing some of the rock constituents. The process is also used to remove mud introduced during drilling. Syn: *acid treatment.*

acid mine drainage   Drainage with a pH of 2.0 to 4.5, issuing from mines and their wastes. The process is initiated with the oxidation of sulfides exposed during mining, which produces sulfuric acid and sulfate salts. The quality of the drainage water continues to be lowered as the acid dissolves minerals in the rocks.

aclinic line   *Magnetic equator.*

acme-zone   A biozone consisting of a body of strata in which a particular species or genus of an organism occurs with maximum frequency. *Hemera* is the corresponding unit of geologic time. Cf: *assemblage zone; range-zone.* Syn: *epibole; peak zone.*

acmite   A brown or green mineral of the clinopyroxene group, $NaFe(SiO_3)_2$, found in certain alkali-rich igneous rocks. Syn: *aegirine.*

acoustic log A generic term for a well log that shows any of various measurements of acoustic waves traveling in rocks exposed by the sinking of a borehole. To obtain such a log, certain devices are used to generate an excitation through the adjacent materials in a borehole. The excitation profiles are recorded on a measuring device, and the differences in the inherent acoustical properties of the various materials are determined from this record. See *sonic log; well logging*.

acoustics (underwater) Study of the production, transmission, reception, and utilization of sound. Because sound waves are not absorbed by water, sound is used to probe the ocean's depths, study the nature of sediments, and locate objects in the oceans. Reflection and refraction of sound are used by marine geologists and geophysicists for seismic profiling and echo-sounding; sound-scattering effects are used by marine biologists to determine the location of *deep-scattering layers*. The velocity of sound in the ocean depends on temperature, salinity, and pressure (depth). The empirical formula for velocity in terms of these factors is $C = 1449 + 4.6t - 0.055t^2 + 0.0003t^3 + (1.39 - 0.012t)(S - 35) + 0.017d$, where C is the velocity in meters per second; t is temperature in degrees Celsius; S is salinity in parts per thousand; and d is below-surface depth in meters.

acoustic wave A longitudinal wave. Its use is commonly restricted to fluids, but often includes P waves traveling in the solid part of the earth.

ac-plane *Deformation plane.*

acquired character (Biol.) Any modification in function or structure not inherited but acquired by an organism during its lifetime. Such a characteristic might be the result of environmental factors or of the use or disuse of a particular structure according to the organism's mode of life.

acre yield Average amount of gas, oil, or water recovered from 1 acre of a reservoir. (1 acre = 43,560 square ft.)

acritarch A group of organic microfossils of unknown biological affinity, variously characterized by a smooth or spiny texture. The presence of the remains of these organisms in sedimentary rocks of marine origin indicates they were *planktonic*. Acritarchs range from the Precambrian to the present, but are most abundant in the Precambrian and early Paleozoic.

actinolite A mineral composed of hydrous calcium, magnesium, and iron silicate. It is a green, fibrous *amphibole*, generally found in crystalline schists, and similar to tremolite in chemical composition.

activation 1. Treatment of bentonitic clay with acid to further its bleaching action or to improve its adsorptive properties. 2. The process of rendering a substance radioactive by bombarding it with nuclear particles. See *activation energy*.

activation analysis A method for identifying stable isotopes of elements. If a sample is irradiated with neutrons, charged particles, or gamma rays, the elements in the sample are made radioactive, and can then be identified by their characteristic radiations.

activation energy That energy beyond the ground state required to activate a particle or system of particles so that a particular process might occur. For example, the energy needed by a molecule to take part in a chemical reaction, or by an electron to travel to the conduction band in a semiconductor.

active fault Any type of fault in which there is recurrent movement in the form of continuous creep, small, abrupt displacements, or seismic activity. Cf: *capable fault*. See also: *fault*.

active glacier 1. A glacier with an accumulation area and in which there is a flow of ice. Ant: *dead glacier*. 2. A glacier that moves at a relatively rapid rate. See *glacier*.

active layer 1. Surface layer above the permafrost, alternately frozen in the winter and thawed in the summer. Its thickness ranges from several centimeters to a few meters. 2. (Engineering geol.) Surficial material that undergoes seasonal changes of volume, expanding when wet or frozen and shrinking when dry or thawing.

active permafrost See *permafrost*.

active volcano A volcano in the process of erupting or one that has a history of frequent eruptions in recent times. There is no exact distinction between an active and a dormant volcano. See also: *volcano*.

acute bisectrix See *bisectrix*.

adamantine adj. Applied to a mineral with a brilliant luster resembling that of a diamond.

adamellite *Quartz monzonite*.

adaptation The adjustment or modification by natural selection of an organism, or of its parts or functions, so that it becomes better suited to cope with its environmental conditions. The development of protective coloration among fish is an example.

adaptive radiation Process of natural selection whereby lineages separate physically and genetically within a short interval of geologic time. Syn: *divergence*.

adhesion Molecular attraction between closely contiguous surfaces of unlike substances, e.g., liquid in contact with a solid. Cf: *cohesion*.

adiabatic process A process in which a change of state takes place without the transfer of heat between a system and its environment. For example, when a confined gas or other fluid is compressed without loss or gain of heat, the gas is warmed adiabatically. If a gas is allowed to expand without gain or loss of heat in the system, it is cooled adiabatically. In the lower atmosphere, the cooling of rising air and the warming of descending air are largely due to adiabatic expansion and adiabatic compression of air, respectively.

adit A horizontal passage made from the surface of the earth, usually to intersect a mineral vein or coal seam. Although frequently called a tunnel, an adit, unlike a tunnel, is closed at one end. Adits are commonly horseshoe shaped, but may also be square, round, or elliptical in cross-section. They can be driven only in hilly country, where the lower elevation of the entranceway will provide an adequate slope for the outflow of water, and will facilitate the removal of mineral material.

adobe Spanish word for: 1. sun-dried clay bricks; 2. a structure built from such bricks; 3. the clay soil from which the bricks are made. Adobe is a mixture of clay and silt found in the southwestern United States and Mexico. It dries to a hard, uniform mass, and its use for brickmaking dates back thousands of years.

adsorption Attraction of molecules of gases or molecules in solution to the surfaces of solid bodies with which they are in contact. Solids that can adsorb gases or dissolved substances are called adsorbents; the adsorbed molecules are referred to collectively as the adsorbate. Adsorption can be either physical or chemical. *Physical adsorption* resembles the condensation of gases to liquids. In *chemical adsorption*, gases are bound to a solid surface by chemical forces defined for each surface and each gas. Cf: *absorption*.

adularia A colorless variety of orthoclase feldspar, ranging from transparent to translucent.

advance 1. Gradual seaward progression of a shoreline as a result of accumulation or emergence; also, the net seaward progression during a specific time interval. 2. Forward movement of a glacier front. Ant: *recession*.

advection 1. Horizontal transport of air or atmospheric properties within the earth's atmosphere. 2. The horizontal or vertical movement of sea water as a current. 3. Lateral mass movement of material in the earth's mantle. Cf: *convection*.

aegirine A synonym of *acmite*. Sometimes more specifically applied to acmite containing calcium, aluminum, or magnesium. Syn: *aegirite*.

aegirite *Aegirine.*

aeolian *Eolian.*

aeon *Eon.*

aerate To supply or fill with air.

aeration Process of supplying air to the pores in a soil, or to waste water in sewage treatment. See also: *zone of aeration*.

aerial Said of an activity, phenomenon, or object, either animate or inanimate, related to air or the earth's atmosphere. Ex: *aerial currents; aerial reconnaissance; aerial creatures*. Not to be confused with *areal*.

aerial magnetometer *Airborne magnetometer.*

aerial photograph A photograph of earth's surface taken from the air by the use of cameras mounted on aircraft, spacecraft, or orbiting satellites. Usually taken in overlapping series from an aircraft flying in a systematic pattern at a given altitude. Aerial photographs are used for mapping land divisions and to provide information on geology, vegetation, soils, hydrology, etc.

aerobic Said of organisms (e.g., bacteria) that can exist only in the presence of free oxygen, or of processes and activities that require free oxygen. Cf: *anaerobic*.

aerolites Stony meteorites composed mainly of silicate minerals.

aeromagnetic Pertaining to observations and studies made with an *airborne magnetometer*.

aerosol A colloidal system in which the dispersion medium is a gas (usually air), and the dispersed phase consists of liquid droplets or solid particles; e.g., haze, mist, most smoke, some fog.

aff. Abbreviation of *affinity*. Connotes less specific similarity than does Cf.

affine 1. A homogeneous deformation; i.e., one in which initially straight lines remain straight after deformation. 2. A transformation that maps parallel lines to parallel lines and finite points to finite points.

affinity A term used in biology to indicate relationship but not specific identity. Abbrev: aff.

affinity of elements A classification of elements according to their preference, or affinity, for different types of chemical environments. Such preference is the consequence of differing bonding characteristics among the atoms of different elements. 1. *Siderophile* elements, e.g. iron, cobalt, nickel, have an affinity for iron, and are readily soluble in it; they are most concentrated in the core of the earth. 2. *Chalcophile* elements tend to concentrate in sulfide minerals and ores, and so occur in sulfide ore deposits. Examples include sulfur, copper, and zinc. 3. *Lithophile* elements (e.g., lithium, oxygen, sodium, silicon) prefer the silicate phase and are most concentrated in the earth's crust.

AFM diagram   Triangular diagram representing the simplified compositional character of a metamorphosed pelitic rock; constructed by plotting the molecular properties of the three components: A = Al_2O_3; F = FeO; and M = MgO. Cf: *ACF diagram; A'KF diagram.*

aftershock   An earthquake following a larger earthquake and originating at or near the focus of the larger one. Generally, major shallow earthquakes are succeeded by several aftershocks. Although these decrease in number as time goes on, they may continue for days or even months. See also: *seismic focus; tsunami.*

Aftonian   The first interglacial stage of the Pleistocene Epoch (Lower Pleistocene) in North America. It followed the Nebraskan and precedes the Kansan glacial stages.

agate   A fine-grained variety of quartz composed of varicolored bands of *chalcedony.* Agate usually occurs within rock cavities; its natural color generally ranges from white through brown, red, and gray to black. Cf: *moss agate.*

age   1. Unit of geologic time longer than a *subage* and shorter than an *epoch*, during which rocks of a particular *stage* were formed. 2. Frequently used term for a geologic time interval within which were formed the rocks of any stratigraphic unit. 3. Time during which a particular event occurred, or a time characterized by unusual physical conditions; e.g., the Ice Age. 4. A division of earth history, unspecified in length, marked by a dominant life form; e.g., the "age of reptiles." 5. Position of any feature relative to the geologic time scale; e.g., "rocks of Eocene age."

age equation   The relationship between geologic time and radioactive decay, expressed mathematically as $t = 1/\lambda \ln(1 + d/p)$, where t is the age of the sample rock or mineral, λ is the decay constant of the particular radioactive series used for the calculation, *ln* is the logarithm to base e, and d/p is the present ratio of radiogenic daughter atoms to the parent isotope.

age of amphibians   Informal specification of the late Paleozoic (Carboniferous and Permian); i.e., the geologic time span during which amphibians appeared and proliferated.

age of fishes   Informal specification of the Silurian and Devonian, i.e., the geologic period characterized by a great expansion of fishes.

age of mammals   Informal specification of the Cenozoic; i.e., the geologic time span during which mammals became prominent.

age of marine invertebrates   Informal specification of the Cambrian and Ordovician; i.e., the geologic time span during which marine invertebrates (trilobites, mollusks, echinoderms) predominated.

age of reptiles   Informal designation of the Mesozoic, the era during which reptiles predominated. Dinosaurs first appeared in the Triassic and became extinct by the end of the Cretaceous.

age ratio   The ratio of daughter element to parent isotope, with which age is determined. Its validity requires a system that has remained closed since the time of its solidification, sedimentation, or metamorphism; in addition, the sample must be representative of the rock from which it came, and the decay constant must be known.

agglomerate   Pyroclastic rock of consolidated volcanic fragments, rounded and angular, in a finer-grained matrix, in which the rounded fragments predominate. Agglomerates are not well-sorted and are common in volcanic necks. The well-rounded fragments are *volcanic bombs* and the angular fragments are *blocks*. As the proportion of angular fragments increases, agglomerates grade to *volcanic breccias*. See *pyroclastic material; pyroclastic rocks*.

agglutinate A welded pyroclastic deposit in which the pyroclasts are held together by glassy material.

aggradation A gradational process that contributes to the general leveling of the earth's surface by means of deposition that builds up. Some possible agents of this process are running water, wind, waves, and glaciers. Ant: *degradation*.

aggrading stream A stream that progressively builds up its channel or flood plain by receiving more load than it can transport. A stream can be caused to aggrade rather than deepen its valley by any change in the adjacent land resulting in alluviation that causes the stream's gradient to become too low for transport of its load; e.g., sinking of the surrounding land.

aggregate 1. Mass of rock particles, mineral grains, or combination of both. 2. Any hard, inert material (e.g., gravel, sand, blast furnace slag, clinkers) used for mixing with cement, gypsum or other adhesive to form concrete. The aggregate gives volume, stability, and resistance to erosion to the finished product. *Fine aggregate*, usually consisting of sand or crushed stone, is used in making thin concrete slabs or smooth surfaces. The particle size is less than 4.76 mm. *Coarse aggregate* consisting of gravel, slag or other coarse fragments of approx. 4.70 mm is used for more massive construction. *Lightweight aggregate* consists of materials with a low specific gravity; e.g., pumice, volcanic cinders, vermiculite.

aggressive intrusion *Forcible intrusion*.

aging The process by which a young lake becomes an old lake as a result of nutrient saturation, vegetal encroachment or invasion, and *eutrophication*.

Agnatha A class of primitive vertebrates, the jawless fish. Included are the lampreys, hagfish, and some extinct groups. Range, Ordovician to present.

agonic line The line through all points on the earth's surface at which the magnetic declination is zero; it is the locus of all points at which true north and magnetic north coincide. See *isogonic line*.

A horizon Approx. syn: *topsoil*. See *soil profile*.

aiguilles Etymol: Fr. "needles." 1. Pointed granitic rocks. 2. (Physical geography) Peaks of mountains. 3. Needle-shaped peaks, especially certain peaks or clusters of needlelike rocks near Mont Blanc in the French Alps. 4. Another name for *volcanic spine*, particularly in the West Indies.

air Syn: *atmosphere*.

airborne magnetometer An aircraft-transported instrument that measures variations in the earth's magnetic field. Syn: *flying magnetometer*.

air drilling *Rotary drilling* that uses high-velocity air instead of conventional *drilling mud*; unsuitable where appreciable quantities of water may be encountered. See *rotary drilling*.

air gun Energy source frequently used in marine seismic studies; the explosive release of highly compressed air generates a shockwave. Also modified for use in borehole velocity surveys.

air shooting Detonation of an explosive charge above the surface of the earth, so as to produce a seismic pulse. Also used in geophysical exploration.

air wave   The acoustic energy pulse transmitted through the air as a result of a seismic shot. See *seismic shooting.*

Airy hypothesis   A concept of balance for the earth's solid outer crust, which assumes that the crust has uniform density throughout and floats on a more liquid substratum of greater density. It is one explanation of the principle of crustal flotation. Since the thickness of the crustal layer is not uniform, this theory supposes that the thicker parts of the crust, such as mountains, sink deeper into the substratum. This is analogous to an iceberg, the greater part of which is beneath the water. See also: *Pratt hypothesis; isostasy.*

A'KF diagram   Triangular diagram used to show the simplified componential character of a metamorphic rock. The molecular quantities of the following rock components are plotted: $A' = Al_2O_3 + Fe_2O_3 - (Na_2O + K_2O + CaO)$; $K = K_2O$; and $F = FeO + MgO + MnO$. $A' + K + F$ (in mols) are recalculated to 100%. When it is necessary to represent K minerals, this diagram is used in addition to the *ACF diagram*. Cf: *AFM diagram.*

alabaster   Fine-grained, massive gypsum, normally white and often translucent. It is used for statuary, vessels, etc.

alaskite   Plutonic rock consisting chiefly of oligoclase, microcline, and quartz; mafic constituents are few or absent. A commercial source of feldspar.

A layer   See *seismic regions.*

albedo   The ratio of the amount of solar radiation reflected from an object to the total amount incident upon it. An albedo of 1.0 indicates a perfectly reflecting surface, while a value of 0.0 indicates a totally black surface that absorbs all incident light. The albedo of earth is calculated as 0.39, more than half of which is due to reflection from clouds. The greater reflection from new snow,

relative to that of old snow, is an important factor in accounting for the melting of glaciers and snow.

Albers projection   A conic, equal-area projection with two standard parallels that are concentric circles whose center is the point of intersection of the meridians. The meridians are straight lines meeting at a common point beyond the map range. It is one of the best projections for a map of the United States. See also: *map projection.*

Albertan   Term used for the Middle Cambrian of North America, taken from the Canadian Rockies of Alberta, where it is well represented. "Acadian" is the obsolete name.

albite   1. A white or colorless feldspar mineral, $NaAlSi_3O_8$, that forms triclinic crystals. A variety of plagioclase that occurs commonly in metamorphic and igneous rocks. 2. In the plagioclase feldspar series, albite represents the pure sodium end-member. Cf: *potassium feldspar.*

albite-epidote-hornfels facies   A mineral assemblage sometimes appearing in the outer fringes of contact aureoles. It is a low-pressure facies characterized by imperfect recrystallization and the presence of unstable relict phases from the premetamorphic condition.

albitite   A porphyritic igneous rock composed almost entirely of albite phenocrysts in an albite ground mass. Garnet, muscovite, apatite, and quartz are common accessory minerals.

Alexandrian   The Lower Silurian of North America. Obsolete syn: *Medinan.*

alexandrite   Transparent variety of chrysoberyl used as a gemstone. Its color is green in daylight and deep red by artificial light.

algae   A widely diverse group of

photosynthetic plants, almost exclusively aqueous, that includes the seaweeds and freshwater forms. They range in size from unicellular forms to the giant kelps tens of meters in length. Although considered primitive plants, since they lack true leaves, roots, stems, and vascular systems, they manifest extremely varied and complex life cycles and life processes. Algae range from the Precambrian to the present. Singular: *alga*.

algal bloom A relatively sudden increase in algal growth on the surface of a lake, pond, or stream. Phosphate or other nutrient enrichment of the waters stimulates the growth of algal bloom.

algal limestone See *limestone*.

algal structure Calcareous sedimentary structure formed in oceans or lakes by carbonate-depositing colonial algae. Its forms include crusts, cabbage-head shapes, and laminated structures such as *stromatolites*. Syn: *algal reefs*.

Algoman orogeny Orogeny and granitic emplacement evidenced in Precambrian rocks of northern Minnesota and the Algoma district of western Ontario. Dated about 2400 million y.b.p. Synonymous with the Kenoran orogeny of the Canadian Shield.

Algonkian *Proterozoic*.

alidade An instrument used in topographic surveying and mapping by the plane table method; also, any sighting device or indicator used for angular measurement. A surveying alidade has a telescope with an attached vertical circle mounted on a flat, moveable base that carries a straight edge. An alidade is also a component of *stadia* surveying, the method usually used to measure the distance and vertical height differences of distant points. See also: *Gale alidade*.

alkali 1. Carbonate of sodium or potassium, or any bitter-tasting salt, occurring at or near the surface in arid and semiarid regions. 2. A strong base (hydroxide); e.g., NaOH or KOH. 3. adj. Minerals rich in sodium or potassium; e.g., *alkali feldspar*.

alkali-calcic series Those series of igneous rocks with alkali-lime indices in the 51 to 55 range.

alkalic igneous rocks Those igneous rocks in which the sodium and potassium content exceeds the average for the group of rocks to which they belong, or exceeds the amount required to form feldspar with the available silica. Also, those igneous rocks having an *alkali-lime index* below 51, and those belonging to the *Atlantic suite*.

alkali feldspar Feldspar rich in sodium or potassium; e.g., albite, microcline, orthoclase, or sanidine.

alkali flat A plain or a level area encrusted with alkali salts as a result of evaporation and poor drainage, found in arid or semiarid regions; a *salt* (saline) *flat*. See also: *playa*.

alkali lake A *salt lake* often found in arid regions, the waters of which contain large amounts of sodium and potassium carbonates in solution, as well as NaCl and other alkaline compounds; ex: Lake Magadi in the Eastern Rift Valley of Kenya. See also: *soda lake*.

alkali-lime index Percent of silica (by weight) in a sequence of igneous rocks as shown on a variation diagram, where the weight percentages of CaO and those of $K_2O + Na_2O$ are equal; i.e., the intersection point of the curves for CaO and ($K_2O + Na_2O$).

alkali metal Any metal of the alkali group; Group Ia of the periodic classification: lithium, sodium, potassium, rubidium, cesium, or francium.

alkaline 1. Showing the qualities of a base; basic. 2. Sometimes used instead of *alkalic* in discussing igneous rocks.

alkalinity   1. In a lake, the quantity and kinds of compounds that collectively shift the pH to the alkaline end of its range. 2. In seawater, the number of millequivalents of hydrogen ion neutralized by 1 liter of seawater at 20° C. Has no bearing on the hydroxyl content of seawater.

Alkemade line   A straight line in a ternary phase diagram that joins the composition points of two primary phases bounded by an interface with adjacent phases.

Alleghenian   Stratigraphic time unit that represents the Lower Middle Pennsylvanian of eastern North America.

Allegheny orogeny   A mountain-building event, most pronounced in the central and southern Appalachians, produced by the folding and faulting of the Ridge and Valley province. Most of the orogeny probably occurred in the Late Paleozoic, but some phases may have endured into the Early Triassic.

Alling grade scale   A metric scale of grain size for thin and polished sections of sedimentary rocks on which measurements can be taken only in two dimensions. Major divisions (boulder, cobble, gravel, sand, silt, clay, colloid) are in the constant ratio of 10; minor divisions are in the ratio of the fourth root of 10.

allochem   The carbonate particles that form the framework in most mechanically deposited limestones, as distinguished from micrite matrix. Silt-, sand-, and gravel-sized interclasts are some important allochems. Also included are *oöliths*, *fossils*, and *pellets*.

allochemical metamorphism   Metamorphism accompanied by the removal or addition of material, such that the chemical composition of the greater mass of the rock is altered.

allochthon   A term originally proposed for sedimentary rocks whose constituents have been transported to and deposited in locations some distance from their origin. The term is presently applied to rock masses displaced considerable distances by tectonic processes such as overthrusting. Ant: *autochthonous*. adj. *Allochthonous*, said of a rock or material formed in a location other than where it is found.

allogene   Allogenic mineral or rock constituent such as a pebble in a conglomerate or xenolith in an igneous rock.

allogenic   1. (Geol.) Minerals and rock constituents derived from pre-existing rocks transported from their original location; also, a stream fed by water from distant terrain. Ant: *authigenic*. 2. (Ecol.) Successive ecologic conditions or occurrences that originate outside the natural community and alter its habitat. Cf: *autogenic*.

allotriomorphic   *Xenomorphic*.

allotropic   Said of chemical elements that may exist in two or more forms that have differing atomic arrangements as crystalline solids, or of molecules that contain different numbers of atoms. For example, both graphite and diamond are forms of the element carbon.

alluvial   1. Composed or of pertaining to *alluvium*, or deposited by running water. 2. Applied to a placer formed by running water, or to the mineral associated with it.

alluvial cone   A sharply inclined alluvial deposit formed where a stream emerges onto a lowland after its descent from a steep upland area. Alluvial cones are a continuous series of landforms with intervening gradations: *talus sheet, talus cone, alluvial cone, alluvial fan*. Talus forms are deposited largely by the action of gravity, and alluvial forms by alluvial processes. Alluvial cones are steeper than alluvial fans and show a greater average particle size. Fans are characterized by a much greater degree of stratification and by a higher incidence of mudflows.

alluvial fan Outspread mass of alluvium deposited by flowing water where it debouches from a steep, narrow canyon onto a plain or valley floor. The abrupt change of gradient eventually reduces the transport of sediment by the issuing stream. Viewed from above, the deposits are generally fan-shaped; they are especially prominent in arid regions, but may occur anywhere. Cf: *bajada.*

alluvial plain General name for a plain produced by the deposition of alluvium from the action of rivers; e.g., flood plain; delta plain; alluvial fan.

alluviation Deposition of *alluvium* along streamways; *aggradation*. Also, the filling of a depression or covering of a surface with alluvium.

alluvium General term for detrital deposits made by rivers or streams or found on *alluvial fans*, flood plains, etc. Alluvium consists of gravel, sand, silt, and clay, and often contains organic matter that makes it a fertile soil. It does not include the subaqueous sediments of lakes and seas.

almandine The iron-aluminum end-member of the *garnet* group, $Fe_3Al_2(SiO_4)_3$. It characteristically occurs in mica schists with colors ranging from light red to brown red. Its presence in rocks indicates the grade of metamorphism. Deep-red crystals are valued as gemstones. Syn: *almandite.*

alp 1. A high, rugged, and steep mountain resembling those of the European Alps. 2. General term for topographic features resembling those of the European Alps; e.g., the *alpine* topography of parts of New Zealand.

alpha decay Radioactive decay due to the loss of an *alpha particle* from the nucleus of an atom. Mass of the atom decreases by four and the atomic number by two.

alpha particle The nucleus of a helium atom, also called *alpha ray.* It consists of two protons and two neutrons; no electrons are present, and hence the particle charge is $+2$. Alpha particles are highly ionizing and have a short range through matter. Their energy is always uniform for a particular reaction. See *alpha decay.*

alpha quartz Also spelled α-quartz; the polymorph of quartz stable below 573° C. It has a higher refractive index and birefringence than *beta quartz*, the form into which it passes at temperatures above 573° C. Common in igneous, metamorphic, and sedimentary rocks, and in veins and geodes. Syn: *low quartz.*

Alpides The long east-west orogenic belt that includes the European Alps and the Himalayas.

alpine 1. Pertaining to the European Alps. 2. Term applied to topographic features that resemble the form of the European Alps; e.g., the *alpine* topography of parts of *New Zealand.*

alpine glacier See *glacier.*

Alpine orogeny A series of diastrophic movements in southern Europe and the Mediterranean regions associated with the collision of Africa with Europe. The Alpine orogeny perhaps began late in the Triassic and continued well into the Cenozoic Era. Cf: *Laramide orogeny.* It was responsible for metamorphism, folding, faulting accompanied by uplift, elevation of the present Alps, and the uplifting of plateaus in the Balkans. Corsica and Sardinia, England, France, and Iceland experienced volcanic activity during this time.

alteration Changes in the chemical or mineralogical composition of a rock, usually produced by hydrothermal solutions or weathering.

alternation of generations Alternation of sexual and asexual phases in the life cycle of an organism. The phases are often morphologically dissimilar, and sometimes chromosomally distinct.

altimeter An instrument that measures elevation or altitude. The two main types are the *pressure altimeter* (an *aneroid barometer*) and the *radio altimeter*, which measures the time required for a radio pulse to traverse the distance from an object in the atmosphere to the ground and back.

altiplanation The development of terrace-like surfaces or flattened summits of *solifluction* and related mass movements. Altiplanation is most common at high elevations and at latitudes in which periglacial conditions or processes predominate. Cf: *cryoplanation; equiplanation.*

altiplano Etymol: Span., "high plains." A tableland plateau at high elevation; specifically, the Altiplano of western Bolivia and southwestern Peru, which is a series of intermontane basins.

altithermal n. A period of elevated temperature, esp. the postglacial thermal optimum. adj. Pertaining to a climate characterized by high or rising temperatures. See *glaciers*.

altitude 1. Vertical distance measured between a point and some reference surface, usually mean sea level. 2. Vertical angle measured between the plane of the horizon and a reference line at some higher point such as a summit peak. See also: *elevation.*

alum 1. A mineral, $KAl(SO_4)_2 \cdot 12H_2O$. Also known as potassium alum or potash alum. 2. A group of minerals usually consisting of aluminum sulfate, water of hydration, and the sulfate of another element.

alumina Aluminum oxide, Al_2O_3.

alum shale An argillaceous, and often carbonaceous, rock impregnated with *alum*. It derives from a shale originally containing iron sulfide which, upon decomposition, forms sulfuric acid that reacts with the aluminas and potassic materials of the rock to produce aluminum sulfates.

alunite A rhombohedral mineral, $KAl_3(SO_4)_2(OH)_6$, used in producing alum. It usually occurs as gray, white, or pink masses in hydrothermally altered feldspathic rocks. The introduction of alunite as a replacement mineral is called *alunitization.*

amalgam 1. A naturally occurring alloy of silver; gold and palladium amalgams are also known. 2. An alloy of mercury and one or more metals. Amalgams are generally crystalline in structure, but those with a high mercury content are liquid.

amazonite (amazonstone) A green or blue-green variety of microcline feldspar, used as ornamental and gem material. Syn: *amazonstone.*

amber Fossil tree resin that has achieved a stable state after ground burial, through chemical change and the loss of volatile constituents. Amber has been found throughout the world, but the most extensive deposits, which represent extinct flora of the Tertiary (40 to 60 m.y.a.), occur along the Baltic Sea. Some specimens of amber are opaque white, but it most often occurs in shades from yellow to brown. Fossil insects and plants may be found as inclusions. The hardness of amber is 2 to 3 on the Mohs scale, and its specific gravity is between 1.00 and 1.09. Optically, amber is mainly isotropic; its average index of refraction is 1.54 to 1.55. Isolation and identification of many of the resin components are still under investigation. Amber was believed to be completely amorphous, but X-ray diffraction studies have revealed crystalline components in certain fossil resins.

amblygonite   A triclinic mineral, $(Li,Na)Al(PO_4)$ (F,OH). Found in pegmatites as white or greenish masses. Used as a source of lithium.

amethyst   A transparent purple or blue-violet variety of quartz, SiO_2, valued as a semiprecious gem.

ammonite   See *ammonoid.*

ammonoid   Any of a group of cephalopods evolved from the nautiloids of the Silurian to Lower Devonian and extinct by the end of the Cretaceous Period; related to the modern pearly nautilus. Ammonoids are characterized and distinguished from nautiloids by a complex suture structure. Their shells are typically coiled in a plane, but some forms are spirally coiled and others are essentially straight. They are very important as index fossils because of their rapid evolution and wide geographic distribution in shallow marine waters.

Three groups of ammonoids succeeded one another, each group showing a more complex suture pattern. *Goniatites*, characterized by a relatively simple combination of a curved and angular suture pattern, were the dominant ammonoids of the Paleozoic (Devonian to Permian). *Ceratites*, characterized by a more highly folded suture, replaced goniatites in the Permian, and were most abundant in the Triassic Period. The *ammonite* suture, composed of wrinkled saddles and lobes, appeared first in the Permian, although some Late Carboniferous forms had suture patterns transitional between ceratitic and ammonitic.

The ammonoids became almost extinct at the end of the Permian and again at the close of the Triassic. Families that survived evolved into uncoiled forms in the Jurassic, and into partly coiled or

straight shells, or helical forms (turrilites) in the Cretaceous.

amorphous   A term (literally "without form") applied to rocks and minerals that show no definite crystalline structure; e.g., obsidian. Ant: *crystalline.*

amphibian   A cold-blooded, four-footed vertebrate, belonging to a class that is midway in evolutionary development between fishes and reptiles. Amphibians breathe by means of gills in the early stages of life (a larval tadpole stage) and by means of lungs in the adult stage. Ex: *frogs, toads, salamanders.*

amphiboles   A group of common rock-forming minerals that occur most frequently in igneous and metamorphic rocks. They have the general formula $A_2B_5(Si,Al)_8O_{22}(OH)_2$, where A is chiefly Mg, Fe, Ca, or Na, and B is mainly Mg, Fe^{+2}, Al, and Fe^{+3}. The amphiboles have a structure of silicate tetrahedra connected in double chains or bonds, as distinguished from the single chain of *pyroxenes.* They are characterized by good prismatic cleavage in two directions intersecting at angles of 56° and 124°. The most common amphibole minerals are hornblende, the tremolite-actinolite series, and the cummingtonite-grunerite series. One of the varieties, nephrite, is a form of jade.

amphibolite   A metamorphic rock consisting mainly of amphibole and plagioclase, little or no quartz, and having *crystalloblastic* texture. Amphibolite grades into hornblende-plagioclase gneiss as the content of quartz increases.

amphibolite facies   One of the major divisions of metamorphic mineral assemblages, the rocks of which form under conditions of moderate to high temperature and pressure. Less intense temperature and pressure conditions form rocks of the *epidote-amphibolite facies*, and more intense conditions form rocks of the *granulite facies*. (See *Reef.*) The amphibolite facies is typically represented by amphibole, diopside, epidote, and plagioclase. An increase in metamorphic intensity is characteristically marked by the disappearance of epidote and the formation of more calcic plagioclase. Amphibolite facies rocks are widely distributed in Precambrian gneiss; this is construed as an indication that the rocks formed in the deeper parts of folded mountain belts.

amphineuran   A bilaterally symmetrical marine mollusk of the class Amphineura. They are usually oval and have a segmented, univalve shell in the form of eight overlapping plates. A common form is the chiton. Fossil forms date back to the Cambrian Period, these differing from Mesozoic to recent forms in that the shell plates are not articulated. Amphineurans have little geological importance.

amphoteric   A term referring to a substance that acts as an acid in the presence of a base, and as a base in the presence of an acid. Water is an example.

amplitude   1. (Physics) The maximum displacement from rest position of a point on a vibrating body or wave. As measured, it is equal to one half the length of the vibration path; i.e., the distance between zero position and trough, or zero position and crest. 2. (Geol.) (a) In a symmetrical fold, half the orthogonal distance between antiformal crust and synformal trough. (b) In meandering channels, half the wavelength of a cycle; i.e., one crest and one trough.

amygdale   *Amygdule.*

amygdaloid, amygdaloidal   1. adj. Describing the structure of rocks containing *amygdules*. 2. n. Amygdaloid. Igneous rock in which rounded cavities have become filled with mineral deposits.

amygdule   Lit. "almond-shaped." Deposits of *secondary minerals* in elongated, rounded, or almond-shaped vesicles of igneous rocks, esp. basalt.

They range in size from 1 mm to 30 cm. The vesicles, bubbles formed as a result of dissolved gas, are generally larger in basaltic lava flows than in other extrusions. Their size appears to be a function of the lower viscosity of the melt and higher temperatures of the basaltic extrusions. Among the secondary minerals that may fill these vesicles are quartz, carbonates, and zeolites, all of which are precipitates from percolating ground water. See also: *druse; vug.*

anaerobic   Lit. "without air." Organisms that are anaerobic do not require free atmospheric or dissolved oxygen.

anaerobic sediment   Sediment that occurs in basins where the limited circulation of water results in the absence or near absence of oxygen at the sediment surface. Water at the bottom of such sediment is rich in hydrogen sulfide, H_2S.

analcime   A zeolite mineral; $NaAlSi_2O_6 \cdot H_2O$, common in diabase granite and basalt. Syn: *analcite.*

analog   Referring to the representation of a magnitude by some continuously varying parameter. An *analog computer* is one that operates with continuously varying amplitudes. Cf: *digital.*

analytic group   A rock-stratigraphic unit formerly called a *formation* but later classified as a *group* because subdivisions of the unit are themselves considered to be formations. Cf: *synthetic group.*

analyzer   A device (esp. the *Nicol prism*) that is placed in the eyepiece of a polariscope or similar instrument, and which transmits only plane-polarized light.

anamorphism   A deep-seated constructive metamorphic process by which new complex minerals of greater density are formed from pre-existing simpler minerals of low density through silication, decarbonization, deoxidation, and dehydration. Cf: *katamorphism*.

anastomosing 1. A pattern of branching and recombining, as in a braided stream, which is referred to as an *anastomosing stream*. 2. Interveined, as in leaves showing netlike vein patterns.

anatase A mineral, TiO_2, trimorphous with rutile and brooklite. Syn: *octahedrite*.

anatexis The partial melting of pre-existing rock, which generates variable magmas that can subsequently be further diversified by mixing, contamination, and magmatic differentiation. Cf: *syntexis*.

anatexite Rock formed by anatexis; also spelled *anatectite*. Cf: *arterite*. See also: *syntectite*.

anauxite A clay composed of kaolin and amorphous silica.

anchor ice Ice formed at the bottom of streams on a submerged object or structure when the temperature of the water is above the freezing point. Radiation of heat from the stream bottom is probably the prime factor in the formation of anchor ice. Syn: *bottom ice*.

anchored dune A sand dune stabilized by vegetal growth. Syn: *stabilized dune*.

andalusite A mineral, Al_2SiO_5, that occurs as orthorhombic prisms in schists and gneisses. It is trimorphous with kyanite and sillimanite.

andesine A mineral of the plagioclase-feldspar group, composed of 70 to 90% albite and 30 to 50% anorthite. It is a primary constituent of intermediate rocks such as andesites and diorites.

andesite A fine-grained, dark-colored volcanic rock, frequently *porphyritic*; the extrusive equivalent of diorite. Phenocrysts of porphyritic andesite consist chiefly of sodic plagioclase (esp. andesine) and one or more ferromagnesian minerals such as biotite or pyroxene; the groundmass is composed generally of the same minerals as the phenocrysts. The andesite rock family is subdivided into *quartz-bearing andesites* (the dacites), *hornblende- and biotite-andesites*, and *pyroxene-andesites* (the most common). It is generally thought that andesitic magma originates from basaltic magma, either by crystal-liquid fractionation or by contamination with sialic material. Andesite takes its name from the Andes Mountains, where it was first applied to a series of lavas.

andesite line A boundary separating the basaltic rocks of the *Atlantic suite* from the chiefly andesitic rocks of the *Pacific suite*; i.e., the island arc and continental regions of andesitic volcanic rock from the typically basaltic volcanic rock of the ocean regions. In the west, it extends from Alaska to Japan and south to the Marianas. The eastern boundary is less clearly traced, but probably runs along the coast of North and South America.

andradite The calcium-iron end-member of the garnet group, $Ca_3Fe_2(SiO_4)_3$, which characteristically occurs in contact-metamorphosed limestones. Its color may be in various shades of yellow, green, or brown.

aneroid barometer A barometer consisting basically of a partially evacuated short, hollow cylinder with a flexible diaphragm at one end and a recording point at the other end. Atmospheric pressure changes are registered by movements of the diaphragm.

Angaraland A small shield of exposed Precambrian rocks in north-central Siberia, once believed to have formed the nucleus about which all other tectonic structures of Asia were assembled.

angiosperm Any member of the

Angiospermae; the entire range of plants with true flowers having seeds enclosed in an ovary. Such plants originated in the Early Cretaceous, or perhaps earlier. Syn: *flowering plants.* Cf: *gymnosperm.*

angle of emergence The angle formed between a ray of energy—acoustic, optic, or electromagnetic—and the horizontal; complement of the *angle of incidence.*

angle of incidence The angle which a ray of energy—acoustic, optic, or electromagnetic—makes with the normal to the surface on which it impinges; complement of the *angle of emergence.* See also: *critical angle.*

angle of repose The maximum slope or angle at which loose earth materials, such as soil or sand, remain stable. It varies greatly with the composition, grain size, and shape of the material, as well as with its water content. For dry sand, the angle of repose is between 32° and 34°. *Talus* slopes and the slopes of cinder cones are commonly at the angle of repose.

anglesite A white, orthorhombic lead sulfate mineral, $PbSO_4$, a common secondary mineral that is a minor ore of lead; usually formed by the oxidation of galena.

Angström unit Unit of length equal to 10^{-8} cm, used chiefly in measuring wavelengths of light. Also called an *angstrom* and abbreviated as Å or A.

angular Having sharp angles or edges, such as a particle or fragment evidencing little or no abrasion.

angular momentum The product of the linear momentum of a point on the surface of a rotating planet or globe and the perpendicular distance of that point from the axis of rotation. It is a vector quantity, $M = \vec{r} \times m\vec{v}$ where M is the angular momentum about a point, $\vec{r}$ is the position vector from the axis to the particle, m the mass of the particle, and $\vec{v}$ the linear velocity of the particle. The angular momentum of the earth is greatest at the equator, where the position vector is at a maximum, and decreases with increasing north and south latitudes until its value becomes zero at the poles.

angular unconformity See *unconformity.*

angular velocity The rate at which a rotating body turns, expressed in revolutions per unit of time, radians, or degrees. The earth's angular velocity is the speed of its rotation about the geographic pole and not with reference to a point in space. The angular velocity of the earth is approximately 15.0411 seconds of arc per second of time.

anhedral Referring to mineral crystals not bounded by their typical crystal faces (*rational facies*). Cf: *euhedral; subhedral.*

anhydrite An orthorhombic mineral, anhydrous calcium sulfate, $CaSO_4$. It is common in evaporite deposits, and alters to gypsum in humid conditions.

anhydrous Totally or essentially without water; e.g., an anhydrous mineral.

anion A negatively charged ion that moves toward the positive electrode (anode) during electrolysis. Cf: *cation.*

anisometric Said of crystals with unequal dimensions. Ant: *equant.*

anisotropic Said of material having some physical property that varies with direction. In the case of crystals, these variations are relative to the crystal axes. All crystals except those of the isometric system are anisotropic with regard to some property; e.g., elasticity, conductivity, refractive index. However, the term usually refers to optical properties. Ant: *isotropic.*

anisotropy Condition or state of having different properties in different directions. In geologic strata, sound waves may travel at one velocity in the vertical direction and at a different velocity in the horizontal.

ankerite Variety of iron-containing dolomite, $Ca(Fe,Mg,Mn)(CO_3)_2$. In the natural species, considerable magnesium or manganese has replaced the iron.

annelid Any wormlike invertebrate belonging to the phylum *Annelida,* characterized by a segmented body with a distinct head and appendages. Since they lack skeletal structures (hard parts), their fossil evidence is usually only in the form of trails and burrows.

annual layer Any sedimentary or precipitated layer of material of variable composition and thickness, presumed to have been deposited during the course of one year. Annual layers are most frequently utilized in the study of *varved* deposits laid down in glacially fed lakes. Certain nonglacial deposits also exhibit annual layers; e.g., the alternating laminae of anhydrite and calcite in a salt intrusion. It is thought that the anhydrite and calcite represent deposits of dry and humid seasons, respectively.

annual ring In cross-sections of the stems of woody plants, the annual ring is the increment of wood added by a year's growth. The number of annual rings in a tree trunk at its base indicates the age of the tree. Often used in dating ancient wood structures. See also: *dating methods.*

annular drainage pattern See *drainage patterns.*

anomalous lead Lead deposits that give model ages older or younger than

the age of the enclosing rock. Anomalous, or nonconformable, leads contain either too much or too little radiogenic lead. The first type, *B-* or *Bleiberg-type,* gives older model ages, and the second, *J-* or *Joplin-type,* younger ages than the age of the enclosing rock. Bleiberg, Austria and Joplin, Missouri are locations in which notable deposits of such leads are found. The existence of anomalous lead ores can be recognized from a model showing the evolution of common lead isotopes. Although both ordinary and anomalous lead evolved from primordial lead through the addition of radiogenic lead, the anomalous variety, unlike ordinary lead, does not follow the requirement that the radiogenic fraction be produced in a region having but a single source of uranium and thorium. See also: *dating methods.*

anomaly 1. A deviation from the common or normal type, form, or rule. 2. *Gravity anomaly.* 3. *Magnetic anomaly.*

anorogenic 1. Said of features not related to tectonic disturbance. 2. Said of a region that is crustally inactive.

anorthite A feldspar mineral, $CaAl_2Si_2O_8$, found as white or grayish triclinic crystals in basic igneous rocks. It is the most calcic member of the plagioclase series. Also, the pure calcium-feldspar end member of the plagioclase series. Syn: *calcium feldspar.*

anorthoclase A mineral of the alkali feldspar group, $(Na,K) AlSi_3O_8$. It is a sodium-rich feldspar with constituents $Or_{40}Ab_{60}$ to $Or_{10}Ab_{90}$ (Or = Orthoclase, Ab = Albite).

anorthosite Intrusive igneous rock composed almost entirely of plagioclase feldspar. Although less abundant than basalt or granite, anorthosite often occurs in massive complexes, usually *lopolithic* or *laccolithic.*

antarctic n. The area within the

Antarctic Circle; the region about the South Pole. adj. Pertaining to climate, features, animals, and vegetation typically found in the South Polar region.

antecedent stream   A stream established before the onset of some remolding event, such as uplift, and maintaining its course unaffected by that event.

anthophyllite   An amphibole mineral, $(Mg,Fe)_7Si_8O_{22}(OH)_2$, a variety of asbestos. Commonly produced by regional metamorphism of ultrabasic rocks.

anthozoan   Any marine benthonic coelenterate of the class Anthozoa, characterized by an external skeletons having a stonelike or leathery texture. Sea anemones and the reef-building corals are both included in Anthozoa, but only the corals have geological importance. Range, Ordovician to present.

anthracite   The highest rank of coal and the only one that can be considered a metamorphic rock. It is jet black with a somewhat metallic luster, and shows semi-conchoidal fracture. Anthracite has the lowest percentage of volatile matter (less than 8%), the highest percentage of fixed carbon (over 90%, on a dry mineral-matter-free basis), and a calorific value of 9000 to 16,000 BTU/lb. It is a clean coal that burns with a long-lasting, short blue flame without smoke. Syn: *hard coal*. See also: *coal; meta-anthracite.*

anthraxolite   A hard, black *asphaltite* found in sedimentary rocks, especially in association with oil shales, but not restricted to such locations.

anthraxylon   Etymol: Greek, *anthrax*, "coal" and *xylon*, "wood." Composite term for a coal component (*maceral*) that has a vitreous luster. In *banded coal* it forms bands interlayered with opaque or dull *attritus*. Anthraxylon is derived from woody tissues of plants whose structures remain filled with humic matter or resin,

following the coalification process. See also: *coal.*

anticlinal axis   See *anticline.*

anticlinal theory   A theory, originated in 1885, to account for the close association of petroleum and natural gas with anticlines. It was based on the fact that oil and gas accumulate as close as possible to the top of a reservoir, and was reasoned that fracturing of the strata, a requirement for oil accumulation, is a maximum where bending is a maximum; i.e., the anticlinal axis. For many years domes and anticlines were used almost exclusively when drilling for oil, but in the 1930s the importance of other structural features came to be recognized. See: *Oil trap.*

anticlinal valley   See *anticline.*

anticline   A fold, generally of inverted U-shape, which, unless overturned, contains stratigraphically older rocks toward the center of curvature. An *anticlinorium* is a large composite anticline composed of lesser folds. Cf: *synclinorium.* The *anticlinal axis* is that line which, if displaced parallel to itself, generates the form of an *anticline.* Valleys coinciding with this axis are referred to as *anticlinal valleys.* Ant: *syncline.* See also: *antiform; fold.*

anticlinorium   See *anticline.*

antidune   A transient accumulation of sediment on a stream bed, comparable to a sand dune, that moves upstream. Also, any marking or bed feature, whether moving or stationary, produced by unidirectional flow, and in phase with surface water waves.

antiform   Inverted U-shaped fold in strata of unknown stratigraphic sequence. Cf: *anticline.* Ant: *synform.*

antigorite   A variety of serpentine, $(Mg,Fe)_3Si_2O_5(OH)_4$, occurring either in corrugated plates, or fibers.

antimony A mineral in which the native element is a silvery-white metal having a crystalline (hexagonal) structure. It occurs chiefly in the mineral stibnite, Sb_2S_3.

antiperthite A variety of alkali feldspar, composed of crystal intergrowths in which the sodium-rich phase (albite, oligoclase, or andesine) seems to be that component of a solid solution from which the potassium-rich phase (usually orthoclase) exsolved. Cf: *perthite*.

antiroot See *Pratt hypothesis*.

antistress mineral A mineral whose formation in metamorphosed rocks is aided by conditions not influenced by shearing stress but rather by thermal action and hydrostatic pressure. Feldspar and pyroxene are examples.

antithetic fault A minor normal fault whose orientation is opposite to that of the major fault with which it is associated. Cf: *synthetic fault*.

Antler orogeny A mountain-building event in the late Devonian and Mississippian time that produced extensively deformed rocks in the Great Basin of Nevada. It was named by R.J. Roberts for relations in Antler Peak quadrangle near Battle Mountain, Nevada. The main evidence for the Antler orogeny consists of widespread coarse *clastic* sediments of Mississippian and Early Pennsylvanian, and a prominent eastward-directed thrust fault, the Roberts Mountain Thrust, in the Early Mississippian.

apalhraun The name used by Icelanders to describe *aa* flows as they occur in Iceland.

apatite A series of hexagonal phosphate minerals composed of calcium phosphate together with chlorine, fluorine, hydroxyl, or carbonate in varying amounts; general formula: $Ca_5(PO_4,CO_3)_3(F,OH,Cl)$; it also includes fluorapatite, chlorapatite, hydroxylapatite, and carbonate- apatite; where no specification is given, the term usually refers to *fluorapatite*, the most important commercial mineral. The apatite series occurs as accessory minerals in igneous and metamorphic rocks and in ore deposits, and is the world's major source of phosphorus. Syn: *calcium phosphate*.

apex 1. The highest point of a mountain or other land form. 2. The crest of an anticline. 3. In mining, the highest point of a vein relative to the surface, regardless of whether or not it appears as an outcrop.

aphanite A rock with *aphanitic* texture.

aphanitic Textural term used to describe igneous rock with crystalline aggregates of such fine grain size that constituent minerals are not visible to the naked eye. These aggregates form by rapid crystallization or, secondarily, by the devitrification of glasses. A rock having such fine texture is designated as aphanitic, regardless of the presence of phenocrysts.

aphotic A term for ocean regions that do not allow the penetration of sunlight. These regions are normally found at depths of 200 m and greater.

API gravity scale A standard scale adopted by the American Petroleum Institute for designating the specific weight of oils. API gravity is expressed as: API° (degrees API) = (141.5/sp.gr. at 60° F) − 131.5. The API scale allows the linear calibration of hydrometers, which are used to measure specific gravity. Cf: *Baumé gravity scale*.

aplite An intrusive igneous rock of simple composition, e.g., a granite composed only of alkali feldspar, quartz, and muscovite mica. Aplites are fine-grained and sugary in texture, and thus the term *aplite* is often used to designate various rocks characterized by a sugary texture and a very small percentage of dark minerals.

apophysis   In geology, a tongue or other direct offshoot of a larger intrusive body, such as a vein or dike.

Appalachian orogeny   A far-reaching disturbance in the southern Appalachian orogenic belt of North America. It probably began in the Late Devonian and continued to the end of the Permian. Frequently called *Appalachian revolution*, it corresponds to the *Variscan orogeny* of Europe. Practically all folds and faults south of New England date from this disturbance. The moving force was from the southeast. Plate tectonics attributes the Appalachian orogeny to the collision of Africa and North America.

apparent dip   The angle of inclination of a fault plane or bedding surface exposed on the face of an outcrop. It is an angular measurement made on a dipping surface other than on a plane perpendicular to the strike. The result is always less than the *true dip* of the planar feature.

apparent movement of a fault   The movement observed on any randomly viewed section across a fault. It is a function of the disrupted strata, attitude of the fault, particular site of observation, and actual slip of the fault.

apparent resistivity   The electrical resistivity of rocks, as measured by a series of voltage and current electrodes positioned on the surface of the earth or in a borehole. If the earth were homogeneous, this value would be equivalent to the actual resistivity; in practice, however, it is the weighted average of the resistivities that are measured. See *resistivity; resistivity method.*

apparent surface velocity   The velocity at which the phase of a seismic wave train appears to travel along the surface of the earth. If the wave train is not traveling parallel to the surface, then the apparent velocity is greater than the actual value.

applanation   All processes that tend to reduce the relief of a region and cause it to become progressively plainlike. High parts are diminished by denudation and low parts are filled in or raised by the addition of material.

apron   A broadly extended deposit of unconsolidated material at the base of a mountain or in front of a glacier.

aquamarine   A transparent, pale blue-green gem variety of beryl. It occurs in pegmatite, in which it may form crystals of tremendous size.

aqueous   1. Pertaining to water. 2. Prepared with water, such as an *aqueous solution.* 3. Said of rocks that have been formed of matter deposited by or in water.

aqueous ripple mark   A ripple mark in sand, soil, etc., made by waves or water currents, in distinction to one made by wind.

aquiclude   A rock body or layer, e.g., a fine-grained silt or clay layer, that will absorb water but not release it fast enough or in amounts sufficient to supply a spring. Cf: *aquifuge; aquitard; aquifer.*

aquifer   A body of rock that contains water and releases it in significant quantities for use. The rock contains water-filled pore spaces that are sufficiently connected to allow the water to flow through the rock matrix to wells and springs. Syn: *water horizon; groundwater reservoir.*

aquifuge   A rock body or rock layer having no interconnected openings or interstices, and which, thus, neither absorbs nor conducts water. Cf: *aquiclude; aquitard.*

aquitard A *confining bed* that retards but does not completely stop the flow of water to or from an adjacent *aquifer*. Although it does not readily release water to wells or springs, it may function as a storage chamber for groundwater. Cf: *aquifuge; aquiclude.*

aragonite An orthorhombic mineral, one form of calcium carbonate, $CaCO_3$; polymorphous with calcite. Aragonite is always found as a low-temperature, near-surface deposit; e.g., in caves as stalactites and near hot springs as sinter deposits. It is the mineral normally found in pearls, and is an important constituent in the shells and tests of many marine invertebrates that secrete this mineral from waters that would ordinarily yield only calcite.

arborescent *Dendritic.*

arch See *natural bridges and arches.*

archaeocyathid Any marine organism of the phylum *Archaeocyatha*, characterized by a tubular or conical skeleton of calcium carbonate. These organisms have been variously classified as sponges, corals, and calcareous algae, but seem to most closely resemble the calcareous sponges. They occur as fossils in marine limestones of the Lower and Middle Cambrian throughout the world. Syn: *pleosponge.*

Archean Said of the rocks of the Archeozoic, the earlier part of Precambrian time. Also *Archaean.*

Archeozoic The earlier part of Precambrian time. Cf: *Proterozoic.*

archipelago Any large body of water interspersed with a group of islands or with many islands. The term is also applied to the island group itself. The West Indies and the islands of Japan are archipelagos.

arctic n. The area within the Arctic Circle; the region about the North Pole.

adj. Pertaining to frigid temperatures or to the climate, features, animals, and vegetation characteristic of the North Polar region.

arcuate Curved or bowed.

areal Pertaining to an area or the nature of an area, e.g., its surface features; should not be confused with *aerial.*

areal geology The geology of a particular area in which the position and distribution of structural features, surface forms, and stratigraphic elements are observed.

areal map A geologic map showing the extent and distribution of surface-exposed rock structures within a certain area.

arenaceous 1. Said of a sediment or sedimentary rock composed wholly or partly of sand-sized fragments, or having a sandy appearance or texture; relating to sand or arenite, or to such a texture. The term connotes no particular composition and should not be used synonymously with *siliceous.* Syn: *sandy.* 2. Said of organisms occupying sandy habitats.

arenite Etymol: Latin, arena, "sand." 1. A general name for consolidated sedimentary rocks composed of sand-sized particles (0.06 to 2 mm in diameter) irrespective of composition; e.g., arkose, sandstone, quartzite, graywacke. If the particles consist of fossil materials, pebbles, and granules of carbonate rock, and if calcium carbonate comprises more than half of the consolidated rock, it is called a *calcarenite.* Syn: *psammite.* Cf: *coquina.* See also: *lutite; rudite.* 2. A well-sorted sandstone ("clean" sandstone) containing little or no matrix material, and having a relatively simple mineralogical composition. The term is used for a major classification of sandstone as apart from *wacke.* adj. *Arenitic.* See also: *sedimentary rock.*

arête Etymol: Fr. "fish bone." A sharp-crested, serrated, or knife-edged ridge separating the heads of abutting valleys (*cirques*) that previously contained *Alpine glaciers*. It may also be a divide between the glaciation of two parallel *valley glaciers*. An arête sometimes culminates in a high triangular *horn* or peak formed by three or more cirques eroding toward the same point. See also: *glacier*.

argentiferous Containing silver.

argentite A silver sulfide mineral, Ag_2S, the most important ore of silver. Argentite is isometric and stable at temperatures above 180° C; below this point it inverts to monoclinic acanthite, which is stable at low temperatures.

argillaceous Term applied to rocks or substances composed of clay minerals, or having a significant amount of clay in their composition. Syn: *clayey.*

argillic 1. Pertaining to clay. 2. Descriptive term for the process "argillic alteration," in which certain minerals are converted to minerals of the clay series.

argillite A rock derived from mudstone or shale that has been altered and indurated by pressure and cementation. Argillites are midway in metamorphism between shale and slate, lacking the fissility of the former and the cleavage of the latter. See also: *sedimentary rock.*

arid Said of a climate or area characterized by extreme dryness, which is defined in various ways, including an annual rainfall of less than 25 cm, an evaporation rate that significantly exceeds the precipitation rate, and insufficient rainfall for plant life or crops without irrigation. Syn: *dry.*

arithmetic mean The total sum of *n* values of some attribute (e.g., weight, length, specific gravity, etc.) divided by *n*. Usually referred to as the *mean* or *average.*

arkose A coarse pink, red, or gray sandstone formed by the disintegration of granite, without appreciable decomposition. Arkose is composed chiefly of quartz and feldspar grains together in a matrix of calcite or, less frequently, of iron oxides or silica. Arkoses are distinguished from normal quartzose sandstones by a high feldspar content (greater than 25%), and from graywacke by their lighter color. Also, arkoses are usually derived from more acidic rocks, and *graywacke* from more basic ones. When arkose consists of granitic detritus it is referred to as *granite wash*. The high feldspar content indicates that arkose deposits are a product of conditions that would retard the normal decomposition of feldspar; e.g., arid or glacial climate, high relief. adj. *Arkosic.* See also: *sandstone.*

arkosic sandstone A sandstone with a high feldspar content (at least 25% of the mineral grains) together with fragments of various other minerals (mica, quartz) in a predominantly clay matrix. *Arkose* is also used imprecisely to include *feldspathic sandstone, subarkose,* and *arkose.*

armored mud ball A spheroidal mass of clay or silt (5 to 10 cm in diameter) in which gravel and sand become embedded as it rolls downstream.

arrival time The initial appearance of seismic energy as shown on a seismic record. At locations close to an earthquake source, P waves and S waves will arrive fairly close together, even though P waves travel significantly faster than S waves. At locations more removed from the source, however, S waves will lag behind the P waves. The difference in arrival time between the two types of waves at any one station is used to calculate the distance of the earthquake from the station.

arroyo Etymol: Span., "mine shaft." Term applied in southwestern U.S. to channels of intermittent streams, characteristic of semi-arid regions.

Arroyos are gullies or gulches with flat bottoms and nearly vertical walls.

arsenate   A mineral compound formed from the arsenate group, AsO_4. Ex: scorodite, $FeAsO_4 \cdot 2H_2O$. Many arsenate minerals form solid solutions with both phosphates and vanadates.

arsenic   Element (As) of the nitrogen family, existing in both gray and yellow crystalline forms, and having both non-metallic and metallic (mineral) forms. The metallic form, which is the more common and more stable, is gray and very brittle.

arsenopyrite   An iron sulfarsenide mineral, FeAsS, that forms metallic-white to steel-gray orthorhombic crystals. It is the principal ore of arsenic and a common mineral with lead and tin ores in ore veins, and in pegmatites, probably having been deposited by the action of both hydrothermal solutions and vapors. Syn: *mispickel*.

arterite   A *migmatite* with artery-like granite intrusions in the metamorphic host rock. Syn: *injection gneiss*. See also: *migmatite*.

artesian   Etymol: Fr., *Artois*; province in France where naturally flowing wells were first drilled. Pertaining to or characteristic of artesian conditions or their utilization and application. Artesian conditions obtain when the hydrostatic pressure exerted on an aquifer is great enough to cause the water to rise above the water table. Artesian conditions depend upon certain requirements: 1. An inclined aquifer whose lower end is buried and upper end is exposed. 2. Impervious layers above and below the aquifer that prevent leakage and allow hydrostatic pressure to develop (an aquifer so specified is an *artesian aquifer*). 3. Precipitation that infiltrates the aquifer at its exposed end. 4. A spring or well that allows the water of the aquifier to discharge.
    An *artesian spring* occurs where artesian water (*confined ground water*)

flows through faults or joints in the overlying impervious rock layer. *An artesian well* is one that is made below the exposed end of the aquifer, with the result that the water rises above the water table at that position; it is called an artesian well regardless of whether or not the water flows out at the land surface. An *artesian basin* is an area, commonly basin shaped, containing an artesian aquifer whose structure and orientation allow easy and wide ranging access to its water. The Great Artesian Basin of Australia, underlying about one-fifth of the continent, is the world's largest area of artesian water.

artesian aquifer   See *artesian*.

artesian basin   See *artesian*.

artesian spring   See *artesian*.

artesian water   See *artesian*.

artesian well   See *artesian*.

arthrodiran   One of an extinct order of armored, jawed fishes, found in fresh water and marine deposits of the Devonian. They reached their peak during the Late Devonian, when some genera attained lengths of 9 m.

arthrophycus   A trace fossil originally described as a fossil seaweed, assigned to the genus *Arthrophycus*. It is a curving, branching, sand-filled form, variously regarded as a feeding burrow or trail produced by an annelid, mollusk, or arthropod.

arthropod   Any of a group of invertebrates of the phylum Arthropoda, characterized by segmented bodies, jointed appendages, and an exoskeleton or carapace of chitin. Arthropoda is the largest phylum in the animal kingdom, with species inhabiting all environments. Some typical members are trilobites, crustaceans, chelicerates, and insects. Range, Lower Cambrian to present.

artificial brine   Brine produced from

an underground deposit of salt or other soluble rock material during *solution mining*.

asbestos A general term applied to varieties of several fibrous minerals. Although different in chemical composition, their heat-resistant fibers allow them to be used for fireproofing and heat-insulating materials. Chrysotile, known as fibrous serpentine, comprises about 95% of commercial asbestos; other types are all amphiboles. All varieties occur in metamorphic rock.

aseismic ridge A submarine ridge that is a fragment of continental crust. It is non-seismic, and is thus distinguished from the seismically active mid-oceanic ridge.

ash 1. *Volcanic ash.* 2. The inorganic residue remaining after coal is burned. See: *pyroclastic material.*

ash fall A shower of volcanic ash that falls from an eruption cloud; also, a deposit of the ash lying on the ground.

ash flow A density current, generally a highly heated mixture of volcanic gases and ash; it rolls down the flanks of a volcano or close along the ground, and is produced by the violent disintegration of lava in a volcanic crater, or by the explosive ejection of gas-laden ash from a fissure. The velocity of these flows is determined by gravity, and sometimes by the explosive force of the eruption. The solid material of a typical ash flow is unsorted and includes *blocks, scoria,* and *pumice,* in addition to ash. Syn: *glowing avalanche.* See: *ash.* See also: *nuée ardente.*

asphalt A dark brown to black bitumen with a consistency varying from viscous liquid to solid. It consists almost entirely of hydrogen and carbon disulfide. Natural asphalt is found in oil-bearing rocks and is believed to be an early stage in the breakdown of organic marine materials into petroleum.

asphalt-base crude Crude oil that contains a high percentage of asphaltic and naphthenic hydrocarbons. Cf: *paraffin-base crude.*

asphaltic sand A naturally occurring mixture of asphalt and sand.

asphaltite Any naturally occurring solid black bitumen that is insoluble in petroleum naphthas. Ex: glance pitch, grahamite.

assay v. To determine the proportions of metal in an ore, or to test an ore or a mineral for composition, weight, purity, etc. n. The analysis itself of such an ore or mineral.

assay foot (inch) In the determination of the *assay value* of an ore body, the assay foot (inch) is the number of feet (inches) along which the sample was taken, multiplied by the *assay grade.*

assay grade Percentage of valuable or useful constituents in an ore, as determined from assay. Cf: *assay value.*

assay limit The limits of an orebody as defined by assay rather than by structural, stratigraphic, or other geologic parameters. Syn: *cutoff limit.*

assay ton Unit of weight used in assaying to represent proportionately the *assay value* of an ore. Equivalent to 29.167 gm.

assay value The quantity of valuable constituents in an ore as determined by the product of its *assay grade* and its dimensions (assay foot, assay inch). For precious metals, the figure is usually given in troy ounces per ton of ore.

assemblage 1. A grouping of similar organisms; in particular, a group of fossils found at the same stratigraphic level. Cf: *association; biocoenosis; community.* 2. The minerals that make up a rock, particularly an igneous or metamorphic rock.

assemblage-zone A biostratigraphic unit defined by a grouping of associated fossils rather than by any single guide fossil. Cf: *biozone*.

assimilation The incorporation of solid or fluid matter from wall rocks, or from physically ingested inclusions, into magma to form a *contaminated* or *hybrid* igneous rock. The degree of assimilation of cool wall rock into a magma is determined by factors such as the structure and fabric of the wall rock and the thermal energy of the magma itself.

association 1. A group of organisms, living or fossil, that occur together because of similar environmental tolerances or requirements. 2. *Rock association*.

asterism A starlike optical phenomenon displayed by some crystals when viewed in reflected light (e.g., star sapphire), or in transmitted light (e.g., some phlogopite mica). It is the result of minute, oriented acicular inclusions.

asteroid 1. A member of the subclass of echinoderms that includes the starfish. Range, Ordovician to Recent. The asteroids have broad, hollow arms that are radially arranged and not separable from the central disc. The number of arms is typically five, but may be in multiples of five. 2. One of several small celestial bodies in orbit around the sun. Most of these have orbits between those of Jupiter and Mars.

asthenosphere The layer or shell of the earth directly below the *lithosphere*. Its upper boundary begins about 100 km below the surface and extends to a depth of about 300 km. The upper limit is marked by an abrupt decrease in the velocity of seismic waves as they enter this region. The temperature and pressure in this zone reduce the strength of rocks, so that they flow plastically. For this reason, the asthenosphere is referred to as the "weak zone." (See lithosphere.) According to one theory, convection currents in the asthenosphere permit the weakened material to flow in the form of closed loops (cells). The motion of these cells effects a drag on the adjacent lithosphere, causing it and the outer crust to move along the earth's surface. At present, this is the most widely accepted mechanism for moving the continents (*continental drift*). Syn: *zone of mobility*. See also: *global tectonics; plate tectonics; sea-floor spreading*.

astrobleme Eroded remains on the earth's surface of an ancient impact structure produced by a cosmic body; in particular, a meteor or comet. Astroblemes are usually characterized by a circular to elliptical outline and highly disturbed rocks. Since features of crater walls are altered by erosion, astroblemes are identified chiefly by the presence of *shatter cones*, which are impact structures that form under the point of impact. See also: *shatter cone*.

astrogeology A science that applies the knowledge and principles of all terrestrial branches of the geological sciences to the study of the origin and history of the condensed matter and gases of other bodies in the solar system. Syn: *extraterrestrial geology*. See also: *planetology*.

asymmetrical Without symmetry or having no axis, center, or plane of symmetry.

asymmetrical fold See *fold*.

Atlantic, Pacific, and Mediterranean suites The division of igneous rocks of the Tertiary and Holocene periods into groups according to their petrographic characteristics. The *Atlantic* and *Pacific suites* were defined on the basis of apparent differences in the rock types surrounding the two ocean basins. The *Pacific suite* was characterized by calcic and calc-alkalic rocks, whereas rocks of the *Atlantic suite* were more alkaline. The *Atlantic suite* was later subdivided with rocks high in sodium called the *Atlantic suite* and those rich in potassium defined as the *Mediterranean suite*. The three suites were also referred to as *series*, and the three ocean basins were called *provinces* with the same names. Further study has found so many exceptions as to almost render these terms obsolete. See also: *andesite line*.

Atlantic-type coastline   A coastline that develops where the structural features of the land, such as mountain ranges, are transverse to the boundary of the ocean basin. Such coastlines are usually irregular with many inlets. Coastlines of this type are seen ideally along Ireland, Brittany, and Newfoundland. See also: *Pacific-type coastline*.

atmophile elements   1. The most representative elements of the atmosphere: H,N,He,C,O,I, and inert gases. 2. Elements that were concentrated in the gaseous primordial atmosphere or that occur in the uncombined state.

atmosphere   1. The gaseous envelope surrounding the earth; it consists (by volume) of 78% nitrogen, 21% oxygen, 0.9% argon, 0.03% carbon dioxide, and small amounts of helium, krypton, neon, and xenon. The mobility of the earth's atmosphere is one 'of the determining factors of earth's climate and wind systems. The atmosphere is a poor conductor of heat but a rather good medium for the transmission of vibrations. Atmospheric density at sea level is 1.225 kg/m^3. This value decreases markedly with elevation, because half the total mass of the atmosphere is less than 5.5 km from the earth's surface. Syn: *air*. 2. A unit of pressure equal to the pressure exerted by a vertical column of mercury 760 mm in height at 0° C., with gravity taken at 980.665 cm/sec^2; equivalent to about 14.7 lb/in^2.

atmospheric pressure   The force per unit area exerted by an atmospheric column; most commonly measured with a mercury or aneroid barometer. Some expressions for standard sea-level pressure are: 760 mm or 29.92 inches of mercury; 1033.3 cm or 33.9 feet of water; 1013.25 x 10^3 dynes/cm^2; 14.66 lb/in^2; and 1013.25 millibars or 1 atmosphere.

atmospheric water   Water present in the atmosphere in a gaseous, liquid, or solid state (water vapor, rain, snow, hail).

Atokan   Major division of Pennsylvanian rocks and time in North America. It designates the Lower Middle Pennsylvanian in North America, and is roughly equivalent to the Upper Carboniferous.

atoll   A ring-shaped coral reef composed of closely spaced coral islets that enclose a shallow lagoon; may be circular to elliptical in plan view. Atolls are most abundant in the Pacific Ocean; they may be as great as 130 km in diameter, but most measure only a few kilometers. They may rise thousands of meters above the sea floor, with the highest portion surrounding the lagoon. The theory of atoll origin that is most compatible with modern data is based on the subsidence of volcanic islands and the simultaneous growth of a coral reef. The reef develops initially as a fringe around a volcanic island; subsidence of the extinct volcano and upward growth of the reef produce a *barrier reef*; progressive subsidence and reef growth eventually leave a lagoon surrounded by an atoll.

atom percent   The percent by atom fraction of a given element, as distinct from reference to the weight or number of molecules, in a substance consisting of two or more elements.

attapulgite   *Palygorskite*.

attenuation   1. A reduction in the magnitude or energy of a signal. 2. A decrease in the amplitude of seismic waves due to *divergence*, reflection and scattering, and *absorption*. 3. That portion of the reduction in seismic or sonar signal strength that depends on the physical characteristics of the transmitting medium and not on geometrical divergence.

attenuation constant A mathematical parameter in an equation that expresses the relationship between the initial value x_0 of some physical quantity and x_1, the value it assumes by virtue of traveling a unit distance through a medium, or by virtue of a unit of elapsed time. An example in geophysics is the relation between the initial amplitude I_0 of a seismic disturbance and its amplitude at a distance $r = I = \frac{I_0 e^{-qr}}{r}$, where q is the attenuation constant.

Atterberg limits system A definition of the plastic properties of soils, including *liquid* and *plastic limits* and *plasticity index*; it is based on a series of tests that delineate the transition between states of consistency. The system is important in civil engineering, where it is used to determine rock and soil properties prior to construction work.

Atterberg scale A decimal scale of grain size, based on the unit value of 2 mm. Each grade stands in a fixed ratio of 10 to each successive grade. Subdivisions are the geometric means of the grade limits.

attitude General term to describe the relation between some directional feature in a rock and a horizontal plane. The attitude of planar features, e.g., joints, bedding, foliation, is defined by the strike and dip. The attitude of a linear feature, e.g., fold axis, lineation, etc., is defined by the strike of the horizontal projection of the linear feature and its plunge.

attrital coal See *attritus*.

attrition Wearing away by friction; in particular, the deterioration that rock fragments in transit, such as pebbles in a stream, undergo through mutual grinding, bumping, and scraping, with a resulting decrease in size and angularity.

attritus Composite term for coal constituents of varying maceral content, ranging from translucent to opaque. Attritus may form the bulk of some coals, or may be interlayered with anthraxylon bands in others. (Syn: *durain*.) *Attrital coal* refers to that coal in which the ratio of anthraxylon to attritus ranges from 1:1 to 1:3, or to the matrix of *banded coal*, in which *vitrain* and frequently *fusain* are embedded. See also: *coal*.

augen Etymol: Ger., "eyes." Large lenticular mineral grains or aggregates in foliated metamorphic rock. Feldspar, quartz, and garnet are common as augen, which are generally associated with the schistose and gneissic varieties of metamorphic rock. Such rocks are usually referred to as *augen schist* and *augen gneiss*, and are said to show *augen structure*.

augite Dark-colored pyroxene mineral $(Ca,Na)(Mg,Fe^2 2Al)_2O_6$. It commonly occurs as dark crystals in many basic igeous rocks.

aureole A zone surrounding an igneous intrusion where contact metamorphism of the host rock has occurred. Aureoles are products of mineralogical and chemical changes produced by heat and the migration of chemically active fluids from the magma into the earlier formed adjacent solid rock. The alteration is greatest along the solid rock-magma contact, and decreases with distance from the contact to a point where no alteration has occurred. Syn: *contact zone*.

auric 1. Pertaining to or containing gold. 2. Pertaining to trivalent gold; e.g., *auric chloride*, $AuCl_3$.

auriferous Containing gold.

austral Pertaining to the south; southern. Cf: *boreal*.

authigenic A term applied to mineral constituents formed in a rock at the site where the rock is found. It is applied chiefly to sedimentary material. Cf: *allogenic*.

autoclastic   A term applied to rocks that have been fragmented *in situ* by mechanical processes such as faulting, or by shrinkage following desiccation. Cf: *crush breccia*. *Autoclastic processes create essentially monolithologic, poorly sorted deposits of angular juvenile clasts.*

autochthon   A massive rock body that has remained essentially at its place of origin, and which may show moderate to extensive deformation. Although continental drift could have moved the entire framework containing the rock body into a different geographical location, the autochthon has not been displaced by "local" tectonic processes, such as overthrust faulting, that transport the allochthon. adj. *Autochthonous*: produced or formed in the location where found; ex: coal. The term refers to whole formations rather than constituents (*authigenic*). Ant: *allochthonous*.

autoecology   The study of the relationship between individual organisms or species and their environment. Cf: *synecology*.

autogenetic   1. Term applied to landforms that have developed within the boundaries of local conditions, without the influence of orogenic movement. 2. Said of a certain drainage system that is determined solely by the land surface over which streams flow. Syn: *autogenic*.

autogenic   1. Said of an ecologic sequence resulting from factors that originated within the natural biotic community and altered its habitat. Cf: *allogenic*. 2. *Autogenetic*.

autogeosyncline   See *geosyncline*.

autointrusion   The injection of the residual liquid fraction of a magma that is differentiating into fissures formed in the crystallized fraction at a late stage by some deformation.

autolith   1. An inclusion in an igneous rock to which it is genetically related. It is an aggregate of early-formed crystals precipitated from the magma itself. Syn: *cognate inclusion*. Cf: *xenolith*. 2. Fe-Mg minerals of uncertain origin that are found in a granitoid rock, and which may have the form of round, oval, or elongate clots.

autometamorphism   1. Pervasive mineralogical and fabric changes that occur in an igneous rock body under conditions of decreasing temperature, and which are attributed to the action of the volatiles of the rock body itself. 2. Alteration of an igneous rock body that proceeds more or less automatically during cooling of the body in the presence of its own aqueous fluids. This process should preferably be called *deuteric* alteration.

autometasomatism   A *deuteric* effect in which newly crystallized igneous rock is altered by its own residual fluids that are held within the rock. See also: *metasomatism*.

automorphic   1. Adjective applied to the holocrystalline texture of an igneous or metamorphic rock that is characterized by crystals bounded by their typical crystal faces. Also applied to a rock with such texture. The texture is fairly rare, since its development requires nearly simultaneous crystallization of all minerals and minimum interference among the grains during the process. Automorphic texture is well developed among lamprophyres. Syn: *idiomorphic*. Cf: *subautomorphic*. 2. Syn. of *euhedral*, generally preferred in European usage, but obsolete in America.

auxiliary fault   A minor fault branching from a major one or abutting it. Syn: *branch fault*.

auxiliary minerals   As defined in the Johannsen classification of igneous rocks, those light-colored, relatively rare minerals such as apatite, corundum, fluorite, muscovite, and topaz.

available relief Vertical distance between upland divide and the mouth of a valley floor. It determines the degree of headwater erosion and the depth to which valleys can be cut.

avalanche A large mass of snow, ice, or rock, or a mixture of these that moves rapidly down a mountain slope under the force of gravity. Avalanche velocities may reach 500 km/hr. An avalanche begins when a mass of material overcomes the frictional resistance of the surface on which it is deposited; this often follows loosening of the foundation by spring rains and warm, dry winds. Vibrations from thunder or artillery fire can also initiate movement. Rapidly moving avalanches of rock or dry snow may generate powerful winds (*avalanche winds*) that sometimes cause widespread destruction.

aven A vertical shaft leading upward from a cave passage, and sometimes connecting with passages higher up.

aventurine A gem mineral that is a translucent variety of quartz or plagioclase feldspar. Its spangled appearance is due to tiny inclusions of such minerals as mica and hematite. adj. Referring to the sparkling appearance of a mineral containing shiny or golden inclusions. Cf: *goldstone*.

axial angle The acute angle between the optic axes of a biaxial crystal, designated as 2V. Syn: *optic angle*.

axial elements See *crystal axes*.

axial plane 1. A crystallographic plane in which two of the crystal axes are included. 2. The plane in which lie the optic axes of an optically biaxial crystal. 3. A planar surface connecting the *hinge lines* of the strata in a fold.

axial-plane cleavage Cleavage that is generally parallel with the axial planes of folds in rock. Although some axial plane cleavage parallels the regional fold

axes, most is related to the minor folds occurring in individual outcrops. Axial plane cleavage is usually *slaty cleavage*. Cf: *bedding-plane cleavage*.

axial-plane folding Axial planes are folded during extensive secondary folding of pre-existing folds in rock. Such secondary changes are the result of movements that were considerably different from those effecting the original folding. See *fold*.

axial ratio See *crystal axes*.

axial surface A surface joining the *hinge lines* of strata in a fold.

axial trace The intersection of the axial plane of a fold with a specified surface; e.g., the surface of the earth.

axis 1. A line passing through a body, on which the body rotates or may be imagined to rotate; a line centrally bisecting a body or system around which the parts are arranged. 2. *Crystal axes*. 3. *Anticlinal axis* or *synclinal axis*.

axis of symmetry An imaginary line through a crystal, about which it may be rotated so as to return to its same spatial position during a complete rotation. Syn: *rotation axis; symmetry axis*.

azimuth 1. (Survey.) The angle of horizontal difference, measured clockwise, of a bearing from a standard direction, as from north or south. 2. A horizontal angle, measured clockwise, between north meridian and the arc of the great circle connecting an earthquake epicenter and the receiver.

azimuth compass A magnetic compass fitted with sights for determining the angle between a line on the earth's surface (or the vertical circle through a heavenly body) and the magnetic meridian.

azimuthal projection See *map projection*.

azonal soil   A soil that lacks a definite profile and resembles its parent material. Stream alluvium and dry sands are examples. Syn: *immature soil*.

azurite   A deep-blue monoclinic mineral, $Cu_3(CO_3)_2(OH)_2$. It is an ore of copper and occurs in the upper (oxidized) zones of copper deposits, where it is usually found with *malachite*. Compact azurite is the source of semiprecious stones used in jewelry and decorative objects.

B

back   Roof or ceiling of an underground mine.

backdeep   *Epieugeosyncline*.

backfill   Any material used to refill a quarry or other excavation; waste rock used to support the roof after removal of ore from a stope.

background   1. The normal radioactivity of the environment, attributed to earth's naturally radioactive substances and cosmic rays. 2. In geochemical prospecting, the normal concentration of a given element in a material under investigation. 3. Amount of pollutants in the surrounding air due to natural sources and processes.

backlimb   That side of an asymmetric anticline having the more gentle slope. Cf: *forelimb*.

back reef   The landward side of a reef, also used in some areas to designate the side of a reef away from the open bar even though no land is immediately near. Cf: *fore reef*.

backset bed   A cross-bed that inclines against the flow direction of a depository wind current.

backset eddy   A small current swirling in the direction opposite to those of the major oceanic eddies; often observed between the main current and the coastline.

backshore   1. Upper zone of a beach or shore, bounded by the high-water line of mean spring tides and the upper limit of shore-zone processes. Only unusually big tides or severe storms affect this area. Cf: *foreshore*. 2. The area at the base of a sea cliff. 3. *berm*.

backsight   A sight or bearing calculated on an already established survey point. A comparison between backsighting and foresighting is made for the purposes of calibrating and verifying the precision of surveying instruments. Cf: *foresight*.

back slope   1. *dip slope*. 2. The gentler slope of a fault block or *cuesta*; it is not necessarily related to the dip of the underlying rocks.

back thrusting   Thrust faulting that occurs toward the interior of an orogenic belt; its direction of displacement is opposite the general direction of tectonic transport.

backwash   See *breaker*.

bacteria   Unicellular microorganisms or organisms that lack a nucleus. May be aerobic or anaerobic. They have been present on earth since the Precambrian.

bacteriogenic   Term applied to mineral deposits formed by the action of certain bacteria. Many structureless limestones of geological periods dating back as far as the early Paleozoic are associated with bacteria which, by their activity, are responsible for the deposition of carbonates. See also: *iron bacteria; sulfur bacteria*.

badlands Derived from a term used by the pioneers of the American West to describe a complex, stream-dissected topography. Badlands (e.g., those of South Dakota) are characterized by a very fine drainage network and short steep slopes with little or no vegetative cover. They develop in regions of erodible sediments where vegetation has been destroyed; e.g., by cattle grazing, or where vegetation was lacking.

bailer A cylindrical steel container used in cable-tool drilling for the removal of water, rock cuttings, and mud from the bottom of a well.

bajada Etymol: Span., "slope." A broad, gently sloping detrital surface formed in a basin, most commonly under arid or semi-arid conditions. A bajada is formed by the lateral coalescence of several alluvial fans, or built up by material washed down from the fans. It is a depositional feature, as distinct from a *pediment*, but distinguishing between the two is difficult in most cases. Bajadas are common in the Basin and Range Province of the southwestern United States. Also, "bahada." Syn: *apron; alluvial plain; compound alluvial fan; piedmont alluvial plain.*

balance The difference between accumulation and ablation of a glacier over a specific time interval, determined as a value at a point, average over an area, or the total mass change of the glacier. Changes in the balance of a glacier determine whether it is in a state of advance, recession, or stagnation. Syn: *regimen.* Cf: *net balance.*

bald-headed anticline A term often used in petroleum geology for an anticline whose crest has been eroded and upon which an unconformable sedimentary layer has been deposited.

ballas See *industrial diamond.*

ballast 1. Gravel, broken stone, slag, or similar material used in the roadbed of a railroad. It forms a bed for the ties and functions to hold the track in line, distribute the load, and provide drainage.

ball clay An extremely plastic clay, commonly containing organic matter, used as a bonding substance of ceramic wares. Its unfired colors range from light buff to various shades of gray. Also called *pipe clay.*

banco Term used in Texas for an oxbow lake or a cut-off meander.

band 1. A stratum or lamina distinguishable from adjacent layers by color or lithologic difference. 2. A range of frequencies between designated limits; e.g., the ultraviolet band of electromagnetic radiation ranging from 4×10^{-7}m to 5.0×10^{-9}m. 3. A glacier band, such as an *ogive.*

banded Term applied to a vein, sediment, or other deposit showing alternating layers differing in texture or color, and possibly in mineral composition. Ex: *banded iron formation.*

banded agate An agate whose colors are arranged in alternating bands or stripes that are often wavy or zigzag, and sometimes concentric; the bands may be clearly demarcated or may blend into one another. Cf: *onyx.*

banded coal Coal containing bands of varying luster due to the presence of certain ingredients (e.g., durain, vitrain). Although such coal is usually *bituminous*, banding occurs in all ranks of coal. Banded coal consisting of more than 5% anthraxylon and less than 20% opaque matter is called *bright coal.* (Cf: *attrital coal.*) See also: *coal.*

banded iron formation Abbreviated as *bif*; these are sedimentary rocks that are typically bedded or laminated, and composed of at least 25% iron and layers of chert, chalcedony, jasper, or quartz. Bifs, which occur in all continents, are older than 1700 m.y., and have been highly metamorphosed. They contain many iron oxides (e.g., magnetite and hematite) and are commonly used as low-grade iron ore. Because bifs apparently have not formed since Precambrian time, it is thought that special conditions must have existed contemporaneously with their formation. Their origin has been ascribed to various processes including volcanic activity, seasonally rhythmic deposition from iron and silica solutions, and the oxidation of iron-rich sediments at the time of deposition.

banded ore Ore that consists of different minerals, or of the same minerals differing in texture or color.

banded structure A feature of igneous and metamorphic rock, resulting from the alternation of layers or streaks differing clearly in texture and/or mineral composition.

banding 1. The appearance of *banded structure*, due to *layering*, in an outcrop of igneous or metamorphic rock. Cf: *flow layering*. 2. In sedimentary rocks, thin bedding, conspicuous in cross section, produced by different materials in alternating layers. 3. In glacier ice, layered structures due to alternating layers of fine- and coarse-grained ice, or of clear and clouded ice. Syn: *foliation*.

bank 1. A sandy or rocky submerged elevation of the sea floor. It may be a local prominence on continental or island shelves. 2. Along the Atlantic coast of the U.S., a long, narrow island composed of sand forming a barrier between a lagoon and the ocean. 3. A *shoal*. 4. The rising ground edging a stream. 5. A mound or ridgelike deposit of shells formed *in situ* by organisms such as brachiopods and crinoids. Cf: *reef*. 6. The surface of a coal deposit that is being worked.

bankfull stage Position of the water surface of a stream when it is at *channel capacity*. Discharge at this stage is *bankfull discharge*.

bar 1. Any of various types of submerged or emergent elongated embankments of sand and gravel built by waves and currents. 2. A *river bar, channel-mouth bar,* or *mid-bay bar.* 3. A unit of pressure equal to 10^6dynes/cm^2; equivalent to 750.076 mm at 0° C (mercury barometer). It is equal to the mean *atmospheric pressure* at about 100 m above sea level.

barbed drainage pattern See *drainage patterns*.

barchan A crescent-shaped dune with a convex windward side and horns that point downwind. Barchans usually occur in groups or chains that are disposed toward the wind direction that is most effective, their form possibly being modified later by changes in wind direction. Where sand is abundant, barchans are large and densely packed; where sand is less abundant they are smaller and more dispersed. See also: *seif; dune.*

bar finger A long narrow sand mass of lenticular cross-section underlying a *distributary channel.*

barite A mineral, BaSO$_4$, sp. gr. 4.5, having orthorhombic crystals ranging from colorless to white, yellow, gray, and brown. Its major uses include drilling mud, paint manufacture, and as a filler for paper and textiles. It is the principal ore of barium. Syn: *barytes; heavy spar.*

barite rose A flowerlike group of tabular, sand-filled crystals of barite, arranged in radially symmetrical form. Barite roses usually form in sandstone and are commonly reddish brown. Syn: *petrified rose.*

barograph A *barometer*, usually of an aneroid type, that makes a continuous automatic record of fluctuations in atmospheric pressure.

barometer An instrument used to measure atmospheric pressure. A *mercury barometer* is used when highly

Barchan (barchane or barkhan)

Steep concave
leeward slope

Prevailing wind

Horns

Gentle windward slope

A group of barchans

Prevailing wind

Barchans slowly advance at the rate of
a meter or so per year in the direction of
the prevailing winds.

Wind

Eddy

Eddy currents
steepen the
leeward slope

15
to
30m

Cross section

accurate readings are important, and an aneroid barometer when compactness is a factor.

barometric elevation   An elevation, located above mean sea level and established by the use of a barometer.

barometric pressure   Atmospheric pressure as determined by a barometer.

barred basin   *Silled basin.*

barrens   Level or gently rolling land, usually with a sandy soil and sparse tree growth, and relatively infertile; e.g., the Pine Barrens of New Jersey.

barrier   1. A *barrier beach* or *barrier island.* 2. An *ice shelf.* 3. A *groundwater barrier.* 4. (Ecol.) A condition that tends to prevent the free mixing of populations and individuals; e.g., difference of water quality.

barrier beach   See *barrier island.*

barrier flat   The relatively flat area of a

barrier island, often occupied by small, persistent pools of water. A barrier flat lies between the seaward edge of a barrier island and the lagoon.

barrier ice   *Shelf ice.*

barrier island   An elongate coastal island representing a widened *barrier beach*; a sand ridge rising above high-tide level and roughly paralleling the coast but separated from it by a marsh or a lagoon. One explanation for the existence of barrier islands is that they originated as submerged sandbars that subsequently became emergent after being moved by wave action toward the shoreline. Another possibility is that most barrier islands are remnants of sand spits built from the mainland, then severed by inlets cut through them during storms. Most geologists subscribe to a third theory, which is that dune lines on the mainland were slowly surrounded by rising seas at the conclusion of the last glacial epoch.

barrier reef   A long, narrow reef running roughly parallel to a shore and separated from it by a body of water, e.g., a lagoon, usually too deep to permit coral growth. It may include a volcanic island, either completely or in part. The Great Barrier Reef off the coast of northeast Australia extends for more than 1600 km with only minor breaks in it. A barrier reef is usually pierced by many channels (*passes*), giving access to the lagoon and the coastline beyond it. Barrier reefs may change laterally into *fringing reefs* by joining a coast. See also: *atoll; coral reef.*

Barrovian metamorphism A progressive regional metamorphism that can be noted in the mineralogical and textural changes of a sequence of zones, each one of which is named after an index mineral.

bar theory A theory to account for the origin of large marine deposits of salt, gypsum, and other evaporites. It assumes a standing body of water, e.g., a lagoon, isolated from the ocean by a bar in an arid or semi-arid climate. As water in the lagoon evaporates, water of normal salinity continues to flow in from the ocean. If the rate of evaporation exceeds the rate of inflow over a prolonged period, the concentration of dissolved salts in the water increases and an evaporite is deposited by precipitation. The bar theory, advanced by Ochsenius in 1877, has been well-substantiated with laboratory models.

barysphere *Centrosphere.*

basal cleavage See *cleavage.*

basal conglomerate A well-sorted, homogeneous conglomerate at the base of a sedimentary sequence resting on a surface of erosion. It forms the initial stratigraphic unit in a marine series, and marks an unconformity. Basal conglomerate originates as a thin layer of coarse material deposited by an encroaching sea.

basalt A medium-gray to black igneous rock, more commonly extrusive than intrusive, composed principally of pyroxene and plagioclase with lesser amounts of some combination of olivine, apatite, magnetite, and glass; it is the fine-grained (aphanitic) equivalent of *gabbro.* Most basalts are non-porphyritic, but some contain phenocrysts of olivine and plagioclase. Basalt is the world's most abundant lava and is the chief constituent of isolated oceanic islands.

Because basalts comprise a spectrum of chemical composition, they are, for simplicity and convenience, divided into two types—alkali olivine basalt and tholeiitic basalt—that belong to the alkalic and subalkalic suites, respectively. Without a chemical analysis, classification can be uncertain; however, if olivine is clearly evident in the groundmass, the rock is probably an alkali olivine basalt, and if orthopyroxene is present, it is tholeiitic.

The nomenclature for volcaniclastic rocks is applied to basaltic rocks with clastic fabric; e.g., basalt breccia, basalt tuff. In the U.S., the name *diabase* is used for rocks transitional in grain size between aphanitic basalt and phaneritic gabbro. Cf: *dolerite.* See also: *lunar basalt.*

basaltic layer See *granitic and basaltic layers.*

basal till See *till.*

basanite A feldspathoid olivine basalt composed of calcic plagioclase, augite, olivine, and a feldspathoid (nepheline, analcime, or leucite).

base 1. A substance which, when dissolved in water, forms hydroxyl ions. 2. A nontechnical term for the dominant hydrocarbon series in a crude oil; e.g., asphalt-base crude.

base correction An adjustment of geophysical measurements to express them relative to the values of a referential *base station.*

base exchange In geology, a property that enables certain minerals, notably the zeolites, to exchange cations adsorbed on the surface for cations in the surrounding solution, e.g., the exchange of Ca for Na, when in a suitable environment. Syn: *cation exchange.*

base level The theoretical limit toward which the vertical erosion of the earth's surface progresses but rarely, if ever, reaches. The general base level (*ultimate base level*) is the lowest possible base level; for land surface, it is about 60 m below sea level, because of the reducing effect of wave action over a long period of time; for a stream, it is the plane surface of the sea projected inland as an imaginary surface beneath the stream. A region is at the *temporary base level* whenever it is graded toward some level, other than sea level, below which the land area cannot be reduced for the time being by erosion. Ex: a level locally controlled by a resistant stratum in a stream bed.

base line Any line established with precision instruments for the purpose of making additional azimuthal measurements.

base map Any map that shows the basic outlines required for adequate geological reference, and on which supplementary or specialized information is plotted for some particular purpose.

basement 1. Geological basement is the surface beneath which sedimentary rocks are not found; the igneous, metamorphic, granitized, or highly deformed rock underlying sedimentary rocks. Syn: *basement complex.* 2. Petroleum economic basement, the surface below which there is no current interest in exploration, even though some sedimentary units may lie deeper. 3. Magnetic basement, the upper surface of igneous or metamorphic rocks that are magnetized more than the overlying sedimentary rocks. 4. Electrical basement, the surface below which resistivity is so high that its variations, as noted in electrical survey results, are not significant.

basement complex *Basement.*

base metal 1. Any of the more common chemically active metals, e.g., copper, lead, tin, zinc, etc. Cf: *noble metal.* 2. The principal metal of an alloy, e.g., the copper in bronze.

base of weathering In seismic studies, the boundary between the surface layer over which seismic waves travel with low velocity and an underlying layer through which such waves travel with appreciably higher velocity. The base of weathering, which may correspond to geological weathering or to the water table, is used to derive time corrections for seismic records. See also: *seismic waves.*

base station In geophysical surveys, a referential observation point to which measurements at additional points can be compared. See also: *base correction.*

basic 1. A classification applied to igneous rocks having relatively low silica content, about 45 to 50%; e.g., basalt, gabbro. Basic rocks contain relatively high amounts of iron, magnesium, and/or calcium, and so include most mafic rocks as well as other types. Basic is one of four subdivisions (*acid, intermediate, basic, ultrabasic*) of a commonly used system for classifying igneous rocks by their silica content. Cf: *femic.* See *silica concentration.* 2. Generally said of any igneous rock compound composed mainly of dark-colored minerals. Cf: *silicic; mafic.*

basification Enrichment of a rock in elements such as calcium, iron, magnesium, and manganese.

basin-and-range landscape The Basin and Range physiographic province extends southward from the Columbia Plateau, including most of Nevada and parts of Oregon, Idaho, Utah, California, Arizona, and New Mexico. The topography is characterized by isolated, nearly parallel mountain ranges with intervening valleys or basins composed of sediments derived from the mountains. In cross-profile the ranges are generally asymmetrical, the steeper slope commonly being fairly straight. Both sides of a range may be bounded by steep slopes, but generally only one side is so bounded. The origin of basin-and-range topography is related to the structure of the underlying bedrock. At present, there seems to be general acceptance that it originated with block faulting, followed by modification of the fault blocks by erosion.

bastnaesite   A mineral, $(Ce, La)CO_3(F, OH)$, wax-yellow to reddish-brown in color. It occurs in alkaline igneous rocks, esp. carbonatite, and is the chief U.S. source of rare-earth elements.

batholith   Typically a large, cross-cutting plutonic mass with more than 100 km^2 of surface exposure and with no visible or clearly inferred floor, hence showing no direct evidence of emplacement by simple injection into the earth's crust. Deep-earth temperature and density measurements suggest that the floors of some batholiths are as much as 30 km beneath earth's present surface. Although the origin and mechanism of emplacement of batholiths are still controversial, their formation is generally believed to involve magmatic processes. The most common of these plutons are granitic or granodioritic in composition and are usually, but not always, associated with orogenic belts. One of the greatest batholiths forms a virtually continuous mass extending from southwestern Alaska through British Columbia to the Washington Cascades. Cf: *stock.* See *intrusion.*

bathyal zone   The marine ecological zone that is deeper than the continental shelf but shallower than the deep ocean floor. This range is usually given as 200 to 3000 m. Fluctuations in temperature, oxygen concentration, and sedimentation in the bathyal zone result in conditions that are much more variable than in the abyssal region. The bathyal zone is almost entirely *aphotic*, with general

temperatures between 5° and 15° C. Its salinity may range from 34 to 36%, and current movement in this region is essentially geostrophic. See *continental shelf.*

bathymetric chart   A topographic record of the ocean floor or bottom of some other body of water; contour lines called *isobaths* indicate points of equal elevation.

bathymetry   The measurement and charting of ocean depths. A series of echo soundings are made and the depths are plotted on a chart called an *echogram.* Echograms are then used to establish contour maps and physiographic conditions for the area under investigation. See *echogram; echo sounding.*

bathypelagic   Pertaining to the part of open sea or ocean at bathyal depth.

bathyscaph   A manned submarine vessel for deep-sea exploration; the first such unit was built between 1946 and 1948. It supplanted the *bathysphere,* exceeding it in safety and maneuverability.

bathysphere   A manned spherical steel vessel that is lowered into ocean depths by cable for observations. The first bathysphere dive was made in 1930.

battery ore   Manganese oxide ore of a grade suitable for dry cells.

Baumé gravity scale   A scale originally developed for the purpose of hydrometer calibration, given by the relationship: Baumé° (degrees Baumé) = (140/sp. gr. at 60° F) − 130. Although still used in parts of Europe, the Baumé scale has been largely replaced by the *API gravity scale.*

bauxite   A gray, yellow, or reddish-brown rock largely composed of a mixture of various aluminum oxides and hydroxides, along with free silica, silt, iron hydroxides, and clay minerals. It is the principal commercial source of aluminum. Bauxite is formed by the weathering of many different rocks, and varies physically according to the origin and geological history of deposits; its forms may be concretionary, earthy, pisolitic, or oölitic.

bauxitization The formation of bauxite from either primary aluminum silicates or secondary clay minerals under conditions of thorough tropical or subtropical weathering.

b-axis 1. The crystallographic axis that is oriented horizontally, right to left, and perpendicular to the c-axis. 2. In deformed rocks, the direction in the plane of movement perpendicular to the direction of tectonic transport; b is located on a slickensided surface at right angles to the striae.

bay 1. A body of water forming a recess or indentation of the shoreline, smaller than a gulf but larger than a cove. 2. Any landform resembling a bay of the sea, such as a recess of land partially surrounded by hills. Cf: *bight, embayment*.

bay bar *Baymouth bar.*

bayhead bar A bar formed at the head of a bay not far from shore. Wave refraction on bay shores produces currents that tend to move detritus toward the bayhead, where it accumulates.

baymouth bar A ridge, commonly composed of sand, extending partially or completely across the mouth of a bay from one headland to another; usually formed by a *spit* growing from one or both of the headlands. As the bar develops, the bay becomes a lagoon. Complete closure is made only if shore currents are stronger than currents entering and leaving the lagoon.

bayou 1. A name given (chiefly in the southern states of North America) to marshy offshoots and overflows of rivers and lakes. 2. General term for a stagnant inlet or outlet of a bay or lake. 3. A stagnant or abandoned course of a meandering river. 4. An *oxbow lake*.

beach The sloping shore of a body of water, especially those parts on which sand, gravel, pebbles, shells, or other hard parts of marine organisms are deposited by waves or tides. The configuration of a beach and the nature of the deposits covering it are related to coastal profile, types of debris available, wave pattern, and the rate of deposit. On mountainous coasts, beaches are narrow and broken, and often composed of poorly sorted debris; along plain coasts, beaches may extend laterally for hundreds of kilometers and are usually composed of well-sorted sand. Beach sands on coasts of high relief may be rich in minerals such as feldspar; beach sands on plain coasts consist of quartz and minerals such as rutile and garnet. Some tropical beaches are composed almost exclusively of shell fragments.

beach berm *Berm.*

beach cusp An arcuate seaward projection of sand or pebbles formed on the foreshore of a beach by the action of waves. The distance between cusps generally increases with an increase in wave height.

beach face The section of a beach that is normally subjected to the action of the wave surge. Also, the *foreshore* of a beach.

beach placer A *placer* deposit of valuable minerals on a present-day or ancient beach, or along a coastline. The minerals are formed as a lag deposit by virtue of their greater hardness or density.

beach ridge Near-parallel ridges of sand, shell, or pebbles varying in height from a few centimeters to several meters. Depressions between them are called *swales*. Ridges are found behind the present shore, and each ridge marks the position of a previous shoreline.

beaded drainage   An arrangement of short streams joining small pools, characteristic of areas underlain by permafrost.

Beaufort wind scale   A system of estimating wind characteristics, used at sea, over large land areas, or where there are no wind instruments. Code numbers and descriptive terms are assigned to various ranges; e.g., a wind velocity of 12 to 18 mph (11 to 16 knots) is Beaufort number 4 and is termed "a moderate breeze." The scale is an adaptation of that devised by Admiral Francis Beaufort in 1805.

Becke test   A comparison of the refractive indices of two contiguous minerals (or of a mineral and a mounting medium or immersion liquid) in a thin section, using a microscope of moderate to high magnification. The difference in refractive index of the two substances is indicated by the *Becke line*, a bright line that separates the substances.

bed   1. The smallest *lithostratigraphic unit* that is distinguishable from beds beneath and above it. The term is usually applied to sedimentary strata, but may be used for other types. Syn: *layer; stratum.* 2. The bottom of a body of water; e.g., stream bed.

bedded   A formation that is arranged in layers (esp. sedimentary rock) is said to be bedded. Also, any mineral deposit aligned with the bedding in a sedimentary rock is described as bedded.

bedding   A stratified or layered feature, esp. associated with sedimentary rocks, but also found in igneous and metamorphic rocks. Bedding results from variations in the grain size and mineral composition of materials during deposition, most of which is effected by water directly, or associated with it. *Current bedding* may form in any fluid (water, air, etc.) carrying a traction load; it may form in air, streams, or in the sea. Water or air currents of variable direction are responsible for *cross-bedding*. High-angle cross-bedding is produced chiefly by avalanching of sand down the downslope of such structures as ripples and dunes. Low-angle cross-bedding is produced during accretionary deposition on sloping surfaces, such as on the upstream slope of wavelike structures. *Scour-and-fill* structures are commonly associated with cross-bedding. *Graded bedding* consists of layers, each with a sharply distinct base, on which the coarsest grains of the bed lie. In each layer, the coarse material progresses gradually upward to increasingly finer material, which is followed abruptly by the base of the next layer. The layers are commonly interbedded with others having parallel laminations. Graded bedding may form under conditions similar to those of a *turbidity current*, i.e., where the velocity of the prevailing current declines in gradual measure.

bedding cleavage   *Bedding-plane cleavage.*

bedding fissility   See *fissility.*

bedding plane   The dividing plane that separates each layer in a sedimentary or stratified rock from layers below and above it. It ordinarily delineates a noticeable change in color, texture, or composition.

bedding-plane cleavage   Cleavage that is parallel with the bedding plane. Syn: *bedding cleavage; parallel cleavage.* Cf: *axial-plane cleavage.*

bed form   Any topographic deviation from a flat stream bed produced by changes in the discharge of streams flowing in sandy channels.

bed load   The part of a load moved by a stream along or near its bed because this fraction of the load consists of particles too large or heavy (boulders, pebbles, gravel) to be carried in suspension. Movement of the bed load is a function of the stream gradient, velocity, and friction of the grains between each other and between grains and bed. Syn: *bottom load; traction load.*

bedrock The solid rock lying beneath superficial material such as gravel or soil.

beds A general or informal term for strata that are lithologically similar or contain substances of economic importance, e.g., "beds of Devonian age," "coal beds."

beheading 1. The cutting off of the upper section of a stream and diversion of the headwaters into a new course by *stream capture.* 2. Removal by wave erosion of the upper part of the drainage area of a stream. 3. Truncation of spurs by glacial action. See *stream capture.*

beidellite Member of the smectite group of clay minerals. Commonly found in soils and in certain clay deposits; e.g., *metabentonite.*

belemnite An extinct type of cephalopod related to the modern squid and octopus but with a long, bulletlike internal shell. The shell was straight in most forms, but loosely coiled in others. Belemnites probably originated from more primitive nautiloid cephalopods, and serve well as index fossils. Range, Mississippian to Cretaceous.

belted coastal plain A series of *cuestas* that develop on a coastal plain, in more or less parallel order, with intervening lowlands. The Gulf Coastal Plain through Alabama and Mississippi is an example.

bench mark Markers that are firmly implanted in stable ground, from which local subsidiary measurements originate, and which are used as a reference in topographic surveys and tidal readings.

beneficiation Upgrading of an ore by some process such as flotation, milling, gravity concentration, or sintering.

Benioff zone A plane of seismic activity that dips beneath a continent or continental margin at an angle of about 45°. It is held that lithospheric plates plunge into the upper mantle through this zone and cause earthquakes; earthquake foci tend to occur along these regions. The zone is named for Hugo Benioff, an American seismologist who located it. Syn: *Benioff seismic zone.* See also: *plate tectonics.*

benthos 1. Those organisms that live on the sea floor, including mobile forms such as crabs and starfish, and nonmobile forms such as barnacles and hydroids. adj. *benthonic.* 2. Deepest part of the sea or ocean. Beyond the littoral zone, the benthic zone is usually divided into archibenthic or bathyal, which is on the continental slope, and the *abyssal benthic,* which is beyond a depth of some 1000 m. adj. *benthic; benthal.* See *continental shelf.*

bentonite A variety of mudrock composed almost totally of montmorillonite and colloidal silica, produced as the alteration product of volcanic debris, usually tuff or ash. Although bentonite is usually of the calcium variety, sodium and potassium bentonites are also known; these contain sanidine and quartz grains, in addition to the montmorillonite. Paleozoic bentonites are generally composed of illite-smectite layers. Certain Ordovician bentonites of the Appalachian region and Upper Mississippi River valley are called *metabentonite.* A bed of bentonite is the result of a single volcanic eruption or of several eruptions within a very brief interval. The thickness of the bed and the size of unaltered fragments in it decrease with distance from the volcanic vent; the fragments also show in-bed grading, with lighter ones near the top and heavier ones toward the bottom.

Because of certain properties, bentonites have some economic importance. Calcium bentonites are highly absorbent and also useful as bleaching and decolorizing agents. The fact that sodium bentonites will absorb large quantities of water makes them useful as *drilling muds*, but also causes them to form very unstable slopes in outcrop.

bergschrund Etymol: Ger., *berg*, "mountain" and *schrund*, "crack." A crevasse that may form at the head of a *mountain* (alpine) *glacier*, and which separates the mobile snow and ice of the glacier from the relatively stationary snow and ice adhering to the rocky headwall. It develops when a glacier moves downvalley, away from bedrock or a stationary snowfield. See also: *cirque*.

berg till Term for a glacial till deposited by grounded icebergs into water (salt or fresh) bordering an ice sheet. It may also refer to a marine or lacustrine clay containing rock debris dropped into it by melting icebergs. Syn: *floe till*.

berm A low shelf or terracelike feature on the backshore of a beach, one side of which is generally bounded by a beach ridge or beach scarp. A *beach berm* is formed of material deposited by storm waves. Syn: beach berm. The seaward limit and usually highest point of a berm on a beach is called the *berm crest*.

Bernoulli's theorem For fluids (liquid or gas) whose compressibility and viscosity are negligible, and whose flow is laminar, Bernoulli's theorem states that the total mechanical energy of the flowing fluid, which comprises the kinetic energy of the fluid motion, the potential energy of elevation and the energy associated with fluid pressure, remains constant. In equation form,
$$v^2/2 + gz + p/d = \text{constant}$$
where v is the velocity of the fluid, g the gravitational acceleration, z the total vertical distance moved, p the pressure, and d the density. Bernoulli's theorem has been used to describe flows in rivers and in the Gulf Stream, and in meteorology to describe air flow over mountains. See also: *laminar flow*.

Bertrand lens Removable lens in a petrographic microscope, used with convergent light to form interference figures.

beryl A hexagonal mineral, $Be_3Al_2Si_6O_{18}$, from which beryllium is extracted; its principal occurrence is in granite pegmatites. Beryl is a major gemstone; when green, it is emerald, when blue or bluish-green, aquamarine, and when pink, morganite.

beta decay Radioactive decay due to the loss of a beta particle (an electron) from one of the neutrons in the nucleus, thus changing the neutron into a proton. The atomic mass remains the same, but the atomic number increases by one.

beta particle An electron that has been ejected from an atom. Beta particles, sometimes called *beta rays*, carry a negative unit charge and are moderately ionizing. They have a longer range than alpha particles and can penetrate light metals. See *beta decay*.

beta quartz Also spelled β-*quartz*. The polymorph of quartz that is stable from 573° to 870° C; it has a lower refractive index and birefringence than *alpha quartz*. It occurs as phenocrysts in quartz porphyries and granite pegmatites. Syn: *high quartz*.

beveling Erosional planing or shearing off of the outcropped edges of strata. Cf: *truncation*.

B horizon Approx. syn: *subsoil*. See *soil profile*.

biaxial Said of crystals having two optic axes, i.e., directions showing no double refraction of light. Crystals of the orthorhombic, monoclinic, and triclinic systems can exhibit this property. Cf: *uniaxial*.

bicarbonate A salt containing the HCO_3 radical; e.g., $NaHCO_3$.

bif *Banded iron formation*.

bight 1. A bend or curve in the shoreline of a sea or river. 2. A body of water formed by such a curve.

bilateral symmetry A pattern of symmetry of an organism in which the individual parts are arranged along opposite sides of a median axis, so that one and only one plane can divide the whole into essentially identical halves. Cf: *radial symmetry*.

binary system See *phase diagram*.

biochemical oxygen demand The amount of oxygen removed from aquatic environments by aerobic microorganisms. Abbrev: *BOD*. Syn: *biological oxygen demand*. Cf: *chemical oxygen demand*.

biochron The length of time represented by a *biozone*.

biochronology That branch of *geochronology* based on the relative dating of geological occurrence by biostratigraphic or paleontological evidence.

bioclastic rock 1. A biochemical sedimentary rock consisting of fragmental remains of organisms, such as limestone composed of shell fragments. Cf: *biogenic rock*. 2. A rock composed mostly of fragments detached from pre-existing rocks, or which are granulated by the action of living organisms, such as plant roots, earthworms, or man.

biocoenosis Etymol: Greek, *bios*, "mode of life" and *koinos*, "common, general." 1. An assemblage of fossil remains found in the same location where the animals lived. Syn: *life assemblage*. Cf: *thanatocoenoses*. 2. A group of organisms forming a natural ecological unit. Cf: *community*. Var: biocenosis; biocoenose; biocenose.

biodegradable Said of a substance, esp. *manmade*, that is subject to decomposition by microorganisms.

bioecology The branch of ecology dealing with the interactions between plants and animals in a shared environment.

biofacies A term connoting the biological features or fossil character of a stratigraphic facies. It is also used in the sense of a rock or sediment formation differentiated from surrounding bodies purely on the basis of its fossil content. An ecological association of fossils may also be called a biofacies. Syn: *biologic facies*.

biogenesis 1. Formation by the activity or processes of living organisms. 2. The tenet that all life must derive from previously living organisms.

biogenetic law The principle that ontogeny repeats phylogeny.

biogenic rock An *organic rock* that is produced directly by the life activities or processes of organisms, e.g., coral reefs, shelly limestone, pelagic ooze, coal. Cf: *bioclastic rock*.

biogeochemistry A branch of geochemistry that is concerned with the interrelationships between organisms and the distribution and fixation of chemical elements in the biosphere, esp. the soil. (Cf: *hydrogeochemistry; lithogeochemistry*.) A buried deposit may impart to the soil an abnormal amount of the metal it contains, and the soil, in turn, may provide some amount of the same metal to the plants above. Analysis of such plants for the purpose of detecting an orebody is called *biogeochemical prospecting*. Cf: *geobotanical prospecting*.

bioherm A moundlike mass of rock built by sedentary organisms such as colonial corals, stromatoporoids, or calcareous algae. Voids in the superstructure built by these organisms are filled with skeletons of other organisms as well as organic and inorganic detritus. Cf: *biostrome*.

biohorizon A surface of particular biostratigraphic character or showing some biostratigraphic change. Usually a biozone boundary. Theoretically, it is a surface or interface, but in practice it may be a very thin bed with clearly distinguishable biostratigraphic features. Cf: *chronohorizon; lithohorizon.*

biologic facies *Biofacies.*

biology The study of life and living matter, esp. with regard to origin, structure, growth, reproduction, and interrelationships. Although this includes *neontology* and *paleontology*, the term is most often used to imply neontology alone.

biomass The total mass of living organisms within a specified habitat expressed in terms of weight of organisms per unit area of habitat, or as the volume of organisms per unit volume of habitat.

biome 1. (Ecol.) A complex of communities characterized by distinctive vegetation and maintained within the limitations of climatic conditions of the region. 2. The largest land community region recognized by ecologists; e.g., savanna, tundra, desert, grassland.

biomechanical rock *Bioclastic rock.*

biometrics The statistical analysis of biologic observations and phenomena.

biomicrite A limestone composed of fossil skeletal remains and carbonate mud in varying proportions. When the term is used, the predominant organism should be specified; e.g., "gastropod biomicrite." Cf: *micrite.*

biophile An element that most typically occurs in organisms and organic material, or is concentrated in and by living plants and animals.

biosome A mass of sediment deposited under uniform biological conditions. It is the biostratigraphic equivalent of *lithosome*, not to be confused with *biostrome.*

biosparite A limestone composed of skeletal remains and clear calcite (spar) in variable proportions. When the term is used, the predominant organism is usually specified; e.g., "crinoidal biosparite." Cf: *sparite.*

biosphere 1. The zone at or near the earth's surface occupied by living organisms. It includes parts of the lithosphere, hydrosphere, and atmosphere. Cf: *ecosphere.* 2. A reference term for all living organisms of the earth.

biostratigraphic unit See *stratigraphic unit.*

biostratigraphic zone *Biozone.*

biostratigraphy Stratigraphy based on the paleontological features of rocks; also, the differentiation of rock units by means of the fossils they contain.

biostrome A clearly bedded, sheetlike mass of rock built by sedentary organisms and composed chiefly of their remains; e.g., a bed of shells. Not to be confused with *biosome.* Cf: *bioherm.*

biota The collective flora and fauna of a region.

biotic community *Community.*

biotite A mineral of the mica group, $K(Mg,Fe^{+2})_3(Al, Fe^{+3})Si_3O_{10}(OH)_2$, widely distributed in igneous rocks as black crystals. Biotite is black as an in-hand specimen, greenish brown in thin section, and has perfect basal cleavage.

biotope 1. An area characterized by uniform ecology and organic adaptation. 2. An environment in which an assemblage of flora and fauna lives or has lived.

bioturbation The stirring of a sediment by organisms, often destroying its structure. For example, the feeding by burrowing of many sea-floor organisms through freshly deposited sediment destroys bedding and lamination.

biozone A general term for any type of

biostratigraphic unit. The fundamental unit in biostratigraphic classification, and usually the smallest such unit on which intercontinental correlation can be established. In oceanography, a depth division within the ocean, characterized by particular conditions and organisms. Syn: *biostratigraphic zone.* Cf: *acme-zone; range-zone.*

bird-foot delta See *delta.*

birefringence The separation of an ordinary light beam by certain crystals into two unequally refracted beams. The measure of the birefringence of a crystal is the difference between the greatest and smallest refractive indices for light passing through it. Crystals having this property are said to be *birefringent,* since they have more than one *index of refraction.* The property is characteristic of crystals in which the velocity of light rays is not the same in all directions, i.e., all crystals but those of the isometric systems.

biscuit-board topography Glaciated landscape in which initial or incomplete erosion in cirques may leave gently rolling sections of the proglacial upland surface. The cirques on the side of the upland resemble bits removed by a biscuit cutter at the edge of the dough. An example may be found in the Presidential Upland in the White Mountains of New Hampshire.

bisector A line or plane of symmetry.

bisectrix An imaginary line bisecting either of the angles between the two optic axes of a biaxial crystal. The *acute bisectrix* bisects the acute angle, and the *obtuse bisectrix* bisects the obtuse angle.

bittern A bitter liquid remaining in salt-making after seawater or brine has evaporated and the salt has crystallized out; used as a source of bromides, iodides, and certain other salts.

bitumen Any of various solid and semisolid substances consisting mainly of hydrocarbons; e.g., petroleum, asphalt.

bituminous coal A dark brown to black coal, usually banded. It is the most abundant rank of coal, much of it dating to the Carboniferous, and is formed through the further coalification of lignite. Bituminous coal, which burns with a smoky, yellowish flame, has more than 14% volatiles on a dry, ash-free basis, and a calorific value of about 12,000 BTU/lb. (moist, mineral-matter free). Coals that are of gradations between lignite and bituminous are called *subbituminous.* Syn: *soft coal.* See also: *coal.*

bivalve An animal having a shell composed of two distinct parts that open and shut. Specif: a mollusk of the class Bivalvia (Pelecypoda). Clams, oysters, scallops, and mussels are examples. See also: *pelecypod.*

blackdamp A nonexplosive coal-mine gas consisting of about 85% nitrogen and 15% carbon dioxide. Because it causes choking, it is also called *chokedamp.* Cf: *firedamp.*

black diamond 1. *Carbonado.* 2. *Coal.* 3. A black gem diamond. 4. A dense, dark hematite that can be highly polished.

black earth *Chernozem soil.*

black granite A diabase, diorite, gabbro, or other rock that is dark gray to black when polished, and which is used as a commercial "granite."

blackjack A dark-colored, iron-rich variety of sphalerite.

black light An ultraviolet or infrared lamp used to detect mineral fluorescence.

black mud A marine-derived mud that acquires its color from black sulfides of iron and from organic matter. Black mud forms in bodies of water characterized by limited circulation and anaerobic conditions; e.g., lagoons, bays of water with weak tides, deep holes in neritic regions. The waters support few organisms. See also: *euxinic.*

black sand An alluvial or beach sand containing heavy, dark minerals (magnetite, ilmenite, rutile) that have been transported by rivers and concentrated by wave and wind action.

blanket deposit 1. A sedimentary deposit of relatively uniform thickness covering a large area. 2. A flat deposit of ore, relatively thin as compared with its length and breadth.

blanket sand A *blanket deposit* of sand or sandstone covering an unusually great area; typically deposited by a transgressive sea. Syn: *sheet sand; sand sheet.*

blastoid A class of echinoderms of Middle Ordivician to Late Permian occurrence, possibly evolved from cystoids. They were attached to the sea floor by a stalk of circular plates, surmounted by a calyx or cup, and commonly had fivefold radial symmetry. Some blastoids are useful as *index fossils.*

B layer See *seismic regions.*

bleaching clay Any clay having the capacity of removing coloring matter from oil through adsorption.

bleach spot A yellowish to greenish area in a red rock, formed by the chemical reduction of ferric oxide around an organic particle. Syn: *deoxidation sphere.*

bleb An informal term for a small, rounded inclusion of one mineral in another; e.g., a bleb of olivine poikilitically enclosed in pyroxene.

Bleiberg-type lead See *anomalous lead.*

blende *Sphalerite.*

blind valley See *karst.*

block 1. Angular rock fragment with a diameter greater than 256 mm. It shows little alteration as a result of transporting agents. Cf: *boulder.* 2. Angular

pyroclastic particle of diameter greater than 64 mm, ejected in a solid state. Cf: *volcanic bomb.* 3. A *fault block.*

block caving A method of mining in which blocks of ore are caused to cave in or collapse by undercutting and blasting away the supporting pillars.

block diagram A plane figure illustrating the three-dimensional geologic structure of an area. A rectangular block of the earth's crust is drawn in three-dimensional perspective to show a top surface and at least two vertical cross-sections. If an orthorhombic projection is drawn, each of the three sides of a rectangular block is foreshortened equally. In isometric block diagrams, equal parts of the three faces of a cube are seen.

block faulting Type of normal or gravity faulting in which the crust is separated into structural units (fault blocks) of different orientations and elevations. *Fault-block mountains* are formed by this process.

block field Accumulation of coarse detritus on level mountaintop surfaces or gently sloping mountain sides; also termed *felsenmeer* or *blockmeer.* Block fields consist mainly of local rocks, often angular, broken by frost shattering of the bedrock, the degree of angularity depending greatly on the rock type. The time required for the formation of block fields varies with rock type and climate, and especially with the number of times the freezing point is passed. Syn: *stone field.* Cf: *block stream.* See also: *frost action.*

block lava See *lava flow.*

blockmeer Another name for *block field.*

block mountain *Fault-block mountain.*

block stream Accumulation of rock

debris, usually at the head of a ravine and following valley floors. Block streams may begin from summit *block fields* extending downslope, or from large talus accumulations. They are regarded as being of periglacial origin, but the mechanism responsible for their formation and possible movement is still open to question. Syn: *rock stream.* Cf: *block field.*

block stripe   See *patterned ground.*

blocky lava   See *aa.*

bloom   1. An *efflorescence;* also the oxidized or decomposed part of a mineral vein or bed that has been exposed. 2. The sudden proliferation of organisms in bodies of fresh or marine water; e.g., *algal bloom; plankton bloom.*

blowhole   1. A vertical, cylindrical fissure in a sea cliff, extending from the bottom of the inner end of a *sea cave* upward to the surface. A blowhole is formed if a joint, which is a line of weakness, extends from the tunnel end to the top of the cliff. A fountainlike effect is produced as waves and tides force water and air into it. 2. An opening through a snow bridge into a crevasse, the presence of which is usually indicated by a current of air. 3. A small vent on the surface of a lava flow.

blowout   1. General term for various hollows formed by wind action in regions of sand or light soil. Also applied to the accumulation of a material derived from the hollow; e.g., a *blowout dune.* 2. The abrupt and uncontrollable escape of gas, oil, or water from a well. 3. The uncontrolled escape of oil and gas from an oil well, possibly because the *drilling mud* was not sufficiently dense.

blue asbestos   Crocidolite.

blue mud   See *terrigenous deposits.*

bluestone   Commercial name for a bluish or blue-gray argillaceous

sandstone used as building or paving stone.

blue vitriol   Chalcanthite.

bluff   A headland, hill, or cliff having a broad, steep face.

BM   Bench mark.

BOD   Biochemical oxygen demand.

body waves   See *seismic waves.*

boehmite   An orthorhombic mineral, AlO(OH), a dimorph of diaspore; its colors may be gray, brown, or reddish. It is an important constituent of some bauxites.

bog   A type of wetland characterized by spongy, poorly drained, peaty soil. Bogs are often divided into: 1. Those of cool regions, dominated by sphagnum and heather; bogs of boreal regions, e.g., Canada, with trees such as tamarack and black spruce on them, are called *muskegs.* 2. Fens (esp. England) are dominated by grasses, sedges, and reeds. 3. *Tropical tree bogs,* in which the peat is composed almost entirely of tree remains.

Bogs form in depressions created by glacial ice, and in small lakes in glaciated regions. Colonization by sphagnum and subsequent poor drainage contribute to a process that may eventually fill the body of water with vegetation. At the stage where surface vegetation is still floating and not coherent, the bog is called a *quaking bog* because of the surface instability.

Peat bogs are not generally found in lowland tropical areas because the high temperatures facilitate rapid decay of organic matter. Tropical peat bogs may develop, however, in areas of very high rainfall and with groundwater of very low mineral content. The peat from such bogs is composed of the remains of seed plants rather than sphagnum. Cf: *marsh; swamp.*

bog burst   The rupture of a bog that

has become swollen because of water retention by fallen vegetation. Conditions of great rainfall rapidly increase the pressure of the swelling, and the bog may "burst," emitting black organic matter that spreads over its surroundings.

boghead coal A nonlayered sapropelic coal similar to cannel coal in its physical properties, e.g., having a high content of volatiles, but composed mainly of algal matter rather than spores. Cf: *torbanite*.

bog iron ore A general term for a poor-quality iron ore found as layered or nodular deposits of hydrous iron oxides at the bottom of swamps and bogs. The deposits, consisting chiefly of goethite, are formed by precipitation of iron from the ambient water, and by the oxidizing action of iron bacteria and algae, or by atmospheric oxidation.

bog manganese *Wad.*

bolson A term applied in the desert areas of the southwestern U.S. and northern Mexico to an extensive basin more or less rimmed by mountains from which drainage flows into the depression. As a result of centripetal drainage, which is characteristic of a bolson, there is usually at or near its center, which is the lowest part, a level plain or *playa*.

bomb *Volcanic bomb.*

bonanza A miner's or mining term in the U.S. for an unusually rich body of ore. The word was taken from the Spanish bonanza, "a calm and fair sea," thus signifying good fortune.

bone beds Any sedimentary stratum, usually a thin bed of limestone, sandstone, or gravel, in which fossil bones, teeth, or scales are abundant. Bone beds suggest unusual features of deposition, such as mass extinction due to salinity changes, earthquakes, or outbursts (blooms) of certain organisms.

bone phosphate of lime Tricalcium phosphate, $Ca_3(PO_4)_2$.

book structure In ore deposits, a formation showing an alternation or interleaving with gangue.

boomer A marine seismic energy source in which capacitors are charged to high voltage and then discharged in a body of water through a transducer, a component of which consists of two metal plates. Abrupt separation of the plates produces a low-pressure region between them, into which water rushes, thus generating a pressure wave.

booming sand See *sounding sand*.

borate A mineral type containing the borate radical, $(BO_3)^{-3}$. Ex: borax, colemanite. Cf: *carbonate; nitrate*.

borax A mineral $Na_2B_4O_7 \cdot 10H_2O$, mined as an ore of boron. Found in arid regions, where it forms as an evaporite in playa deposits and as a surface efflorescence. In a dry environment its clear, transparent crystals lose water and become chalky white *tincalconite*.

bore 1. Abrupt and violent wall-like wave of water produced as an incoming tide rushes up a shallow, narrowing bay or estuary. Syn: *tidal bore*. A famous one occurs where the Bay of Fundy, located between New Brunswick and Nova Scotia, enters the St. John River. 2. Submarine sand ridge in very shallow water that may reach intertidal level. 3. Syn: *borehole* or boring.

boreal Pertaining to the north; northern. Cf: *austral*.

borehole Any exploratory hole, usually a deep hole of small diameter, made by drilling into the earth. The purpose can range from oil, mineral, or water prospecting to determining emplacement sites for dynamite. See also: *Mohole Project*.

bornhardt A large *inselberg.* Bornhardts are dome-shaped monoliths, examples of which are found on every continent but Antarctica. They are best developed in the tropics.

bornite Copper-iron sulfide, Cu_5FeS_4, an ore of copper, occasionally occurring with chalcopyrite and chalcocite. Bornite is a fairly common mineral, but crystals of it (isometric) are rare. Freshly exposed surfaces of the mineral show a bronzelike color that quickly tarnishes in air to a purplish iridescence, giving rise to the name "peacock ore."

bort See *industrial diamond.*

boss 1. A mass of plutonic igneous rock that occurs at the surface of the earth, and which may be knoblike, circular, or elliptical. Its contacts are vertical or steeply inclined. 2. A knoblike excrescence or protuberance on some organ of a plant or animal.

botryoidal Etymol: Greek, "bunch of grapes." Said of certain minerals, e.g., smithsonite, having the form of a bunch of grapes. Cf: *reniform.*

bottom ice Anchor ice.

bottom load Bed load.

bottomset bed See *delta.*

boudinage Etymol: Fr., *boudin,* "sausage." A structure found in greatly deformed sedimentary and metamorphic rocks, in which an originally continuous rigid (competent) band or layer between relatively plastic (less competent) layers is stretched and thinned until rupture occurs; the rigid band, thus separated into long cylindrical pieces, resembles a string of sausages. In the pinched parts, or points of separation, quartz or calcite is usually crystallized. Syn: *sausage structure.*

Bouguer anomaly See *gravity anomaly.*

boulder A separated rock mass larger than a cobble, having a diameter greater than 256 mm. It is rounded in form or shaped by abrasion. Boulders are the largest rock fragments recognized by sedimentologists.

boulder clay See *till.*

boulder fan See *indicator boulder.*

boulder pavement 1. A surface of till consisting largely of boulders, and worn down by glacier movement. 2. An accumulation of boulders, once part of a glacial moraine, remaining after finer materials have been removed by wind, waves, and currents. 3. A *desert pavement* composed of boulders.

boulder rampart A type of *beach-ridge system* composed of boulders thrown up along the seaward side of a reef.

boulder train See *indicator boulder.*

boundary monument A pile of stones or other objects placed at or near a boundary line to identify the location of the line.

boundary stratotype That point in a particular sequence of rock strata that is used as the standard for recognition and definition of a stratigraphic boundary. Cf: *stratotype.*

boundary waves A mode of wave propagation along the interface between media of different properties. Where the interface or boundary is a free surface, such as an earth-to-air surface, the mode is a *surface wave.*

Bowen's reaction series *Reaction series.*

box fold See *fold.*

boxwork A skeletal framework of intersecting plates composed of a hydrated iron oxide (any of several are possible) that has been deposited along fractures and in cavities from which intervening material has been dissolved by groundwater-related processes. Boxwork is common in the oxidized zone of sulfide ores (see gossan), and on the ceilings of caves. In such sites of occurrence, several different boxwork structures may exist separately or become joined together in a complex.

b.p.   *Before present.*

BPL   Bone phosphate of lime.

brachiopod   Any marine invertebrate of the phylum *Brachiopoda*. The shells are of unequal size, bilaterally symmetrical, and may be held together only by muscle or by muscle and hinge structure. Range, Lower Cambrian to present. Used as zonal indices in the Ordovician, Silurian, Carboniferous, and Cretaceous. Syn: *brach; lamp shell.*

brachygenesis   See *acceleration* (Biol.).

brackish   Term applied to water with salinity intermediate between that of normal fresh water and normal sea water.

bradygenesis   *Bradytely.*

bradytely   A slowing down in the development of a group of organisms, possibly causing particular members to lag behind the normal rate of progress in certain or all of their characteristics.

braided stream   A stream that divides into numerous channelways that branch, separate, and rejoin, thus becoming a tangle of channels, islands, and sandbars resembling a complex braid. Braided streams occur under conditions of abundant bed load and highly variable discharge. These conditions are provided where a river in a semi-arid region receives most of its discharge from precipitous headwaters, e.g., the Platte River, Nebraska. Braiding also occurs where glaciers are melting and the debris

supply is consistently large, whereas the discharge varies as the ice melts and refreezes. Examples of this type of braiding are found wherever melting glaciers occur; e.g., Iceland. Information concerning ancient environments is often based on recognition of braided stream deposits in older sediments. Syn: *anastomosing stream.*

branch fault   *Auxiliary fault.*

Bravais lattice   Syn. for *crystal lattice.*

breached anticline   An anticline whose center has been eroded to the extent that its limbs form erosional scarps that flank the breach and face upward. Cf: *bald-headed anticline.*

breadcrust bomb   A *volcanic bomb* with a cracked exterior resembling breadcrust. Its appearance is due to expansion of the interior after surface solidification.

break   1. A general term for a sudden change in a slope. 2. An abrupt change in the lithology or faunal content of a stratigraphic sequence. 3. *Arrival time.* 4. An irregular piece of land that gives the impression of being discontinuous, or interrupted; often used in the plural. See *breaks.*

breaker   A wave in which the velocity of the crest exceeds the rate at which the body of the wave is moving forward, thereby causing a curl to form at the top of the wave front, which plunges forward and collapses. Conditions for breakers (also called *breaking waves*) are set as waves decrease in velocity when moving into shallow water; as a consequence of this velocity change, there is a crowding together of the waves and a steepening of their fronts. *Breaker depth* is the still-water depth taken at the point of wave break. *Breaker height* is the average height from trough to crest of breaking waves. Following the breaking of a wave, a surge of water, or *swash*, rushes onto the shore. After the swash reaches its maximum forward position, the water returns down the seaward slope as *backwash*, leaving behind a thin line or ridge of debris (sand, seaweed) called a *swash mark*. See also: *surf*.

breaks 1. A term used esp. in the western U.S. for an area of land that is extensively cut by gullies or ravines. 2. Any sudden topographic change, such as from a plain to hilly country.

breccia Etymol: Ital., "broken stones." A coarse-grained (rudaceous) clastic rock composed of broken, angular rock fragments enclosed in a fine-grained matrix or held together by a mineral cement. Unlike *conglomerates*, in which the fragments are rounded, breccias consist of fragments that were not worn by abrasion prior to their embedment in a matrix. Sedimentary breccias are relatively rare, but may indicate terrestrial mudflows or submarine landslides. Tectonic or fault breccias are composed of fragments produced by rock fracturing during faulting or other crustal deformation. See also: *volcanic breccia*.

bridal-veil fall A cataract characterized by a small volume of water falling from such a great height that it becomes a veil-like mist or spray before reaching bottom level. Bridalveil Fall in Yosemite Valley, Calif., is the type example.

bridge See *natural bridges and arches*.

bright coal See *banded coal*.

bright spot In seismic prospecting, an unusually strong signal on a seismic profile, frequently indicating a pocket of natural gas.

brimstone Common or commercial name for sulfur.

brine Salt water, esp. a highly concentrated aqueous solution of common salt (NaCl). Natural brines occur in salt lakes, underground, or as seawater, and have commercial importance as sources of NaCl and other salts such as chlorides and sulfites of potassium and magnesium.

British thermal unit The amount of heat required to raise the temperature of one pound of water one Fahrenheit degree (usually from 60° to 61°). Equivalent to 252 calories. Abbrev: BTU.

brittleness A solid material that ruptures easily with little or no plastic flow is said to be characterized by brittleness. Cf: *ductility*. The adjective, *brittle*, is applied to a rock that fractures at less than 3 to 5% strain or deformation. Cf: *ductile*.

bromoform Tribromomethane, $CHBr_3$, sp. gr. 2.9. Used as a *heavy liquid*.

bronzite A brown or green orthopyroxene, intermediate in composition between enstatite and hypersthene. It contains iron and often has a distinctive, bronzelike luster.

brookite A brown or reddish lustrous orthorhombic mineral, TiO_2. It is trimorphous with anatase and rutile, and occurs as *druse* in veins and cavities.

brown coal A substance intermediate in composition between peat and lignite, in which the original plant structures can be seen. The term is generally used in Europe, where it is also used as a synonym for lignite. Cf: *lignite*.

brown iron ore *Limonite*.

brownstone A brown or reddish-brown sandstone, the grains of which are usually coated with iron oxide. The brownstone that was once quarried in the Connecticut River valley for use as building stone was a ferruginous quartz sandstone of Triassic age.

brucite A hexagonal mineral, $Mg(OH_2)$ that commonly occurs as thin plates and as fibrous masses in serpentine and metamorphosed dolomite.

Brunhes normal epoch See *constant polarity epoch.*

Brunton compass A pocket instrument consisting of a compass, mirror, folding open sights, and spirit-level clinometer. Used in preliminary topographic and geological surveys for leveling, reading horizontal and vertical angles, and for determining the magnetic bearing of a line. In field mapping of bedrock, it is used to determine strike and dip. Usually called a "Brunton." Syn: *pocket transit.*

bryophyte A small, nonvascular plant without true roots, but which may have differentiated stems and leaves. Mosses and liverworts are bryophytes. Cf: *thallophyte; pteridophyte.*

bryozoan Any aquatic invertebrate of the phylum Bryozoa, characterized mainly by colonial growth and an encrusting, branching, or fanlike structure. The organism's skeletal material may be calcium carbonate or chitin, but only the calcareous forms are found as fossils. Range, Ordovician to present; some are useful in strata correlation. Syn: *moss animal; polyzoan.*

B-tectonite See *tectonite.*

BTU *British thermal unit.*

B-type lead See *anomalous lead.*

bubble point The point on a diagram representing a state of equilibrium between a relatively large quantity of liquid and the last vestige (or bubble) of vapor.

bubble trend The distribution, either linear or planar, of bubbles in glacier ice.

Bubnoff unit A standard unit of measure of geologic time-distance rates, e.g., geologic movements and increments, defined as 1 micron/year (or 1 m/m.y.).

buffalo wallow One of the small, shallow depressions that were once found in great numbers on the Great Plains of the western U.S. Its diameter ranges between 15 and 20 m, and its depth from several centimeters to a few meters. It is generally believed that these depressions have been modified, and were perhaps initially formed, by the wallowing of buffalo herds in the mud and dust.

buhrstone Any of various siliceous rocks suitable for use as a millstone. Also, *burstone; burrstone.* Syn: *millstone.*

building stone General term for any type of rock used in construction. See also: *dimension stone.*

bulk density The density of an object or material, determination of which includes the volume of its pore spaces. Syn: *apparent density.*

bulk modulus A numerical constant that describes the elastic properties of a solid or fluid by relating the change in volume when the substance is subjected to pressure on all surfaces. Thus, bulk modulus = pressure/strain = $\frac{P}{V_o - V_n / V_o}$, where V_o, the original volume of a material, is reduced by an applied pressure P to a new volume V_n; the strain may be expressed as the change in volume, $V_o - V_n$, divided by the original volume, V_o. The bulk modulus is the reciprocal of compressibility. Syn: *volume elasticity; modulus of incompressibility.*

bullion 1. A carbonate or silica concretion found in some coals, often formed about a nucleus of preserved plant structure. Its diameter ranges from several centimeters to a meter. Cf: *coal ball.* 2. A nodule of some material (clay, shale, ironstone) that generally holds a fossil.

Burgess shale The site of an important discovery of fossil fauna in British Columbia. It is a bed of black shale of the Middle Cambrian. The fossils are preserved as carbon films on the *bedding planes* where bedding and cleavage coincide. All are the forms of soft-bodied animals. Among 70 genera and 130 species identified are sponges, jellyfish, annelid worms, and primitive arthropods. See *carbonization; fossil.*

buried hill   A hill covered with sediments that were deposited over resistant, older rock.

burrow   A pipelike cavity in sedimentary rock, made by an animal that lived in the soft sediment. Burrows, often filled with clay or sand, may be along the bedding plane or may penetrate the rock. The porosity in a sedimentary rock that results from animal activity is called *burrow porosity.*

bushveld   See *savanna.*

butane   Either of two odorless, colorless, inflammable paraffinic hydrocarbons, formula C_4H_{10}, that occur in natural gas and crude petroleum. The straight-chain compound is denoted *n*-butane; the branched-chain form is isobutane.

butte   Etymol: Fr., "hillock." A term used in the western U.S. and Canada for an isolated, flat-topped hill with steep slopes, representing an erosional remnant of horizontal sedimentary rocks. A butte resembles a *mesa,* but is smaller, with a more limited summit.

b.y.   *Billion years.*

bysmalith   A roughly vertical and cylindrical body of intrusive igneous rock, cross-cutting adjacent sediments. It is injected by pushing up the overlying strata, and is bounded by steep faults. It has been interpreted as a *laccolith* that has broken through to the surface along marginal faults. Examples are known in Arizona, Iceland, Yellowstone Park, and elsewhere.

bytownite   A bluish to dark-gray member of the plagioclase feldspar series, occurring in basic and ultrabasic igneous

rocks. It consists of 10 to 30% albite, with the remainder being anorthite.

C

cable-tool drilling   The standard and earlier method of drilling for oil. A chisel-like bit is suspended by a cable to a lever at the surface, and the lever alternately lifts and drops the bit, causing it to chip away the rock, which is removed by a bailer. Gushers are a frequent occurrence with this type of drilling. Cf: *rotary drilling.*

cafemic   A mnemonic term for an igneous rock or magma containing calcium, iron, and magnesium.

Cainozoic   *Cenozoic.*

calamine   1. A term used among mining people in the U.S. for *hemimorphite.* 2. In Great Britain, a former term for *smithsonite.* 3. A commercial name for oxidized ores of zinc, as distinguished from the sulfide ores. 4. *Hydrozincite.*

calamite   Giant scouring rush (sphenopsid) with a jointed, ribbed trunk and leaf whorls at the joints. Height about 12 m; Mississippian to Permian. See also: *fossil plants.*

calaverite   A tin-white or bronze-yellow monoclinic mineral, gold telluride, $AuTe_2$. It resembles silver more than gold, but yields a button of gold when the tellurium is burned off by heat.

calc-   A prefix indicating that the substance contains calcium carbonate.

calc-alkalic series   A rock suite comprised of those igneous rocks in which the percentage of silica by weight is between

53 and 61% when the weight percentages of CaO and of K_2O + Na_2O are equal. Cf: *calcic series*. See also: *silica concentration*.

calcarenite button   See *arenite*.

calcareous   Said of a substance containing calcium carbonate. When used with a rock name, it generally implies that as much as 50% of the rock is calcium carbonate.

calcareous algae   Certain algae that precipitate calcium carbonate to build up large globular or irregular masses of organic limestone. Such algae have formed characteristic limy deposits in all ages since early in the Precambrian.

calcareous ooze   See *pelagic deposits*.

calcareous tufa   *Tufa.*

calciclastic   Referring to a clastic carbonate rock.

calcic series   A rock series comprised of those igneous rocks in which the percentage of silica by weight is greater than 61% when the weight percentages of CaO and of K_2O + Na_2O are equal. Cf: *calc-alcalic series*. See also: *silica concentration*.

calcification   Replacement of the original hard parts of organisms by calcium carbonate ($CaCO_3$) in the form of calcite or aragonite. Calcification is common among corals, brachiopods, echinoderms, and some mollusks. The siliceous skeletons of some animals, certain sponges, and radiolarians may also be replaced by calcite. See *fossil; replacement*.

calcify   1. To make hard or strong by the deposition of calcium salts. 2. To become hard or stonelike because of the deposition of calcium salts.

calcilutite   See *lutite*.

calcimicrite   A limestone in which the micrite component is greater than the allochem component, and particles are less than 20 microns in diameter. See also: *micrite*.

calcination   1. The result of prolonged heating of a material at fairly high temperature. 2. The heating of ores to expel water and carbon dioxide.

calcirudite   See *rudite*.

calcisiltite   A sedimentary rock composed mostly of detrital, silt-sized calcite particles that have been consolidated. Cf: *calcilutite*.

calcite   A widely distributed rock-forming mineral, $CaCO_3$, the chief constituent of limestone and most marble, and common in the shells of invertebrates. Its crystals, which have perfect rhombohedral cleavage, are usually white or gray; the transparent variety is *Iceland spar*. It is polymorphous with aragonite, vaterite, and some other exotic forms. Calcite is the stable form of calcium carbonate at all those temperatures and pressures obtaining at or near earth's surface; because of this, it is possible that all other forms convert to it over geologic time.

calcium bentonite   See *bentonite*.

calcium carbonate   A solid, $CaCO_3$, very abundant in nature, and occurring chiefly as chalk, limestone, calcite, and aragonite.

calcrete   1. A limestone precipitated as surface or near-surface crusts and nodules by the evaporation of soil moisture in semi-arid climates; may be combined with sand and gravel. Syn: *caliche*. 2. A calcareous *duricrust*. Cf: *silcrete*.

calc-schist   A metamorphosed, argillaceous limestone showing a schistose structure.

calc-silicate rock   A type of

metamorphic rock in which calcium and silicon are the dominant constituents, and which is derived from quartz-bearing dolomites and limestones, or from carbonate rock metasomatized by siliceous solutions from abutting granitic intrusions. Examples include epidote, calcic plagioclase, and sphene.

calc-sinter   *Travertine.*

caldera   A very large, bowl-shaped volcanic depression whose horizontal dimension is much greater than its vertical dimension. Usually formed by a combination of the explosion and collapse of the top of a volcanic cone or group of cones. Such subsidence of the roof can be caused by displacement of the underlying body of magma. The collapse is often that of a composite cone that rapidly depleted the underlying magma reservoir by great eruptions of pumice. The caldera may later fill up with water; e.g., Crater Lake in Oregon. Other "super depressions" larger than calderas, are found in volcanic areas. However, their angular outlines and angular indentations at the edges indicate that their structure is affected by joints and faults in the underlying older rocks. Therefore, these structures are referred to as *volcanic-tectonic depressions*. The rapid extrusion of large amounts of lava seem at least partly responsible for their collapse. See *volcanic crater; volcano.*

Caledonian orogeny   A name commonly used for the Lower Paleozoic deformation in northwest Europe, extending from The Middle Ordovician to Middle Devonian. It corresponds to the *Taconian* plus *Acadian* disturbances of North America, and produced a mountain range extending from southern Wales, across Scotland, and northeastward through Scandinavia. In Scandinavia the pre-Devonian formations were folded and overthrust with eastward movement. The mountains in Scotland and Wales, although seemingly parallel to those in Norway, show thrusts to the west. While western Europe, in late Silurian time, experienced

widespread orogeny, North America (excepting eastern Greenland) was stable. Plate tectonic theory indicates that the collision of Europe with North America was responsible for the Caledonian orogeny.

Caledonides   The orogenic belt formed by the Caledonian orogeny. It extends from southern Wales across Scotland, and northeastward through Scandinavia.

caliche   1. A layer, chiefly of calcium carbonate at or near ground surface, in certain areas of scant rainfall. It is formed when groundwater containing dissolved calcium carbonate moves upward and evaporates, leaving a crust. 2. Alluvium cemented with sodium salts in the nitrate deposits of northern Chile. 3. In the southwestern U.S. and Mexico, a quarried aggregate of desert debris cemented by porous calcium carbonate; used as a road material. See also: *duricrust; calcrete.*

calorific value   The number of heat units obtained by the complete combination of a unit mass of fuel. The numerical value obtained depends on the units, e.g., for solid and liquid fuels, BTU/lb. or MJ/kg; for gaseous fuels, $BTU/ft.^3$ or MJ/m^3.

calving   The breaking off of large masses of a glacier as it moves into the sea. See *iceberg.*

camber   A superficial feature observed in areas near horizontal and relatively thin rocks that comprise a competent bed and are underlain by weaker strata. The weaker strata will sometimes show a general valleyward increase of dip, to which the competent bed adjusts by sagging. This can be seen along the edges of the outcrop.

Cambrian System   The oldest period of the Paleozoic Era, having a duration of about 70 m.y. and beginning about 570 m.y.a. The Cambrian System is the oldest in which fossil remains are sufficiently

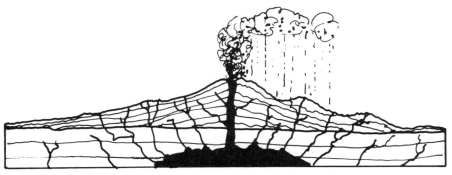

Magma stands high in volcano.

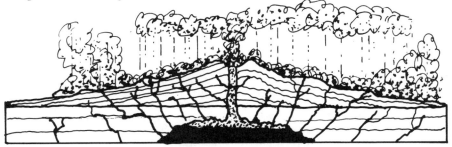

Magma sinks as eruptions of pumice increase.

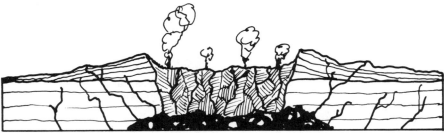

Volcano collapses into its chamber.

Caldera fills with water and Wizard Island arises.

Formation of a caldera

abundant and distinct, because of hard parts, so as to provide reliable geological information. Since the oldest rocks of this series occur in Northern Wales, the series was named "Cambrian," after the ancient English name for Wales. Southern Scandinavia, however, serves as the Cambrian-referred section because northern Wales is poorly fossiliferous. The most complete Cambrian sedimentation and faunal developments are in western North America and Siberia.

Cambrian rocks occur on all continents, the most extensive being those of North America, in the United States. Most are found in varying states of faulting, folding, and metamorphism, and are rich sources of many minerals, e.g., zinc, lead, gold, and silver. The presence of fossil marine organisms in most of the rocks attests to the Cambrian submergence, when fully 30% of North America was submerged.

Cambrian marine sediments are usually grouped into three major facies referred to as *terrigenous*, shallow-water calcareous sediments, and deep-water deposits. Most marine invertebrate phyla were represented in the earliest Cambrian. Hydrozoans, foraminifers, and corals were absent, as were any terrestrial or freshwater organisms. *Trilobites* are the most sufficiently abundant for stratigraphic purposes, with *brachiopods* second in use as index fossils. Graptolites and conodonts are found under special conditions in Middle and Late Cambrian rocks. Marine plants were mainly blue-green, green, or red algae that formed massive calcareous structures. Because most Cambrian regions have significant carbonate accumulations, which proliferate only in temperature regions, it is thought that the Cambrian climate was mild.

The Cambrian Period shows only few records of active volcanism; these are found in northeastern China, parts of the U.S.S.R., and the extreme eastern United States. Evidence of *plate tectonics* indicates that, at least in the Late Cambrian, the relation of Cambrian continents to each other was very different from the present one. A threefold division of the Cambrian System is followed in all parts of the world, though local names differ. In North America, the Lower, Middle, and Upper Cambrian are the *Waucoban, Albertan, and Croixan.*

camptonite A *lamprophyre* composed mainly of plagioclase (usually labradorite) and hornblende (usually barkenite).

Canadian The Lower Ordovician of North America.

Canadian Shield The largest and most studied Precambrian rock area of the world. It occupies about 5,000,000 km^2, covering a large part of northern Canada and extending down into the Lake Superior region of the United States. It is rich in deposits of iron, nickel, copper, silver, and gold, and is extremely complex in structure. About three-fourths of the shield is formed of banded gneiss and granite, the rest consisting of small sections of sedimentary and volcanic rock. Since the sedimentary rock patches are isolated and lacking in sufficient fossil information, their relative ages can be judged only by physical appearance or radioactive dating. See *shield.*

canal 1. An artificial waterway cut through a tract of land for navigation, irrigation, water supply, or drainage. 2. A long narrow arm of a body of water connecting two larger water areas. 3. A partially water-filled passage within a cave.

cannel coal Variety of bituminous coal of fine-grained uniform texture; a common fireplace fuel. It is composed mainly of microspores, and probably developed in lakes where floating spores were deposited. It has a dull or waxy luster and conchoidal fracture. Because of its high content of volatiles, it ignites easily and burns with a bright, smoky flame, the feature responsible for its early name—candle coal. Cannel coal grades into *torbanite* as its petroleum content decreases.

canyon   Also cañon. A deep, steep-walled gorge cut by a river or stream, generally into bedrock. Canyons are most frequently found in arid or semi-arid regions where the effect of stream action greatly outweighs weathering. See also: *submarine canyon*.

capable fault   Three criteria are used by the U.S. Nuclear Regulatory Commission to define a *capable fault*: 1. Movement at or near the ground surface at least once within the last 35,000 years, or recurring movement within the past 500,000 years. 2. Macroseismicity, instrumentally determined with sufficient precision to show a direct relationship with the fault. 3. Existence of a structural relationship with a capable fault (as def. by 1 and 2), such that movement on one could be reasonably expected to be accompanied by movement on the other. The definition was formulated for use in the siting of nuclear power plants. Cf: *active fault*.

capacity   1. The ability of a stream or wind current to transport debris, as gauged by quantity per unit time at a given point. Stream capacity increases with an increase in slope and discharge, and a decrease in width. Cf: *competence*. 2. The yield of a well, reservoir, or pump. 3. Ability of a particular soil to retain water.

cape   A relatively large section of land jutting into the sea or some other large body of water from a larger land mass; e.g., Cape of Good Hope.

capillarity   A manifestation of surface tension by which the portion of the surface of a liquid coming in contact with a solid is depressed or elevated, depending upon the cohesive or adhesive properties of the liquid. It is illustrated by the rising of water in very fine tubes or the drawing up of water between grains of rock.

capillarity conductivity   The ability of an unsaturated soil or rock to transmit fluid. When water is the fluid, the conductivity, which is zero for dry material, increases with moisture content to the maximum value, which is equal to the permeability coefficient.

capillary   1. A mineral *habit* composed of flexible, threadlike crystals; e.g., millerite and the chalcotrichite variety of cuprite. 2. Said of tubes with such small openings that fluids can be retained in them by capillarity.

capillary fringe   A lower portion of the *zone of aeration* immediately above the water table, in which interstices are filled with water. Although the water is at less than atmospheric pressure, being part of the water below the water table, it is held above that level by surface tension. Syn: *zone of capillarity*.

capillary water   Water of the capillary fringe.

cap rock   A hard covering atop the salt covering in a salt dome, consisting mainly of gypsum and anhydrite, some calcite, and sulfur in varying amounts. Large sulfur deposits occur in the cap rock of certain salt domes in the Gulf Coast area of the United States. Experiments imply that cap rock sulfur was formed through the action of *sulfur bacteria*.

capture   Syn: *stream capture*.

carat   A unit of weight for gemstones. The metric carat equals 200 mg. Abbrev: c; ct. Cf: *karat*.

carbide   Any one of a class of chemical compounds in which carbon is combined with a metallic or semimetallic element. As based on their atomic structures, carbides may be saltlike, metallic, or diamantine.

carbon-14   A radioactive isotope of carbon, ^{14}C, having a mass number of 14 and a half-life of 5730 ± 40 years. It occurs in nature as the result of reaction between atmospheric nitrogen, ^{14}N, and neutrons produced by cosmic-ray collisions. Carbon-14 is used in dating substances directly or indirectly associated with the carbon cycle, provided they are not over 50,000 years old. Partial syn: *radiocarbon*.

carbon-14 dating See *dating methods*.

carbonaceous Said of a rock sediment or other material that consists of or contains carbon, or is similar to it in some aspect.

carbonaceous chondrites Stony meteorites, containing material (e.g., amino acids, hydrocarbons) possibly associated with life. Some researchers suppose an extraterrestrial biological origin. Carbonaceous chondrites contain claylike hydrous silicate minerals, instead of the anhydrous silicates found in most chondrites.

carbonado See *industrial diamond*.

carbonate 1. A mineral type containing the carbonate radical, $(CO_3)^{-2}$. Calcite, aragonite, and dolomite represent three groups of carbonate minerals. Cf: *borate; nitrate*. 2. A sediment composed of calcium, magnesium, and/or iron; e.g., limestone, dolomite.

carbonatite A rock of apparently magmatic origin consisting chiefly of the carbonates of Ca, Mg, and Na, with subordinate feldspar, pyroxene, olivine, and other materials.

carbonation 1. A process of chemical weathering during which minerals containing calcium, iron, magnesium, potassium, and sodium are transformed into carbonates of these metals by carbon dioxide. Syn: *carbonization*. 2. Saturation of a fluid with carbon dioxide.

carbon dioxide A colorless, odorless gas, CO_2. Over 99% of terrestrial CO_2 is contained in the oceans, and is exchanged between ocean water and the atmosphere in a continuous process.

Carboniferous The time interval between 345 and 280 million y.b.p. It was defined by Conybeare in 1822 for type strata in Wales, which contained great coal deposits, and the limestone on which the coal had formed. The period is divided into the Lower and Upper Carboniferous, called the Mississippian and Pennsylvanian Periods in North America. The division is based on the presence of coal measures in the upper strata and barren rock beneath. (See *Mississippian-Pennsylvanian boundary*).

Lower Carboniferous denotes different periods in America, Europe, and the Soviet Union. Although the bottom limits are approximately contemporaneous, the upper limits differ. Terrestrial deposits are rare in the Lower Carboniferous. Rocks are predominantly marine sediments of two basic types: limestone strata, much of it fossiliferous, and clastic rocks such as shales and sandstones. Igneous and metamorphic rocks are nowhere dominant. The limestones were thickly and uniformly deposited in the stable shelf regions by shallow shelf seas (ex: Central North America). Orogenic regions were sites of clastic deposits less uniform than the limestone series. Coal is found in some Lower Carboniferous strata, and limestones are quarried on all continents. Oil shales and petroleum are important in various localities.

Flora that had appeared during the Devonian became plentiful and varied during the Lower Carboniferous. Although it is widely believed that the Devonian forests were populated by some true trees, the first indisputable evidence of trees with woody trunks occurs in the Lower Carboniferous. Psilophytes (primitive land plants without leaves) die out in the Carboniferous. *Lycopods* (lower vascular plants) comprise the "coal forests" of the Lower Carboniferous. The first seed plants appear, and gymnosperms emerge as trees for the first time; lime-depositing algae are the most significant algal type. *Goniatites* (cephalopods having characteristic sutures) appear initially.

Brachiopods decline in number and forms; only the productids (spiny brachiopods) expand during this period, and are therefore useful as an index fossil. Crinoids and blastoids (Echinoderms) expand into a great diversity, especially on the North American mid-continent shelf. Crossopterygian-like amphibians inhabit the swamp forests.

The clastic facies of the Lower Carboniferous contains much sparser faunal representation than does the limestone facies. This is due to the low oxygen supply at those depths where the clastic sediments are deposited. Extensive orogenic activity that began near the close of the Lower Carboniferous formed the *Variscan* (Hercynian) range in Europe. In North America, the *Appalachian orogeny* effected a sea withdrawal from the continents. As indicated by the contemporaneity of Lower Carboniferous rocks, the continents were probably in close proximity at this time. In Europe, much of England, Belgium, and Northern France were submerged during the entire period; in North America, the sea covered most of the Mississippi Valley.

Upper Carboniferous is the time interval from about 315 to 280 million y.b.p. The base of the Upper Carboniferous is generally defined by the first occurrence of a particular cephalopod series; there is no generally accepted upper limit definition. Late Carboniferous sediments formed in basins created by orogenic episodes of Lower Carboniferous time. These developments coincided with a relatively abrupt evolution of land plants and accumulations of peats; following decomposition, the compression and petrifaction of the peats produced coals. Gold, silver, lead, copper, zinc, and other metallic ores also occur in these strata, as well as natural gas and oil reserves. There is evidence that the present continental masses formed two great supercontinents during the Upper Carboniferous: *Gondwanaland*, the southern land mass, and *Laurasia*, the northern mass. Within this geographic framework, the Upper

Carboniferous sediments accumulated as marginal deposits in shallow areas along the supercontinental blocks. These are represented by graywackes, limestones, and evaporites. Granites and metamorphic rocks, products of earlier action in the Variscan and Appalachian orogenies, are also found in Upper Carboniferous strata. In Britain and other parts of Europe, the Upper Carboniferous differs from the lower in that freshwater sediments generally predominate and marine layers are only occasional. Columnar sections of the *Pennsylvanian System* in the eastern United States show repeated alternations of sandstone, shale, and coal with thin, interlayered marine limestone. The presence of the marine limestone denotes repeated short incursions of the sea. See *cyclothem*.

Fusulinids (relatively large protozoans) appear for the first time during the Upper Carboniferous; goniatites developed steadily. Both invertebrates are used as Upper Carboniferous *index fossils*. Larger fauna include sharklike fish and large amphibians, and the first reptiles emerge in Late Carboniferous time. Freshwater life (clams, and other shellfish) abounds during this period. Four hundred forms of insects, mainly primitive, are known from the lower and middle Late Carboniferous. More than 20 of these, including cockroaches, exceeded lengths of 10 cm; one dragonfly-like insect had a wingspread of 73 cm. Algae comprised much of certain Upper Carboniferous limestones. Ferns were plentiful and of various kinds; seed ferns were even more common at this time. Other flora included calamites (scouring rushes), *Lepidodendron* and *Sigillaria* (two scale trees), and *cordaites* (forerunners of modern conifers). Although a warm, moist climate is indicated by the abundant Upper Carboniferous plant growth, toward the end of the period, widespread glaciation became established in the southern hemisphere, near the present equator. This climate change was preceded by a rather abrupt change of the magnetic poles.

carbonization The decomposition of organic matter under water or sediment, during which the hydrogen, oxygen, and nitrogen in the original tissues are lost by distillation, so that only a thin film of carbon remains; this residue may retain many features of the original organism. Plants, arthropods and fish have been so preserved. Animals that lack preservable hard parts, e.g., sponges, jellyfish, and annelid worms, may also be preserved in this way. Perhaps the best known and most important carbonized remains are those of the *Burgess Shale* fauna in western Canada. See *fossil.*

carbon monoxide A colorless, odorless, poisonous gas, CO, produced when carbon burns with insufficient air—such as during the incomplete burning of fossil fuels.

carbon ratio 1. Ratio of the fixed carbon content in a coal to the fixed carbon plus volatile matter. 2. Ratio of ^{12}C, the most common carbon isotope, to either ^{13}C or ^{14}C. If unspecified, the term is usually $^{12}C/^{13}C$.

Carborundum Trade name for a manufactured substance (silicon carbide) used as an abrasive.

Carlsbad twin See *twinning.*

carnallite A white orthorhombic mineral, $KMgCl_3 \cdot 6H_2O$, occurring in marine salt deposits, apparently as an alteration product of pre-existing salts. It is a source of potassium for fertilizers.

carnelian Also called *cornelian.* A translucent, reddish semiprecious variety of *chalcedony*; as it becomes brownish, it grades into sard. The color is due to colloidally dispersed hematite.

carnotite A highly radioactive, bright-yellow mineral ore of vanadium and uranium: $K_2(UO_2)_2(VO_4)_2 \cdot 3H_2O$. It is of secondary origin, occurring as loosely coherent masses in sandstone, or as an earthy powder, particularly around petrified wood.

Carolina bays Elliptical-shaped shallow depressions that occur in particular abundance on the coastal plain of the Carolinas, but extend from Florida to New Jersey. All the bays are on terraces of Pleistocene age. The deepest part of a bay is generally at its southeast end, and west of the axial line. The origin of the bays has been attributed to meteorites, shock waves, or the rising of artesian springs through the coastal-plain sediments. Another hypothesis contends that they are basins of extinct lakes.

Carrara marble A general name for marbles quarried in the vicinity of Carrara, Italy. Several pure varieties are found there, including fine-grade statuary marble. Its colors are mostly white to bluish or blue-veined. See also: *marble.*

cartography The art and science of constructing maps or charts. It may involve the superimposition of non-geographical divisions (political, cultural) onto the representation of a geographic area.

cascade 1. A series of shallow, closely spaced waterfalls descending over a steep, rocky surface. 2. A series of descending recumbent folds associated with gravity sliding. They are found when a bed or an inclined surface buckles as it slides downward under the effect of gravity.

Cascadian orogeny A time of regional mountain building and deformation that was especially effective in the Miocene and Pliocene, ending in the Late Pleistocene. It was centralized in the Pacific Northwest as a broad upwarp of the Rocky Mountain regions, which had been *peneplained* by Late Oligocene time. See also: *Laramide orogeny.*

casing-head gas An unprocessed natural gas produced from an oil-containing reservoir. It is so named because it is most often produced under low-pressure conditions through the casing-head of an oil well.

cassiterite   A red, brown, or black tetragonal mineral, SnO_2, the major ore of tin. Significant deposits occur in high-temperature hydrothermal veins and in placers, as well as in granite and pegmatites.

cast   A positive representation of organic remains, derived from different types of molds or the impressions of original forms. It results when any substance, artificial or mineral, fills the void created by the dissolution of the original hard parts of an original form. (Cf: *replacement*.) A natural cast is produced when a natural mold is filled with mineral substance while still in the embedding rock. Numerous fossils are preserved in this manner. It is especially characteristic in the preservation of fossil clams and snails, since their aragonitic shells are easily dissolved. See *fossil; mold*.

cataclasis   A type of metamorphism in which rock deformation is produced by the crushing or shattering of brittle rock, e.g., along faults or fault zones, to the extent that the mineral composition and texture of the original rock are still recognizable. A metamorphic rock produced by such a process is a *cataclasite*; e.g., a *tectonic breccia*. A structure or a material that exhibits the effects of severe mechanical stress during metamorphism is described as *cataclastic*; such features include bending, breaking, and cracking of the minerals. See also: *mortar structure*.

cataclysm   Any geological occurrence that effects sudden and widespread changes of the earth's surface; e.g., an earthquake or volcanic eruption of exceptional magnitude. adj. *Cataclysmal; cataclysmic*. Cf: *catastrophe*.

catanorm   See *depth zone of emplacement*.

cataract   A synonym of *waterfall*, the term implies a large volume of water falling over a long, vertical unbroken drop. Although it is differentiated from *cascade*

(a series of small falls) and from *rapids*, which are not as steep, the term *cataract* is used for series of steep rapids in large rivers; e.g., the Nile.

catastrophe   A sudden and violent disturbance of nature ascribed to unusual or supernatural agents, and affecting both physical conditions and inhabitants of the earth. Cf: *cataclysm*.

catastrophism   An hypothesis that proffered recurrent, violent, worldwide events as the reason for the sudden disappearance of some species and the abrupt rise of new ones. Cuvier, a French biologist, supported the doctrine, explaining most extinctions and landscape changes as consequences of the Biblical deluge. Although catastrophism had wide acceptance in the scientific community, it was short-lived and was supplanted by Hutton's principle of *uniformitarianism*.

catazone   See *depth zone of emplacement*.

catchment area   1. The waterproofed area of a storage reservoir. 2. The *recharge* area and all other areas that contribute water to an aquifer. 3. Drainage basin.

catchment basin   *Drainage basin*.

catena   Etymol: Latin, "a chain." A soil catena is a soil series consisting of a group of soils that developed from similar parent material, but show different profiles because of variations in topography and drainage conditions. Inasmuch as each topographic setting affects the nature of the soil profile, a soil catena holds both lithologic and topographic implications.

cation   A positively charged ion that moves toward the negative electrode (cathode) during electrolysis. Cf: *anion*.

cation exchange   *Base exchange*.

catoctin   A residual hill, knob, or ridge of resistant rock material rising above a peneplained surface and maintaining on its summit a remnant of a still older peneplain. Named after Catoctin Mountain, Maryland and Virginia. Cf: *Monadnock.*

cat's-eye Any of several gemstones displaying *chatoyancy.* Precious, or oriental, cat's-eye is a greenish variety of chrysoberyl; quartz cat's-eye, the most common type, is often called *occidental cat's-eye.*

cauldron subsidence A geological structure formed by the lowering of a large segment of *country rock* into an intruding magma. If the magma breaks loose great portions of the overlying crust, and if these portions extend upward to the earth's surface, the blocks may sink in a pool of magma. Such large-scale subsidence (cf: *stoping*) produces catastrophic eruption of volcanic rocks. Shallow, intrusive bodies formed in this way resemble rings. Successive intrusions, usually younger toward the center, build up roughly circular exposures called *ring complexes* or *ring structures.*

cave A natural cavity or system of chambers beneath the surface of the earth, large enough for a person to enter. If a group of caves are connected, or the same underground river or stream flows through a cave group, it is referred to as a *cave system* or *cavern system.* Caves are formed by the action of waves against cliffs, or may be cut into glacier bottoms by meltwater streams. They may also be formed in lava. (See *lava tunnel.*) The largest caves and caverns, however, are formed mostly in limestone and dolomite. See also: *karst.*

cave breccia Angular fragments of limestone that have broken off the walls and roof of a cave, and are cemented with calcium carbonate. See also: *collapse breccia; solution breccia.*

cave coral Coral-shaped cave deposit of calcite.

cave draperies Sheetlike cave deposits, usually of calcite, formed by water trickling along a slanted ceiling.

cave marble *Cave onyx.*

cave onyx A compact, banded cave deposit of calcite or aragonite. It resembles true onyx and can take a high polish. Syn: *cave marble.* Cf: *onyx.* See also: *dripstone; flowstone; onyx marble; travertine.*

cave pearls Small, spherical concretions of calcite or aragonite formed by the accretion of layers around a nucleus, such as a grain of sand.

caver A person who explores caves as a hobby. Usually synonymous with *spelunker,* although cave explorers tend to avoid the latter term. See *speleologist.*

cavern A synonym of *cave,* esp. one of large size. May also be used for a cave system.

cavernous Characterized by caverns, cells, or large pore spaces, as in certain volcanic rocks and limestones.

c-axis 1. A vertically oriented crystallographic axis used for reference in crystal descriptions. 2. In deformed rocks, such as in simple shear, the c axis lies in the symmetry plane, normal to the plane of movement. See also: *a axis; b axis.*

cay Etymol: Span., cayo, "shoal" or "reef." A small insular bank of sand, rock, mud, or coral, or a range of low-lying reefs or rocks. The term cay for such features is used in the West Indies. Cf: *key.*

Cayugan Upper Silurian of North America.

celestite An orthorhombic mineral, $SrSO_4$, of the barite group, a source of strontium.

Celsius scale A thermometric scale,

formerly called *centigrade scale*, based on 0 degrees as the melting point of ice and 100 degrees as the boiling point of water. The formula $C = 5/9(F-32)$ is used to convert Fahrenheit to Celsius.

cement   1. A manufactured, finely ground powder that, when mixed with water, sets and hardens. Widely used in building and engineering construction. See also: *portland cement; concrete.* 2. Chemically precipitated mineral matter that is part of the *cementation* process. 3. Ore minerals that have replaced, or are part of, mineral cement.

cementation   The process by which clastic sediments are converted into sedimentary rock by precipitation of a mineral *cement* between the sediment grains, forming an integral part of the rock. Silicon is the most common cement, but calcite and other carbonates, as well as iron oxides, also undergo the process. It is not clear how and when the cement is deposited; a part seems to originate within the formation itself, and another part seems to be imported from outside by circulating waters.

cement rock   Any rock containing those ingredients that allow the processed rock to furnish cement, with little or no addition of other material. These basic ingredients are: calcium carbonates (found in limestone), silica, alumina, and iron oxide (found in clay).

cenote   Etymol: Mayan, *dz'onot.* A natural well or reservoir, common in Yucatan, Mexico, formed by the collapse of a limestone surface, exposing water underneath. In ancient Yucatan, precious objects were thrown into cenotes as offerings to the rain gods.

Cenozoic Era   (Cainozoic, Kainozoic) The Cenozoic ("recent life") Era covers the earth's history during the last 70 m.y. It is subdivided into the Tertiary and Quaternary (last 2.5 m.y.) periods. The early Tertiary includes the Paleocene, Eocene, and Oligocene epochs. The Quaternary includes the Pleistocene and Holocene (last 10,000 years) periods. The term "Tertiary" is used for the younger rocks that rest on "Secondary" or Mesozoic strata, and was once used as a synonym for the entire Cenozoic Era. Most Cenozoic formations have not been deeply buried and consist chiefly of thick deposits of marine and terrestrial sedimentary rocks. In North America, tertiary marine strata occur in narrow belts along the Atlantic, Pacific, and Gulf coastal regions; non-marine strata, are found in the western United States and the Great Plains. Tertiary volcanic rocks are exposed in western and northwestern North America. Marine terraces of Quaternary age are found along the Atlantic and Gulf coastal regions of North America.

The Laramide orogeny, typically recorded in the Rocky Mountains of the U.S., began in the Late Cretaceous and continued into the Eocene. The Sierra Nevada and Coast Ranges of British Columbia were the sites of recurrent orogenic activity that began in the Late Jurassic. Tertiary tectonic activity gave rise to the Alps, Andes, and Himalayas. The most intensive Himalayan orogeny occurred in the middle Miocene, but tectonic activity continued until the mid-Pleistocene of the Quaternary Period.

The northern Rocky Mountains experienced active volcanism during the Early Tertiary; in the southern Rockies, volcanism was more common during the Middle (Oligocene) and Late Tertiary. In Central America, volcanism began near the end of the Oligocene and has continued. Volcanism in South America is restricted to the Andean Belt, where most activity occurred during the Middle and Upper Tertiary. Tertiary volcanic belts extended from Greenland across Iceland to Scotland. Volcanic activity and crustal movement continued into the Quaternary.

Evidence favoring continental drift indicates that present continental positions had been almost established by the onset of the Tertiary, although the Atlantic Ocean was still spreading and continents still diverging from the mid-Atlantic Ridge.

The Tethys Sea was closed and the Himalayas were formed as a result of the northward movement of India. In the North Pacific region, there was no seaway through the Bering Strait area to the Arctic until the end of the Miocene, when it opened briefly, and then was closed again until the end of the Pliocene. The Cenozoic was characterized by a general global cooling, interrupted by periods of warming. The mild climate of the Paleocene and Eocene epochs became cooler with the onset of a cooling trend in the Oligocene. A brief warming interval in the Middle Miocene was followed by increasing cold until the onset of continental glaciation in the Pleistocene, more than 2 m.y.a. Antarctica, however, was already glaciated as early as Miocene time, glaciation having begun there about 20 m.y.a. During the Pleistocene Epoch, called the "Great Ice Age," glaciers and ice sheets covered the land masses four times. Longer warm interglacial periods separated these massive ice advances. See *Ice Age.*

Grasses appear in the Eocene, and Compositae in the Paleocene; these plants and Leguminosae undergo rapid evolution, diversification, and dispersal during the Tertiary. Orchidaceae also arise in the Tertiary. Angiosperms, which appeared in the Cretaceous, replace the gymnosperms. Eocene floras from Alaska and Greenland reach North America during the Miocene. (Temperature decrease during the Quaternary fixed the modern zones of vegetation.)

The *Foraminifera* evolve during the Cenozoic, reaching a maximum in the mid-Eocene; several attain a large size (e.g., *nummulites*, 2.5 to 5.0 cm). Arthropods become more significant throughout the Cenozoic, and several groups of land snails evolve during the Tertiary. (Most early Tertiary genera of gastropods are now extinct, the majority of living species having originated in the Middle Tertiary.) Turtles, snakes, and lizards continue into the Cenozoic. Teleost diversity increases, and most living groups of birds appear to have evolved in the Early Tertiary. Mammalian

evolution is the most notable biologic occurrence of the Tertiary, represented by the hoofed animals, carnivores, primates, and Australian marsupials. Some evolutionary branches (seals and whales) return to the sea. The Tertiary-Quaternary boundary is partially defined by *Equus, Elephas,* and *Ursus.* The giant elk, woolly mammoth, and cave bear appear in the Early Holocene. Hominids appeared at the beginning of the Quaternary, and *Homo sapiens* in the Holocene. (More recent findings have set both these appearances back to earlier dates.)

center of gravity   That point in a body or system of particles at which the entire mass of the body or system seems to be concentrated. It is through this point that the resultant attraction of gravity acts when the body or system is in any position.

center of symmetry   A point within a configuration through which any line cuts the configuration in pairs of points equidistant in opposite directions.

centigrade scale   *Celsius scale.*

central vent eruption   Ejection of lava and debris from a ground hole that is fed by a single, deep-lying channel. Ejected material accumulates around the vent and may form a conical volcano or any of several other forms. Central vent eruptions occur along *subduction zones, mid-oceanic ridges,* and certain *rift valleys.* The differentiation between central vent and *fissure eruption* is not always easy. The large volcanoes of the Hawaiian Islands all appear to have central vents, but the summit craters are located on a fissure zone, and most of the lavas are erupted from the fissures. See also: *volcanic eruption, sites of.*

centrifugal force   A body constrained to move along a curved path reacts against the constraint with a "force" directed away from the center of curvature of the path; this reaction is called the *centrifugal force.* See also: *Coriolis force.*

centripetal drainage pattern   See *drainage patterns.*

centrosphere   See *core.*

cephalopod   A member of Cephalopoda, a class of highly organized marine mollusks of which squid, octopus, nautilus, and cuttlefish are representatives. Range, Cambrian to present. Extinct forms outnumber the living; the greatest cephalopod diversity was attained in the late Paleozoic and Mesozoic. See *ammonite; belemnite.* Cephalopods are characterized by a head surrounded by tentacles, 8 or 10 in most forms. The univalve shell may be internal, external, or absent, and variously coiled or straight (fossil forms), and is commonly composed mainly of aragonite. It may be either wholly chambered when external or partially chambered when internal; the chambers are gas-filled and are used for buoyancy.

ceratite   See *ammonoid.*

cerussite   An orthorhombic mineral, $PbCO_3$, of the aragonite group. A common secondary mineral of lead, it is often found in the weathered zone of lead deposits as an oxidation product of galena.

cf.   1. In paleontology, used to indicate close comparability but not identical sameness between a specimen in question and members of a named species; it implies closer similarity than *aff.* 2. Etymol: Latin, *conferre,* "to compare." Used in this dictionary and other works to mean "compare."

chain   1. Any sequence of similar natural features; e.g., a chain of islands, mountains, or lakes. 2. A distance-measuring device, consisting of joined links of equal length, the total length of which is equivalent to 20.13 m. Although the figure is used as the legal unit of length for the survey of public lands of the U.S., the original surveyor's chain, used as a measuring device, has been replaced by a tape.

chain silicate   *Inosilicate.*

chalcanthite   A blue triclinic mineral, $CuSO_4 \cdot 5H_2O$, that occurs in oxidized zones of copper deposits.

chalcedony   A cryptocrystalline variety of quartz, having a compact fibrous structure and waxy luster. It may be translucent or semi-transparent, and occurs in a variety of colors. Chalcedony is often found as a deposit, lining or filling cavities in rocks. See also: *agate.*

chalcocite   A black or dark lead-gray mineral, Cu_2S, with metallic luster. It is commonly massive, but also occurs infrequently as orthorhombic crystals. Chalcocite is an important ore of copper, occurring most frequently as the result of secondary enrichment.

chalcophile   See *affinity of elements.*

chalcopyrite   A brass or golden-yellow tetragonal mineral, $CuFe_2$, that is the most important ore of copper. Crystals are common, but it is usually massive.

chalk   Etymol: Old English, *cealc* from Latin, *calx,* "lime." A soft, earthy, fine-grained white to grayish limestone of marine origin, composed almost entirely of biochemically derived calcite that is formed mainly by shallow-water accumulations of minute plants and animals, in particular, coccoliths and foraminifers. Chalk deposits, such as those exposed in cliffs along both sides of the English channel, are typical rocks of the Cretaceous age.

chalybeate   adj. Containing or impregnated with iron salts; e.g., a mineral spring.

Champlainian   Middle Ordovician of North America.

chance packing   See *packing.*

Chandler wobble   A small continuous

variation of the earth's rigid body motion that causes it to deviate from pure spin. The wobble completes a cycle in about 428 days.

channel 1. The bed of a stream or waterway. 2. The deeper part of a waterway, esp. the part of a body of water deep enough for navigation. 3. A wide strait, as between an island and a continent; e.g., English Channel.

channel bar See *river bar.*

channel capacity The maximum flow volume that a given channel can sustain without overflowing its banks. See *bankfull stage.*

channel-fill deposit A deposit in a stream channel, consisting largely of bed load materials. The deposit occurs in those places where the transport capacity of the stream is not great enough to remove the material it receives.

channel flow The type of flow exhibited by surface runoff as it streams through long, narrow, V-shaped gullies or troughs.

channel-mouth bar A bar built at the place where a stream enters a body of standing water; a decrease in the stream's velocity at this point results in a buildup of sand and gravel.

channel pattern The configuration of a stream or river course, such as braided, meandering, sinuous, or straight. A river may follow a relatively straight course for some distance, and a meandering course at a different location.

channel sample A rock sample, usually selected across the face of a rock body or vein to provide an average value.

channel sand Sand or sandy material deposited in the bed of some channel cut into the underlying rock. Such sand may contain valuable minerals, ore, or gas. See also: *shoestring sand.*

channel storage The volume of water in a stream channel above a given reference point at a given time.

chaos See *megabreccia.*

characteristic fossil A fossil species or genus that is distinctive of a stratigraphic unit. It is either peculiar to that unit or is especially abundant in it. Cf: *index fossil.*

charnockite A granitic rock of crystalloblastic fabric, consisting of feldspar, orthopyroxene, and quartz. It is found most frequently as a granulite facies assemblage. The origin of charnockites, magmatic or metamorphic, is still debated.

chart datum The level of reference to which readings on a chart, e.g., soundings, are related.

chasm A deep cleft or fissure in the earth's surface, such as a gorge.

chatoyancy An optical property by which certain minerals, such as cat's-eye (chrysoberyl), produce in reflected light a band of light resembling the eye of a cat. The phenomenon is due to the reflection of light from aligned fibers or tubular channels. A mineral possessing such a property is said to be *chatoyant.*

chatter marks Small, curved cracks found on glaciated rock surfaces, usually 1 to 5 cm (1/2 to 2 in.) in length, but may be submicroscopic or as long as 47 cm (20 in). They occur chiefly on hard rocks (e.g., granite), and are formed under a glacier by the pressure of irregularly moving boulders. Chatter marks are commonly found in nested arrangements, with the fractures at right angles to the direction of glacial movement. See also: *glacial scouring.*

Chautauquan Uppermost Devonian of North America.

chemical limestone A rock composed mostly of calcite and formed either by direct chemical precipitation or by consolidation of calcareous ooze.

chemical oxygen demand The amount of oxygen required to oxidize organic matter in a water sample or body of water. Abbrev: *COD*. Syn: *Oxygen demand*. Cf: *biochemical oxygen demand*.

chemical potential A function related to variations in composition in the same manner as V and S are related to pressure and temperature variations, respectively. Chemical potential is analogous to gravitational potential energy, in that the lowest potential is the most stable state and, at equilibrium, potentials of neighboring states are equal. For example, in a hydrous magma system at equilibrium, the chemical potentials of water in the complex silicate melt, of water in the associated vapor, and of the water in biotite crystals suspended in the melt must all be equal to one another.

chemical remanent magnetism See *natural remanent magnetism*.

chemical weathering See weathering.

chenier Etymol: Fr., *chêne*, "oak." A wooded beach ridge found on the Mississippi Delta, composed of sand or shell debris swept onto the deltaic plain. Although beach ridges formed by similar processes are found elsewhere, they are often given other names; the presence of trees, including oaks on those of the Louisiana region, is responsible for the name *chenier*.

chernozem soil Etymol: Russ., "black earth." Soil of subhumid climate, characterized by a deep, dark-colored layer, rich in humus and carbonates, that grades downward to a layer of lime accumulation.

chert A dense, extremely hard, microcrystalline or cryptocrystalline, siliceous sedimentary rock, consisting mainly of interlocking quartz crystals less than 30 mm in diameter and sometimes containing opal (amorphous silica). It is typically white, black, or gray, and has a splintery to conchoidal fracture. Chert occurs mainly as nodular or concretionary aggregations in limestone and dolomite, and less frequently as layered deposits (banded chert). It may be an organic deposit (radiolarian chert), an inorganic precipitate (the primary deposit of colloidal silica), or a siliceous replacement of pre-existing rocks. The term *flint* is basically synonymous.

chertification A kind of silification, especially by a fine-grained or cryptocrystalline quartz.

Chesterian Uppermost Mississippian of North America.

chevron fold See *fold*.

chiastolite A variety of andalusite, embedded in dark schist in cigar-shaped crystals, which in cross-section show light and dark areas caused as the growing crystal forces carbon particles into defined areas.

chickenwire anhydrite A gypsum-anhydrite sequence showing polygonal nodules. Such structure is commonly found as displaced growth in carbonate muds of modern supratidal environments; this is widely believed to be evidence of *sabkha* deposition.

chile saltpeter A sodium nitrate, $NaNO_3$, that occurs in caliche in Chile. Syn: soda niter. Cf: *saltpeter*.

chimney 1. A columnar portion of headland detached from the shore line. 2. A pinnacle-like rock body rising well above its surroundings. 3. A channelway through which magma reaches the surface. 4. A cylindrical, relatively vertical, ore body. 5. A vertical opening in a cave.

china clay *Kaolin* that has been processed for the manufacture of chinaware.

chiton A relatively simple marine mollusk of the class Amphineura, with a shell consisting of overlapping calcareous dorsal plates. Also called "sea mice" or "coat-of-mail" shells. Not common as fossils; range, Ordovician to Recent.

chlorapatite Member of the apatite group of minerals, $Ca_5(PO_4)_3Cl$.

chloride A salt of hydrochloric acid consisting of two elements, one of which is chlorine; e.g., sodium chloride, NaCl.

chlorinity The chloride content of seawater, including the chloride content of all the halides. Syn: *chlorine equivalent*.

chlorite A representative of a group of micaceous greenish minerals of the general formula:
$(Mg, Fe^{+2}, Fe^{+3})_6 AlSi_3O_{10}(OH)_8$.
Chlorates are common in low-grade schists, or as alteration products of ferromagnesian minerals.

chloritization Replacement by alteration into or introduction of chlorite.

chloritoid A micaceous mineral, $Fe_2Al_4Si_2O_{10}(OH)_4$, that crystallizes in the monclinic or triclinic systems. It occurs in yellow-green to black scaly aggregates in low-grade, regionally metamophosed sedimentary rocks.

chokedamp *Blackdamp*.

Chondrichthyes The class of vertebrates including fishes whose skeletons are cartilaginous rather than bony; e.g., sharks and rays.

chondrites One of the two divisions of *stony meteorites* characterized by *chondrules*. They are primitive meteorites, having crystallized some 4700 million y.b.p., and make up over 80% of

meteorite fills. They are divided into three groups according to Mg and Fe ratio (2 to greater than 9). Cf: *achondrites*. See also: *carbonaceous chondrites*.

chondrule A small globular body of various materials, though mainly olivine and pyroxene, found as an inclusion in certain stony meteorites (*chondrites*). Chondrules are usually less than 3 mm in diameter, may be porphyritic or glassy in texture, and are clearly visible on polished surfaces.

chonolith An igneous intrusion that cannot be classified (laccolith, dike, sill, etc.) because its form is so irregular.

Chordata A phylum, including the Vertebrata, of those animals having a notochord. *Protochordata* may or may not be included.

C horizon See *soil profile*.

chromate A mineral containing the chromate ion, $(CrO_4)^{-2}$. Crocoite, $PbCrO_4$, is a common chromate.

chromatography Any of several methods for separating components of a sample. The sample is introduced into a medium using adsorption, partitioning, ion exchange, or some other process to separate the components.

chromite A mineral of the spinel group $FeCrO_4$, that occurs as small, black octahedral crystals, usually in compact masses in mafic and ultramafic rocks. It is the main ore for chromium.

chronohorizon A stratigraphy surface, theoretically without thickness, that is the same age at all points. In practice it is a thin and characteristic interval that serves as a good time-correlation or time-reference zone. Examples include coal beds, bentonite beds, and many *biohorizons*. Cf: *lithohorizon*.

chronolithologic unit *Chrono-stratigraphic unit.*

chronostratigraphic unit See *stratigraphic unit.*

chronostratigraphic zone Chronozone.

chronostratigraphy That area of stratigraphy dealing with the age and time relations of strata. Syn: *time-stratigraphy.*

chronotaxy Similarity of time sequence, such as the correlation of stratigraphic or fossil sequences on the basis of age equivalence. Cf: *homotaxy.*

chronozone 1. General term for all rocks formed at any location during the time span of a particular geologic feature or defined interval of rock strata. 2. Formal term for the bottommost division of *chronostratigraphic units.*

chrysoberyl A mineral, $BeAl_2O_4$, usually green, yellow, gray, or brown. Alexandrite, yellow cat's eye, and some colorless varieties are valued as gems.

chrysocolla A mineral, $(Cu,Al)_2H_2Si_2O_5(OH)_4 \cdot nH_2O$, usually found in the oxidized zone of copper and sulfide deposits. Associated with azurite and malachite.

chrysolite A yellow to yellowish-green gem variety of olivine. In the U.S., it is called *peridot* when used as a gem; the chrysolite of French jewelers is really *chrysoberyl.*

chrysoprase A translucent, pale bluish-green or bright green gem variety of chalcedony.

chrysotile A fibrous white, gray or green mineral of the serpentine group, the most important type of asbestos. Syn: *serpentine asbestos.* Cf: *chrysolite.*

chute An inclined channel, e.g., a tube, trough, or shaft. It may also be used to mean a waterfall or a steep descent or rapids in a river.

chute cutoff A cutoff made through one of the marsh areas of a point bar, thus isolating part of the bar as an island between two river channels. Cf: *neck cutoff.*

cienaga Etymol: Span., *cienaga*, "marsh" or "bog." A term, commonly used in arid regions such as the southwestern U.S., for a marshy area whose wetness is due to springs or seepage.

Cincinnatian Upper Ordovician of North America.

cinder Vesicular, dark-colored basaltic to andesitic volcanic fragments, ranging in size from 4 to 32 mm. When ejected, cinders are in a semi-rigid state and fall to the ground as solid matter. Cf: *lapilli.* See *pyroclastic material; pyroclastic rocks.*

cinder cone See *volcanic cone.*

cinnabar A mineral, HgS, the principal ore of mercury. It normally occurs in brilliant red microcrystalline or earthy masses. It was used extensively by the Chinese as the pigment vermillion.

CIPW classification *CIPW normative composition.*

CIPW normative composition A procedure for recalculating the chemical composition of a rock into a hypothetical group of water-free standard minerals. The merit of the normative composition, or *norm*, is that the limited number of standard normative minerals facilitates comparisons between rocks. The norm disregards the effects of geological processes and P-T conditions, focusing only on the source of matter comprising the magmatic rock. The initials represent the names of Cross, Iddings, Pirsson, and Washington, the four men who devised the system.

circulation 1. The upward and downward cell-like movement of water in an area of the ocean as a result of density variations produced by temperature and salinity changes. 2. The complete mixing of lake waters. 3. A term used in *rotary drilling* for the process of pumping drilling mud down the drill pipe and back to the surface.

circum-Pacific belt A belt, generally 300 km wide, that rings the Pacific Ocean and is one of the two principal regional bands containing the world's high mountains. It is a belt of major tectonic activity and, as such, is one of the loci of most continental earthquakes.

cirque Etymol: Fr., *cirque*, "circle" or "ring." A semicircular basin or indentation with steep walls originating from an ordinary valley head. It is situated high on a mountain slope, and is associated with the erosive activity of a mountain glacier. Walls of the cirque are cut back by disintegration of lower rock. Resulting rock material, embedded in the glacier, gouges a concave floor that may contain a small lake (*tarn*) if the glacier disappears. The expansion of neighboring cirques produces *arêtes, horns,* and *cols.*

cirque lake A *tarn.*

citrine A yellow to yellowish-brown crystalline quartz, sometimes resembling topaz in color.

cladogenesis Phyletic branching or splitting, or progressive specialization through evolution.

clan 1. A group of igneous rocks having closely related chemical compositions. 2. A small ecologic community having only a single dominant species.

Clapeyron equation See *Clausius-Clapeyron equation.*

clarain See *lithotype.*

clarke The abundance of an element in the crust of the earth. Named in honor of F.W. Clarke, an American geochemist. In any ore body, the concentration of an ore element is higher than the *clarke* of the element. This concentration is called the *clarke of concentration* of the ore body.

class In the classification of plants and animals, a category between phylum and order.

clast An individual fragment of a larger rock mass removed by the physical disintegration of the larger mass. The term may also refer to a constituent of a *bioclastic rock,* or to a *pyroclast.*

clastation The breaking-up of rock masses by physical or chemical means.

clastic 1. Pertaining to a sediment or rock composed chiefly of fragments derived from pre-existing rocks or minerals; also, the texture of such a rock. 2. *Pyroclastic.* 3. Pertaining to the fragments (clasts) of which a clastic rock is composed. 4. n. A clastic rock; usually used in the plural. 5. Said of a *bioclastic* rock.

clastic dike A tabular body comprised of clastic materials that cuts across the bedding of a sedimentary formation and is derived from overlying or underlying beds; in particular, a *pebble dike* or *sandstone dike.*

clastic ratio The ratio, in a stratigraphic section, of the amount of clastic material (sandstone, conglomerate) to that of nonclastic material (limestone, dolomite). Syn: *detrital ratio.* Cf: *sand-shale ratio.*

clastic rock 1. A sedimentary rock composed mainly of fragments broken loose from parent material and deposited by mechanical transport; e.g., sandstone, shale, conglomerate. See also: *epiclastic* rock. Syn: *fragmental* rock. 2. *Pyroclastic* rock. 3. *Bioclastic* rock. 4. A *cataclastic* rock.

clastic wedge   The sediments of an exogeosyncline derived from the adjacent orthogeosynclinal belt. Cf: *geosynclinal prism.*

Clausius-Clapeyron equation   An equation that expresses the relation between the vapor pressure of a liquid and its temperature, $\frac{dP}{dT} = \frac{\Delta H}{T\Delta V}$, where P is the pressure, T the temperature, ΔH the change in heat content, and ΔV the change in volume.

clay   1. A detrital mineral particle of any composition having a diameter less than 0.004 mm. 2. A smooth, earthy sediment or soft rock composed chiefly of clay-sized or colloidal particles and a significant content of *clay minerals.* Clays may be classified by color, composition, origin, or use. 3. Common term for any wet earth material.

C layer   See *seismic regions.*

clay ironstone   A fine-grained sedimentary rock, gray or brown, consisting of clay and iron carbonate (siderite); it occurs in concretions or thin beds. Clay ironstone is usually associated with carbonaceous strata, in particular, overlying coal seams in coal measures of the U.S. and Great Britain. See also: *ironstone.*

clay minerals   A member of the aluminosilicate mineral group, which forms about 45% of all minerals in sedimentary materials. Most clay minerals belong to the kaolinite, mortmorillonite, and illite groups. Their crystal structure is the same as that of mica, i.e., sheeted layer structures with strong intra- and intersheet bonding but weak interlayer bonding. The type of clay that is dominant in sedimentary material depends, during early stages of weathering, on the mineral composition of the parent rock; but in later stages, it depends completely upon the climate. Cf: *clay.*

claypan   A dense, clayey subsoil layer that is hard when dry but may be plastic when wet. Its clay content is the result of concentration by downward-percolating waters. Cf: *hardpan.*

clay plug   A mass of clay and silt that may fill an *oxbow lake,* converting it into a marsh. It is perhaps more accurate to say that the marsh constitutes a clay plug, inasmuch as it is resistant to incursion of the future stream channel.

clay shale   A totally or mainly argillaceous shale that reverts to clay upon weathering.

claystone   Sedimentary rock of indurated clay-sized silicate materials, having the texture and composition of shale, but lacking its lamination and fissility.

cleavage   1. The property of some minerals to break along planes related to the molecular structure of the mineral and parallel to actual or possible crystal faces. Cleavage is defined by the quality of the break (excellent, good, etc.) and its direction in the mineral, i.e., the name of the form it parallels. Some of the more important types of cleavage are: *basal,* or parallel with the basal plane, e.g., mica; *cubic,* or parallel with the faces of a cube, e.g., galena; *octahedral,* or parallel with the faces of an octahedron, e.g., fluorite; *prismatic,* or parallel with the faces of a prism, e.g., augite; *rhombohedral,* or parallel with the faces of a rhombohedron, e.g., calcite. Cf: *fracture.* 2. The characteristic or tendency of a rock to split along parallel, closely spaced planar surfaces, e.g., *slaty cleavage.* It is produced by deformation or metamorphism and is independent of bedding. Cf: *schistosity.*

cleavlandite   A white, lamellar variety of albite, found as large crystals and in pegmatite veins.

Clerici solution   An aqueous solution of thallium malonate and thallium formate that is used as a *heavy liquid.* See also: *bromoform; methylene iodide.*

cliff General term for a high, steep face of a rock formation. Cliffs may be formed by faulting, waves and shore currents, volcanic activity, and glaciation. *Sea cliffs* are formed by wave action.

climate Weather conditions of a specified region, averaged over a long time interval. Factors such as temperature, precipitation, and position relative to land and sea are part of climate classification. A region typified by a particular climate is called a *climatic province*; e.g., dry desert climates. A *climate zone* is one of five major divisions of earth's surface, bounded by lines parallel to the equator, and named according to the prevailing temperature; e.g.,, the North Frigid Zone, Torrid Zone.

climatic optimum A period of milder climate, some 4000 to 6000 years ago, during which most middle-latitude valley glaciers disappeared. According to calculations based on pollen distribution and oxygen-isotope ratios in Greenland ice cores, many regions were 5°F warmer, on the average, than they are today.

climax In ecology, that state in the ecological succession of a plant and animal community that is stable and self-perpetuating.

climbing dune See *dune.*

clinker 1. A mass of fused incombustible matter, commonly found when coal is burned. 2. A jagged fragment of lava.

clinoenstatite Monoclinic magnesium silicate, $MgSiO_3$, a mineral of the clinopyroxene group, $(Mg, Fe) SiO_3$.

clinometer An apparatus used with a geologist's compass for measuring angles of slope; in particular, for determining dip.

clinopyroxene A member of the pyroxene group, sometimes containing significant calcium; crystallizes in the monoclinic system.

Clinton ore An oölitic iron ore, one of the ironstones of the Clinton group, or Clinton Formation (Middle Silurian) in the Appalachian Mountain belt. Most ironstones of the group are oölitic hematites, and fossils contained within them are usually partly replaced by iron minerals. All of the oolitic ironstones are cemented chiefly by calcite and dolomite.

closed basin A region that drains into a depression or lake within its environment, and from which water escapes only by evaporation.

closed fold See *fold.*

closed structure A structure, such as an anticline or syncline, whose map representation shows it to be entirely closed by one or more contour lines.

closed system A system in which there is no transfer of matter or energy into or out of the system during a particular process.

close-grained Said of a rock whose constituent particles are fine and tightly packed together.

close packing See *packing.*

closure On a structure *contour* map of a subsurface anticline, closure is the vertical distance between the anticline's highest point and its lowest closed structure contour. The information is used to estimate gas or oil reserves.

coal A combustible, stratified organic sedimentary rock composed of altered and/or decomposed and reconstituted plant remains of non-marine origin, combined with varying minor amounts of inorganic material. Coal is non-crystalline, brittle, and dull to brilliant in luster. With increasing compaction, the color changes from light brown to black, and the specific gravity increases from 1.0 to 1.7. Coals are classified according to their constituent plant materials (type), ratio of fixed carbon content to volatile matter (rank), and amount of impurities present (grade). Coal occurs in rocks of the early Proterozoic, but did not become widespread until the development of woody land plants in the Devonian. Pre-Devonian coals are composed of algal remains. Sedimentary rocks of the Carboniferous-age hold the greatest abundance of coal-bearing rocks. Large deposits of lignite also occur in rocks of Early Tertiary age. Coal beds are commonly found with shales and fine-grained sandstones.

The formation of coal proceeds in stages, beginning with the accumulation of large masses of plant debris in a humid environment. Such accumulations require anaerobic depositional conditions and a slowly subsiding basin. *Peat*, a spongy mass of partially decayed vegetation, is the first stage of coal formation. When covered with sediment, peat may gradually be converted to *lignite*. The major effect of pressure on coalification occurs during this transition, where overburden pressure reduces porosity and moisture by perhaps 50%. Although pressure is not excluded as a factor affecting the transformation of lignite into *bituminous*, and bituminous into *anthracite*, the main factors involved are temperature and time. Time is critical because a suitable temperature must be maintained for extensive periods before the degree of metamorphism, i.e., rank, can be increased. The rate at which coal rank increases with depth depends on the geothermal gradient and on the heat conductivity of the ambient rocks. See also: *banded coal; lithotype; subbituminous; meta-anthracite.*

coal ball   A concretion of mineral and plant material embedded in coal seams or adjacent rocks, ranging from pea to boulder size.

coal field   An area containing coal deposits.

coal gas   A fuel made by the carbonization of a high-volatile bituminous coal. Approximate composition by volume:

50% hydrogen, 30% methane, 6 to 8% carbon monoxide, 7 to 8% carbon dioxide, nitrogen, and oxygen, and 2 to 4% olefins.

coalification   The physical and chemical changes that take place within a deposit of accumulated plant remains in the formation of coal. During coalification, the carbon percentage increases as water and volatile hydrocarbons are expelled from the deposits. See *coal.*

Coal Measures   A stratigraphic term used in Europe (first used in Great Britain for the *Upper Carboniferous*). It bears an approximate correlation with the *Pennsylvanian* of North America. See also: *coal measures.*

coal measures   A succession of sedimentary strata consisting chiefly of clastic rocks with interlayered beds of coal. See also: *Coal Measures.*

coal plant   A fossil plant found in a coal bed, or a plant whose altered substance contributed to the formation of coal beds; e.g., *Lepidodendron* and *Sigillaria*.

coal seam   A layer or bed of coal.

coal type   A classification of coal according to its constituent plant materials (*macerals*).

coarse aggregate   Said of an aggregate composed of particles greater than 4.76 mm in diameter.

coarse-grained   1. Said of a crystalline rock whose individual minerals have an average diameter greater than 5 mm. Syn: *phaneritic*. 2. A sediment or sedimentary rock in which individual constituents are easily visible to the naked eye. Various limits have been suggested.

coarse sand   Sand composed of grains between 0.5 and 1 mm in diameter.

coarse topography   A land surface characterized by large-scale erosional features.

coast   A zone of rather indeterminate width extending landward from the seashore. In geographic terms, it may be that section of a region that is near the coast, and sometimes includes the *coastal plain*.

coastal plain   A broad plain sloping gradually seaward. It generally represents a strip of emerged sea floor. During the emergence, the shoreline of a coastal plain migrates seaward so that progressively younger rocks are found at decreasing distances from the shore.

coastline   1. The boundary between coast and shore. 2. The boundary between water (esp. sea or ocean water) and land. 3. A broad term for the overall appearance of a coastal land trait. 4. A broad zone of indefinite width extending both seaward and landward from a shoreline. Cf: *shoreline*.

Coast Range orogeny   Major deformation, metamorphism, and volcanic activity in the Coast Mountains of the British Columbia Cordillera during the Jurassic and Early Cretaceous. It generally approximates the *Nevadan orogeny* of the U.S.

cobaltite   A steel-gray to silver-white mineral, CoAsS, the principal ore of cobalt.

cobbing   In a mining operation, the separation of valuable minerals from lumps of rock, usually done with a light hammer.

cobble   A rock fragment, rounded or abraded, between 64 and 256 mm in diameter. It is larger than a pebble and smaller than a boulder.

cobblestone   A rounded stone used in paving or other construction.

coccolithophore   A minute marine planktonic photosynthetic organism found only in warm, low-latitude waters. After death, it disintegrates to form *coccoliths*, which are microscopic calcareous plates of many different shapes, built of calcite or aragonite. Coccoliths are found in chalk and in the deep-sea oozes of tropical and temperate oceans.

COD   *Chemical oxygen demand.*

coelacanth   A member of the crossopterygian fishes. Once believed to have been extinct since the end of the Cretaceous Period, several specimens belonging to *Latimeria*, the only surviving genus, have been found during the past 30 years. Range, Upper Devonian to Recent.

coelenterate   Any multicelled aquatic invertebrate, solitary or colonial, of the phylum *Coelenterata*. The name is derived from the simple organization around a central gut (coelenteron). Members typically have tentacles, nematocysts, and a single body opening for digestion and egestion, and radial or biradial symmetry. Range, Precambrian to present.

coesite   A very dense polymorphic form of quartz, SiO_2, stable at room temperature only at pressures exceeding 20,000 bars. It is found in impact craters.

cogeoid   For a point in the earth's surface, the cogeoid is a surface lying above or below the *geoid* at a distance dV/g, where dV is the gravitational potential at the point considered, and g is the gravitational acceleration. The potential, dV, refers to all matter lying above sea level, plus both the mass defect of the oceans and variations of rock density as prescribed by isostasy. Formerly known as *compensated geoid*. See *geoid*.

cognate inclusion   *Autolith.*

cohesion   1. (Physics) The attraction between molecules of the same substance. It is this attraction in liquids that allows the formation of drops and thin films. Cf: *adhesion*. 2. (Geol.) That component of the *shear strength* of a rock that is independent of interparticle friction.

coke A combustible solid material obtained from bituminous coal by carbonization. As a fuel, it is of high calorific value and is almost smokeless. A coal that is suitable for coke production is called a *coking coal*. A *natural coke* is one that has been naturally carbonized by natural combustion, or by contact or near-contact with an igneous intrusion.

col Etymol: Fr., *col*, "neck." 1. A sharp-edged or saddle-shaped *pass* in a mountain range, formed by the headward erosion of two oppositely orientated *cirques*. 2. A saddle-formed depression in a mountain range. See also: *arête*.

colemanite A colorless to white monoclinic mineral, $Ca_2B_6O_{11} \cdot 5H_2O$; an important source of boron. Usually found as massive crystals in Tertiary lake deposits.

collapse breccia Angular rock fragments produced by collapse of the rock formation above an opening; e.g., the collapse of the roof of a cave. See also: *solution breccia*.

collapse caldera See *caldera*.

collapse structure Any rock structure that is the result of removal of support and consequent collapse; e.g., sink-hole collapse, collapse into mine workings, or gravitational sliding on the limbs of folds.

collimator An optical device for obtaining a parallel or near parallel beam of light, or for testing the focus of a lens at infinity. It is used for testing and adjusting certain optical surveying instruments.

colloform A term sometimes used to include all more-or-less spherical forms; e.g., *botryoidal, reniform, mammillary*.

colloid 1. A substance with particle size less than 0.00024 mm, i.e., smaller than clay size. 2. An extremely fine-grained material either in suspension or that can be readily suspended, commonly having special properties due to its extensive surface area.

collophane Any one of the massive cryptocrystalline varieties of apatite that comprise the bulk of phosphate rock and fossil bone; used as a phosphate source for fertilizer. Syn: *collophanite*.

colluvium A general term for unconsolidated material at the bottom of a cliff or slope, generally moved by gravity alone. It lacks stratification and is usually unsorted; its composition depends upon its rock source, and its fragments range greatly in size. Such deposits include cliff debris and talus. adj. Colluvial. Cf: *slope wash*.

colonial coral A unit composed of attached individual coral organisms that cannot exist as separated animals.

color index As used in petrology, esp. in the classification of igneous rocks, a number representing the percent, by volume, of dark-colored (mafic) mineral in a rock. Thus rocks of color index 0 to 30 are "leucocratic"; 30 to 60, "mesocratic"; 60 to 100, "melanocratic." Syn: *color ratio*.

columbite A black mineral, the Nb end member of the columbite-tantalite series, $(Fe, Mn)(Nb, Ta)_2 O_6$. Occurs in granites and related pegmatites.

columnar jointing Long, parallel columns, polygonal in cross-section, occurring most frequently in basalts, but also in other extrusive and igneous rocks. The cooling of very hot lava or pyroclastic flow decreases its volume, causing cracks to form. Because the surface cools first, cracks penetrate the mass along the directional normal to the surface. The rock usually divides into columnar segments approximately perpendicular to the cooling surface. Because the forms are more or less homogenous laterally, the columns tend to be six-sided—the ideal configuration assumed by a solid mass when it cools away from a plane surface. Magnificent basaltic columns can be seen in the Devil's Postpile, Calif. Syn: *columnar structure; prismatic structure.*

columnar section A graphic depiction of the sequence of rock units in a particular locality. The lithology of a section is indicated by symbols, and thicknesses are drawn to scale. See also: *geologic column.*

columnar structure 1. *Columnar jointing.* 2. A columnar, near-parallel arrangement exhibited by aggregates of long, slender, mineral crystals. 3. A primary sedimentary structure found in some calcareous shales or argillaceous limestone. It consists of columns, oval to polygonal in cross-section, and normal to the direction of bedding.

comagmatic A term applied to igneous rocks that are regarded as having been derived from a common parent magma, because they share a common group of chemical and mineralogical features. See also: *consanguinity.*

Comanchean Lower Cretaceous of North America.

comminution A process, either natural or in manufacture, by which a substance is reduced to a fine powder, or pulverized.

common lead Any lead with a low U/Pb and/or Th/Pb value, such that no significant radiogenic lead has been generated *in situ* from the time that the phase formed. See also: *anomalous lead.*

common salt A colorless or white crystalline rock composed almost completely of *halite*; occurs in nature in *salt domes* as beds, and as encrustations that rim *salt lakes.*

community A group of organisms, living or fossil, that occur together because of the existence of a systematic food chain that operates through different feeding levels. Syn: biotic community. Cf: *assemblage; association; biocoenosis.*

compactibility In this sense, the property of a sedimentary material that allows a decrease in thickness or volume under load. It varies with the size, shape, and hardness of the individual particles.

compaction The decrease in pore space of a fine-grained sedimentary rock, and consequent reduction in volume or thickness. Compaction results from the increasing weight of younger sediment material that is continually being deposited, or of pressures due to earth movements.

compensated geoid *Cogeoid.*

competence The ability of a stream or wind current to carry detritus, as determined by particle size rather than amount. The diameter of the largest particle transported is the competence value. Cf: *capacity.*

competent Said of a sedimentary formation, e.g., a bed or stratum, that is strong and able to transmit compressive force much farther than a weak, *incompetent* formation; i.e., the bed or stratum is able to withstand the pressure of folding without flowage or change in original thickness. Some of the factors that determine whether or not a stratum is competent are crushing strength (resistance to crushing), massiveness of the formation, and ability to "heal" fractures.

complementary 1. Said of rock or rock groups, differentiated from a common magma, whose total composition equals that of the parent magma. See *magmatic differentiation.* 2. Said of seemingly related fractures.

composite cone See *volcanic cone.*

composite fault scarp An escarpment whose height is due in part to differentiated erosion and in part to direct movement along the fault. Either process, erosion or faulting, may precede the other.

composite intrusion A term applied to bodies of igneous rock such as *dikes, sills,* or *laccoliths,* which are composed of two or more intrusions that differ in chemical and mineralological composition. For example, an *acid* phase may be injected into a more *basic* one. Cf: *multiple intrusion.*

composite point See *phase diagram.*

composite profile A plot representing the surface of any relief area as viewed in the horizontal plane of the summit levels from an infinite distance. It consists of the highest points of a set of profiles drawn along regularly spaced parallel lines on a map.

composite topography A landscape in which the topographic features have developed during two or more cycles of erosion.

compound alluvial fan A *bajada.*

compound shoreline A shoreline along which features of both *shoreline of submergence* and *shoreline of emergence* are found.

compound vein A term used for a vein composed of several minerals. Also, a vein comprising many parallel fissures cut by cross fissures.

compressibility The capacity of a body to change in volume and density under hydrostatic pressure. It is the reciprocal of *bulk modulus.* Syn: *modulus of compression.*

compression A system of external forces that tends to shorten a body or decrease its volume.

compressional wave P wave.

compressive stress A stress that tends to push together on opposite sides of a plane, in a direction perpendicular to that plane. The maximum compressive stress that can be applied to a material under specified conditions before rupture or facture occurs is the *compressive strength* of the material. Cf: *tensile stress.*

concentric folding *Parallel folding.*

concentric weathering *Spheroidal weathering.*

conchoidal A descriptive term for a type of mineral or rock fracture that is smooth and curved, rather resembling a conch. Conchoidal fracture is characteristic of obsidian.

concordant 1. Term applied to intrusive igneous bodies whose boundaries are parallel with bedding or foliation of the country rock. 2. Said of strata lying parallel with the bedding structure, i.e., structurally conformable. 3. Radiometric ages that are in agreement, although determined from more than one source or by more than one method. Ant: *discordant.*

concrete A mixture of *cement,* some *aggregate,* and water, which will set and harden.

concretion Lumps or nodules found in shales, sandstones and limestones. These have been formed at the same time as the enclosing sediments, by deposition from concentrated mineral solutions around some central nucleus such as leaves, seeds, or shells. Concretions may be rounded or irregular, very small or up to a few meters in diameter. They are usually harder than the enclosing rock. *Septarian* concretions or nodules are particular masses that develop an irregular polygonal system of internal cracks. Under certain conditions of erosion, concretions can have the appearance of plants or animals, and are often mistaken for fossils. See also: *pseudofossil.*

conductance   The measure of the conductivity of a conductor, which depends on the dimensions of the conductor; it is the reciprocal of the resistance.

conduction   A transfer of energy via some sort of conductor that permits the transfer of molecular activity without overall motion. It is distinct from other energy transfer forms such as radiation and heat convection. The transfer of heat, electricity, and sound involves conduction. *Conductivity* is the conduction potential of a medium. See *conductivity, electrical; conductivity, thermal.*

conductivity, electrical   Also called *specific conductance.* It is the reciprocal of resistivity. The conductivity σ of a conductor of length l, cross-section A, and resistance R, is given by $\sigma = lRA$ where σ is expressed in ohms per centimeter. When seats of electromotive force are absent—a condition that obtains in the earth's oceans and atmosphere—Ohm's law is applied in the form of $J = \sigma E$, where J is the current density at a point of the conduction medium, σ is the conductivity of the medium, and E is the electric field intensity at that point. Electrical conductivity is used to measure the temperature and salinity of seawater, and to determine the *redox potential* in sediments, since it varies with the oxygen content.

conductivity, thermal   The amount of heat conducted per unit of time through any cross-section of a substance depends upon the temperature gradient at that section and the area of the section. Thus the quantity of heat is:

$$q = k \frac{dt}{ds} A$$

where A is the sectional area, dt/ds the temperature gradient, and k is a constant—the *thermal conductivity* of the substance. See *conduction.*

cone   1. *Alluvial cone.* 2. *Talus cone.* 3. A *volcanic cone.* 4. A fan-shaped submarine deposit, esp. a deep-sea fan, associated with a large active delta like that of the Mississippi. See also: *submarine fans.*

cone-in-cone structure   A structure sometimes found in fibrous gypsum and fibrous calcite layers in sediments, and less frequently in coal deposits and ironstone. It resembles a series of concentric cones, and is sometimes mistaken for a fossil. Some of these structures may be concretionary, but special pressure conditions seem to be responsible for most. Many cone-in-cone formations show a spiral structure rather than a nested one. See also: *concretion; pseudofossil.*

Conemaughian   Upper Middle Pennsylvanian of eastern U.S.

cone of depression   A cone-shaped depression that forms around a well from which water is being drawn, thus increasing the hydraulic gradient close to the well. Cf: *drawdown.*

cone sheet   See *ring dikes and cone sheets.*

confining bed   A body of impermeable or clearly less permeable material adjacent to one or more aquifers. Cf: *aquitard; aquifuge; aquiclude.* An aquifer bounded above and below by confining beds is a *confined aquifer.* Groundwater that is under sufficient pressure to rise above well level, and whose upper surface is an impermeable bed, is called *confined groundwater;* also referred to as *artesian water.*

confining pressure   Any pressure that is exerted equally on all sides, such as *geostatic pressure* or *hydrostatic pressure.*

confluence   The meeting point of two glaciers or two streams.

conformability   The state of being conformable; *conformity.*

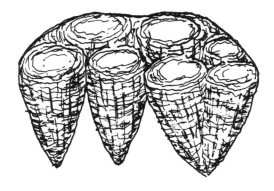

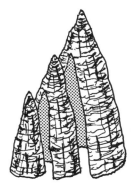

Multiple cone-in-cone   Single cone

conformable 1. Said of rock strata showing a continuous sequence of layers that were deposited without interruption in parallel order, one above the other; also said of the contact area between such strata. Cf: *unconformable.* 2. The contact of an intrusive body is said to be conformable if it is in alignment with the internal structure of the body. Cf: *concordant.*

conformal projection A map projection in which angles formed by lines are preserved, with every small triangle represented by a similar triangle, thus with the effect that any very small area of the surface that is mapped is kept unaltered on the map. See also: *map projection.*

conformity 1. A term indicating true stratigraphic continuity of adjacent sedimentary strata; i.e., they have been deposited in orderly series without evident time lapse. Syn: *conformability.* 2. A surface separating older strata from younger, along which neither erosion nor non-deposition is evident, and with no significant hiatus. Cf: *unconformity.*

congelifraction Rock fragmentation due to the pressure exerted by the freezing of water contained in pores, cracks, or fissures. Syn: *frost splitting; frost weathering; frost wedging.*

congeliturbation The disturbance (stirring or churning) of soil by frost action; e.g., heaving, solifluction. It produces *patterned ground.* Syn: *cryoturbation.*

conglomerate A coarse-grained clastic sedimentary rock, composed of more or less rounded fragments or particles at least 2 mm in diameter (granules, pebbles, cobbles, boulders), set in a fine-textured matrix of sand or silt, and commonly cemented by calcium carbonate, silica, iron oxide, or hardened clay. Syn: *puddingstone.* Cf: *breccia.* See also: *fanglomerate; rudite.*

congruent melting point The temperature at a given pressure at which a solid phase becomes a liquid phase of the same composition. Most pure minerals of fixed composition, such as albite, diopside, and quartz, melt congruently.

conical fold A fold configuration that can be described by a line that is fixed at one end and rotated about that end. Cf: *cylindrical fold.*

conic projection See *map projection*.

conjugate A term for a joint system, the sets of which share the same origin of deformation. Also, faults of the same age and depositional development are said to be conjugate.

connate water Water trapped in the pore space of sediment at the time of its deposition. Other water that may have entered the interstices of a rock subsequent to its deposition is not connate. Syn: *fossil water; native water*.

conodont Etymol: Latin, *conus*, "cone" + Greek, *odons*, "tooth." A small, amber-colored fossil element of phosphatic composition. Conodonts are toothlike in form but not in function. Small marine animals of undetermined affinity produced these elements in bilaterally paired, serial arrangement. Range, Cambrian to Upper Triassic.

Conrad discontinuity A discontinuity in some areas of the earth's crust where the velocities of compressional (P) seismic waves increase from ~6.1 km/sec to 6.4 to 6.7 km/sec. The depth of those locations where this abrupt change is noted is usually at 17 to 20 km. It is believed that the discontinuity indicates a boundary between upper and lower *continental crust*.

consanguinity A term for the relationship between igneous rocks that are presumed to derive from the same parent magma. Such rocks are close in age and location, and usually have comparable chemical and mineralogical features. See also: comagmatic.

consequent adj. Said of a geological or topographic feature whose origin was a result of pre-existing features. For example, a valley or stream whose course was determined by the initial slope of the land is referred to as a *consequent valley* or a *consequent stream*. See also: *subsequent*.

consolidation 1. Any process whereby soft, loose, or liquid earth materials become firm; e.g., the solidification of lava upon cooling, or the cementation of sand. 2. The reaction of a soil or sediment to increased surface load.

constant polarity epoch Long intervals of geologic time during which the polarity of earth's magnetic field was constant. These epochs have been named for geophysicists who have contributed to the study of magnetism.

Brunhes normal epoch extended from about 0.7 m.y.a. to the present. The preceding *Matuyama reversed epoch* extended from 2.5 to 0.7 m.y.a. The *Gauss normal epoch* (3.36 to 2.5 m.y.a.) preceded the Matuyama. The *Gilbert reversed epoch* was earlier than the Gauss, but is of uncertain duration. See also: *geomagnetic polarity reversal*.

contact metamorphism See *metamorphism*.

contact metasomatism See *metasomatism*.

contact twin See *twinning*.

contact zone *Aureole*.

contemporaneous deformation Deformation that takes place in sediments during their deposition or immediately after. Small folds and faults may develop in soft sediments that slide down gentle slopes. Syn: *penecontemporaneous*.

continent One of earth's main land masses usually reckoned as seven in number and, at present, constituting about one-third of the earth's surface. The term includes continental shelves as well as dry land.

continental Formed or generated on land, as opposed to in the ocean or sea. Deposits originating in lakes, swamps, or streams, or moved by land winds, are all continental.

continental apron   *Continental rise.*

continental borderland   A term proposed to describe the type of shelf found off the coast of California, where a series of basins and ridges form a topography that is more complex than the *continental shelf* found off the Atlantic coast.

continental crust   The crustal rocks that underlie the continents; its thickness ranges from 25 to 60 km. The upper continental crust, consisting of rocks rich in silica and alumina, is equivalent to the *sial*; the lower crust, the *sima*, consists largely of silicon and magnesium. Cf: *oceanic crust.* See also: *crust.*

continental deposit   Sedimentary material deposited on land or in bodies of water not directly connected with the ocean, as distinguised from a marine deposit. The term thus includes lacustrine, glacial, fluvial, or eolian deposits originating in a non-marine environment. See also: *terrestrial deposit.*

continental   displacement *continental drift.*

continental divide   A drainage divide separating river systems flowing toward opposite sides of a continent. In North America, the *Continental Divide* is the line of Rocky Mountain summits separating streams flowing toward the Pacific from those flowing toward the Gulf of Mexico and Hudson Bay.

continental drift   The concept that the continents have undergone large-scale horizontal displacement during one or more episodes of geologic time. For more than 300 years the apparent fit of the bulge of eastern South America into the indentation of Africa, which can be seen on any map, has caused scientists to contemplate the movement of continents.

The first detailed and comprehensive theory of continental drift was proposed in 1912 by Alfred Wegener, a German meteorologist. Wegener postulated that throughout most of geological time there was only one continent, which he called *Pangaea*; at some time during the Jurassic Period, Pangaea fragmented and the parts moved away from each other. In 1937, Alexander DuToit, a South African geologist, modified the Wegener hypothesis by supposing two primordial continents: *Gondwanaland* in the south and *Laurasia* in the north. The Wegener hypothesis, when presented, aroused much interest, debate, support, and opposition for some years. It was weakened by its lack of a suitable driving mechanism for continental movement, and by 1926 was no longer a subject of serious investigation.

The proponents of continental drift amassed impressive amounts of data in the matching of geological provinces of different continents. In the 1950s, interest increased as paleomagnetic data accumulated. These data included evidence of *polar wandering*, continental displacement and rotation, and of *geomagnetic polarity reversals.* During the early 1960s, the U.S. geophysicist Harry Hess presented the concept of *sea-floor spreading*, in which new oceanic crust is continually being generated by igneous activity at the crests of *mid-oceanic ridges.* By the late 1960s, several U.S. scientists integrated the notion of sea-floor spreading with that of continental drift and developed the theory of *plate tectonics.* See *plate tectonics; Pangaea.*

continental glacier   An *ice sheet* covering a vast area. See *glacier.*

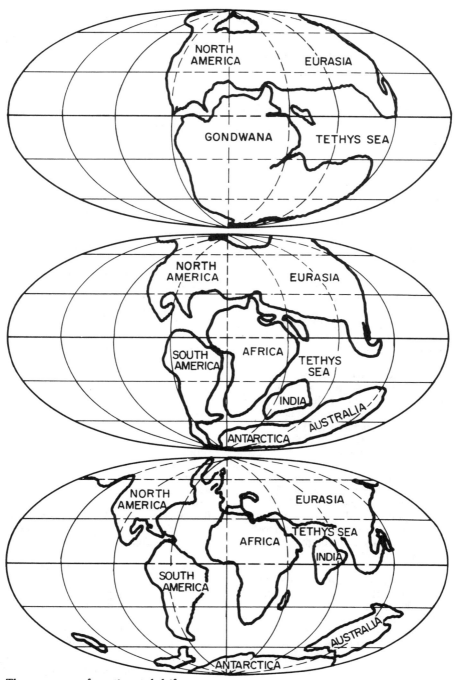

The progress of continental drift

continental margin The totality of the various divisions between the shoreline and abyssal ocean floor. It includes the continental shelf, *continental borderland, continental slope* and *continental rise.*

continental nucleus shield *Shield.*

continental platform *Continental shelf.*

continental rise A geomorphic feature of the lower continental margin where the vertical-to-horizontal ratio ranges from 1:50 to 1:800. It lies between the *continental slope* and the *abyssal plain or abyssal hills*, and its width ranges between 0 and 600 km. Depending on its location, the rise may be divided into "upper" and "lower" steps, or a series of steps resembling a staircase. In some regions the rise is very broad, and in others nonexistent. It appears to consist of fans of sediment derived from terrigenous silts and clays of the continental shelf. See *continental shelf.*

continental shelf A gently sloping, shallow-water platform extending from the coast to a point (the shelf break) where there begins a comparatively sharp descent (*continental slope*) to the ocean floor. The average width of the continental shelves of the world is 70 km. The average depth of shelf termination is about 150 m, although this may vary greatly. Shelves may have parallel ridges and troughs, or may be very flat, such as those found in the Bering Sea. Shelves off glaciated land masses are characteristically deep basins and troughs, and hence very irregular.

continental slope The relatively steep portion of the sea floor extending from the outer edge of the continental shelf to the upper limit of the continental rise. Its average slope is 4°, its width is usually 20 to 100 km, and its range of depth is from 100 to 200 m to 1400 to 3200 m. It may be terraced or smooth. Continental slopes, which are found

throughout the world, may be the products of sedimentational or volcanic processes, or structures that were subjected to faulting and folding. See *continental shelf.*

continuous deformation Deformation by process of flow rather than by fracture. Cf: *discontinuous deformation.*

continuous reaction series See *reaction series.*

contour A line on a map or chart that joins points of equal value; also called a *contour line.* A *contour map* represents the configuration of a land surface through the use of contour lines. In particular, on a topographic map, contours represent lines of equal elevation above sea level; a *structure contour map* is equivalent to a topographic map made of an underground surface, e.g., a coal bed, if all the overburden were first removed. Such maps are almost indispensable in mining and oil-field development.

contraction crack *Frost crack.*

convection 1. (Meteorol.) The transfer of heat by vertical movement in the atmosphere as a result of heat-produced density differences at lower levels. 2. The flow of water through and around heated regions of nearby plutons; water circulation is initiated by thermal gradients within these regions. 3. (Oceanog.) The movement and mixing of oceanic water masses as a result of density differences. 4. (Tectonics) A concept of the mass movement of mantle material, either laterally or in convection cells. It is thought that the decay of radioactive isotopes within the mantle provides the heat energy required. Geophysical studies show that given the pressures at such depths, mantle rock begins to creep at temperatures exceeding 1000° C. Thermal gradients cause hotter, less dense material to rise, displace cooler, denser material, and to flow

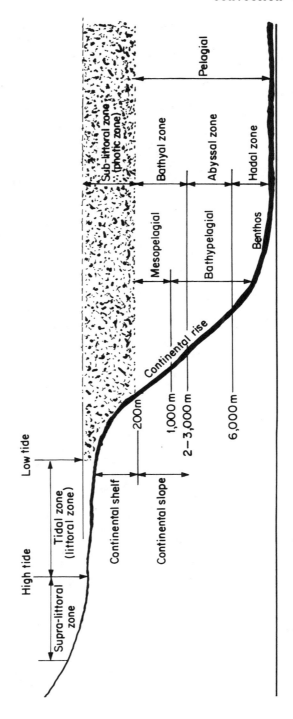

Areas of sea and shore

parallel to the earth's surface while cooling. When sufficiently cooled, the material descends. Model studies indicate that the rising limb of a convection cell is associated with positive gravitational anomaly areas, whereas the sinking portion of the cell coincides with a negative anomaly.

There are two theories of *mantle convection*. One proposes that convection cells are limited to the upper 700 km of the mantle. The other holds that the cells circulate throughout the entire 2900 km-depth of the mantle. Evidence in support of upper mantle convection is stronger: no earthquakes are recorded at depths greater than 700 km; secondly, the observed ratio of heat loss to heat generation at the surface of the earth would be inconsistent with the highly efficient heat transport of the entire mantle convection. The concept of mantle convection helps explain such things as irregular distribution of heat flow within the earth's crust, and the uniformity of basalts along the mid-ocean ridges throughout the world. At present, it offers one of the most acceptable mechanisms for sea-floor spreading, deep-sea trenches, and plate motion. See also: *convection cell; plate tectonics*.

convection cells   In tectonics, a convection cell is a pattern of mass movement of mantle material in which the outer area is downflowing and the central area is uprising due to heat differences. These cells may be of various sizes and shapes. They are not necessarily continuous, and may change velocity or direction, or may stop completely. See also: *convection*.

convergence   1. (Meteorol.) The term used to describe atmospheric motion in which there is a net flow of air into a given volume of the atmosphere; i.e., an increase of mass within that given volume. Mathematically stated, convergence is the contraction of a vector field which, in this case, refers to the atmospheric velocity vector. Cyclones or low-pressure areas are typical regions of convergence at lower levels of the atmosphere. Con-

vergence is normally associated with an increase in cyclonic vorticity. Cf: *divergence*. 2. A meeting of ocean currents or water masses that results in the sinking of the colder, denser, or more saline water. Cf: *divergence*. 3. (Petrol.) *Metamorphic convergence* is a term used when two rocks of different original characteristics become closely similar following metamorphosis. 4. *Stratigraphic convergence* is the gradual decrease in the vertical distance between two sedimentary divisions as a result of narrowing of the intervening strata. 5. (Paleontol.) *Evolutionary convergence* is the appearance of ostensibly similar structures in organisms of different lines of descent; also called *convergent evolution*. Cf: *divergence*.

convergence map   *Isochore map*.

convergent evolution   See *convergence*.

converted wave   A seismic wave that has changed from a P wave to an S wave or vice versa by refraction or reflection at an interface. See also: *seismic waves*.

convolute lamination   Heavily and intricately creased, crumpled, or folded laminae that are contained within an undeformed layer. Parallel, uncontorted layers lie above and below. The structure is thought to be the result of deformation coincident with deposition.

copper   A reddish or orange-red mineral that is the native metallic element, Cu. Small amounts of native copper are found in oxidized zones overlying copper sulfide deposits. It was formerly mined extensively as an ore. Copper is a good conductor of heat and electricity, and is ductile and malleable.

coprolite   A fossilized fecal pellet or casting of animal droppings (fish, reptiles, birds, or mammals). Usually modular, tubular, or pellet shaped, and phosphatic in chemical composition, coprolites provide information about the kinds of organisms that produced them, and these organisms' habits. See *fossils*.

coquina Etymol: Sp., "cockleshell." A limestone formed almost entirely of sorted and cemented fossil debris, usually shells and shell fragments, *Microcoquina* refers to similar rocks composed of finer particles 2 mm or less in diameter. Common among the finer types are those composed of the discs and plates of crinoids. A distinction is drawn between coquina, which is a detrital rock, and coquenoid limestone, which is formed *in situ* and composed of coarse, shelly materials in a fine-grained matrix.

coral Any of a group of bottom-dwelling, attached marine coelenterates of the class Anthozoa. They produce skeletons of calcium carbonate, which may exist individually on sea bottom, or form extensive accumulations (coral reefs). Coral is common in warm seas, and its fossil forms provide important records. Range, Ordovician to present.

coralgal A term applied to the material that makes up the preponderance of a coral reef. It consists of coral fragments and other calcareous organisms, e.g., mollusks, foraminifers, bound together with algal growths.

coralline alga A marine, calcareous red alga commonly associated with coral reefs. It may be a branching form or a massive encrusting type.

coral reefs Wave-resistant ecological units composed predominantly of hermatypic corals and calcareous algae. These reefs fringe Pacific islands and continents where the temperature is suitable (about 18° C, minimum). Dolomite is found at depth in some coral islands. In areas of low rainfall, coral rock, by reaction with phosphates from guano, is altered to phosphate rock composed of apatite. See also: *reef*.

cordaites Gymnosperms ancestral to modern conifers, which they resembled in their soft-wood trunks and parallel-veined leaves. The leaves, however, were bladelike rather than needlelike, as in true conifers. Cordaites grew to heights of 30 to 40 m and were prominent during the Pennsylvanian. Their fossils are found in Late Devonian to Late Permian strata. See also: *fossil plants*.

cordierite An orthorhombic silicate mineral, $(Mg,Fe)_2Al_4Si_5O_{18}$, found commonly in contact metamorphic rocks and low-pressure regional metamorphic rocks. It displays pleochroism visible to the naked eye, appearing blue or violet when viewed parallel to the prism base, and colorless when viewed vertically. When transparent, it is faceted as the gem dichroite.

cordillera Etymol: Span., "chain or range of mountains." A chain of mountains that is usually the dominant mountain system of an extensive mass. As a proper noun, usually plural, the Cordilleras are a mountain system in western North America, including the Sierra Nevada, Cascade Range, and Rocky Mountains. In western South America, it is the mountain system comprised of the Andes and its component ranges. *Cordilleras* is also the term used for the entire chain of mountain ranges that parallel the Pacific coast from Cape Horn to Alaska.

core The core, or innermost part of the earth, is marked by a seismic discontinuity (*Gutenberg discontinuity*) near a depth of 2900 km, at which P-wave velocity decreases from 13.7 to 8 km/sec, and below which S waves are not transmitted. The nontransmission of S waves implies a liquid composition, an inference that has been reinforced by tidal investigations. A corresponding density discontinuity accompanies the seismic discontinuity at 2900 km. It is believed that the source of earth's magnetic field lies in the liquid region of the core. Experimental analysis has shown that in the presence of an initial magnetic field, fluid motions within the core may produce a self-sustaining dynamo action that could induce a system of electrical currents, capable of generating a magnetic field. In general terms, the processes involved in this entail convection in an electrically conducting fluid, with the result that the core acts as a dynamo that maintains and regenerates the magnetic field.

Different mechanisms have been proposed for driving the convective flow within the outer core. A thermally driven flow might be activated by the decay of radioactive isotopes. A compositionally driven flow might occur as an unmixing of dense and less dense regions. One of the more recently proposed concepts states that if the inner core is freezing out of the surrounding liquid, there could be enough heat from the latent heat of crystallization to drive the convective flow. Although the precise energy source is still equivocal, the instability basic to propelling thermal convection is that less dense fluid underlies denser fluid. The high density of the outer core, together with the extremely high electrical conductivity implied by the dynamo theory of earth's magnetic field, suggests that the core may be composed chiefly of iron. Studies of meteorites indicate that the core also contains about 10% metallic nickel. In addition, shock-wave studies at core pressures suggest the presence of sulfur or elementary silicon.

The boundary of the inner core, at about 5100 km, is marked by a P-wave velocity increase of about 10%. As indicated by shock-wave data and calculations, the inner core is probably solid. It may be regarded as the product of the freezing of iron under very high pressure (3000 to 3600 kb). Detailed seismic studies show the presence of a complex structure at the boundary between the outer and inner cores; the same studies also reveal that at least two seismic discontinuities may occur within the transition zone. See also: *crust; mantle.*

core test　A continuous section or core of rock obtained by drilling with a hollow cylinder for the purpose of geological information.

Coriolis Effect　The apparent deflection of a body moving with respect to the earth, as seen by an earthbound observer. This deflection is attributed to a hypothetical force, the *Coriolis force,* but actually is caused by the earth's rotation. It appears as a deflection to the right in the Northern Hemisphere and to the left in the Southern Hemisphere. Aircraft and projectile motion, as well as the motion of a freely falling body from a stationary position, are all influenced by the Coriolis effect. See *centrifugal force; Coriolis force.*

Coriolis force　A hypothetical force that is also known as *deflecting force* and *compound centrifugal force.* It is introduced into the analysis of motions that are measured with respect to rotating coordinate systems, and is one of the equation terms that represents acceleration values as would be observed from the inertial frame of reference. It is measured as $-2\,\vec{\Omega} \times \vec{v}$, where $\vec{\Omega}$ is the angular velocity vector of the rotating coordinate frame and $\vec{v}$ the vector velocity of the motion relative to the rotating system. The Coriolis force is of the same nature as "centrifugal force," in that both are fictitious forces. The former appears to affect moving objects and the latter seems to act on stationary objects on the earth's surface. See *Coriolis effect; centrifugal force.*

corona　A zone of mineralization, often showing radial arrangement, surrounding another mineral. The term has been applied to corrosion rims and reaction rims.

corrasion　1. *Abrasion.* 2. A term used as a syn. of *attrition.* Verb: *corrade.*

correlation　1. Indication of the lithologic or chronologic equivalence of geological phenomena in different regions. 2. In seismology, demonstration of the identity of phases on different seismic records.

corrie Etymol: Gaelic, *coire*, "kettle." Term used in Scotland and England as a synonym of *cirque*. Also spelled *corry*.

corrosion 1. Erosion of rocks by chemical processes; e.g., hydration, oxidation, solution, hydrolysis. The rate of corrosion of a particular area is largely determined by climate, geology, topography, and vegetation. v. *Corrode*. Cf: *corrasion*. 2. The modification (partial resorption, fusion, dissolution) of the outer parts of early-formed crystals or foreign inclusions by the solvent action of the magma in which they occur.

corrosion border One of a series of borders composed of one or more secondary minerals surrounding an early-formed crystal. Such a border indicates the modification of a phenocryst by the corrosive action of the magma in which it occurs. As seen in section, a corrosion border is termed a *corrosion rim*.

corundum A mineral, Al_2O_3, occurring as masses or as variously colored rhombohedral crystals, including ruby and sapphire. It is often found as a primary constituent in igneous rocks containing feldspathoids, and also results from the metamorphism of aluminum-rich rocks. The extreme hardness of corundum, 9 on the Mohs scale, permits it to take a long-lasting, high polish. See also: *emery*.

cosmic dust Dust that exists between the planets, stars, and galaxies. *Primordial dust* is that which remained as residue after their formation. A secondary cosmic dust is constantly produced in the solar system by the disruption of comets that come too close to the sun.

cosmic erosion The wearing away or destruction of rocks on a planetary surface, a cumulative effect of impacts of hypervelocity particles from outer space.

cosmic radiation High-energy subatomic particles from outer space, which strike earth's atmosphere from all directions. *Primary cosmic rays* consist of nuclei of the most abundant elements; upon entering earth's atmosphere, most of the primary rays collide with atomic nuclei in the atmosphere, producing secondary cosmic rays, which consist mainly of elementary particles; e.g., neutrons, mesons.

cosmogeny The study of the age and origin of the universe; in particular, the earth and solar system.

cosmogony The study of scientific hypotheses and cultural myths regarding the origin of the universe.

cosmology The branch of astronomy concerned with both theoretical and observational aspects of the universe as a systematized whole.

coteau Etymol: Canadian Fr., "slight hill." A term used in the U.S. for features such as a side of a valley, a hilly upland region, or a prominence forming the edge of a plateau.

cotectic Conditions of temperature, pressure, and composition under which two or more solid phases crystallize simultaneously from a single liquid over a finite interval of decreasing temperature are said to be *cotectic*; e.g., crystallization of magma on a phase diagram. On a phase diagram, the line or surface representing the corresponding phase boundary on the liquids is called the *cotectic boundary line*.

cotton ball *Ulexite.*

cotylosaur Any member of the reptile order *Cotylosaurea*. Lizard or turtlelike in structure. Range, Lower Pennsylvanian to Upper Triassic.

coulee Etymol: Fr., "to flow." In the U.S. and Canada, an extensive system of drainage trenches that are the abandoned sluiceways of glacial meltwaters. It may also be a long, trenchlike gorge that once carried meltwater, e.g., Grand Coulee in Washington state. A tonguelike flow of lava flowing from a crater or the flank of a cone is called a *volcanic coulee*. When a lava flow dams a stream valley, the confined portion of the stream is called a *coulee lake*.

couloir   Etymol: Fr., "passage," "corridor." A steep valley or gorge on a mountainside, esp. the Alps. It is also used in mountain-climbing parlance for a passage in a cave.

country rock   A general term for the older rock containing an igneous intrusion or penetrated by mineral veins. Also called *host rock*. Country rock may be sedimentary, metamorphic, or older igneous rock. See *intrusion*.

cove   1. One of the lesser indentations of a coast, often the result of differentiated marine erosions. 2. A sheltered hollow or recess in a mountain. 3. A sheltered area between hills or woods.

covellite   A mineral, CuS, found in the zone of sulfide enrichment; forms as an alteration product of other copper minerals; e.g., chalcocite, chalcopyrite. It occurs most often as slaty or compact indigo-blue masses with pronounced iridescence.

crag and tail   A streamlined ridge or hill resulting from glaciation. It consists of a knob (the "crag") of resistant bedrock, with an elongate body (the "tail") or more erodible bedrock on its lee side. Cf: *roche moutonnée*.

crater   1. A *volcanic crater*. 2. A saucer-shaped depression on the earth's surface resulting from a meteorite impact or bomb explosion. 3. A *lunar crater*.

crater lake   A lake formed by precipitation, and the accumulation of groundwater in a volcanic crater or *caldera*.

craton   A part of the earth's crust that has been stable (no orogenic activity) for at least 1500 m.y. Cratons are typically formed of Lower to Middle Precambrian igneous rocks and metamorphosed sedimentary and volcanic rocks, and overlain by generally flat-lying Upper Precambrian or younger sedimentary rocks. They consist of former geosynclinal deposits that have been added to earth's continental nuclei. *Shield* areas are the portion of cratons that are generally exposed, even during times of maximum continental submergence.

creek   1. An informal term used generally in the U.S. and some parts of Canada for a watercourse larger than a brook but smaller than a river. 2. A channel or streamway in a coastal marsh.

creep   1. Slow, imperceptible gravity movement of soil and/or broken rock material from higher to lower levels, or the material itself that has so moved. Accompanying shear stresses may produce permanent deformation, but are too small to cause shear failure, as in landslides. 2. Slow deformation produced by long-acting stresses below the elastic limit of the material subjected to stress. Part of the deformation is elastic (non-permanent) and part is permanent. Cf: *landslides*.

creep recovery   The gradual recovery of elastic strain after stress is released.

crenulation   Small-scale folding superimposed on folding of a larger scale. Crenulation may occur along cleavage planes of a deformed rock.

crescentic fracture   A lunate-shaped mark on a glaciated rock surface, convex in the direction from which the ice moved. It is larger than a *chattermark* and consists of a single mark with no removal of rock material. Syn: *crescentic crack*. Cf: *crescentic gouge*.

crescentic gouge   A crescent-shaped groove resulting from *glacial plucking* on a bedrock surface. It is concave in the direction from which the glacier moved, and is comprised of two fractures from which rock material has been gouged. Syn: *gouge mark*. Cf: *crescentic fracture*.

crest   1. The highest part of a mountain or hill. 2. A ridge or ridgelike formation. 3. The highest part of an anticline fold. The line connecting the highest points on a given stratum in an infinite number of cross-sections is the *crest line*. The surface formed by all crest lines is the *crest surface* or *crestal plane*. 4. The highest part of a wave.

Cretaceous   The interval of geologic time that began about 136 million y.b.p., and lasted about 71 m.y. It is the final period of the Mesozoic Era, and precedes the Tertiary Period. Its name, proposed by Omalius d'Halloy in 1822, referred to chalk (*creta* in Latin), the characteristic rocks found during the period; e.g., the White Cliffs of Dover, England. Rocks of the Cretaceous are distributed throughout the world and vary in thickness and lithographic character. Large-scale marine inundation during this period resulted in extensive covering of lower areas by chalk, clay, and marl. Clastic sediments such as sandstones, shales, and tuffs accumulated in certain geosynclinal areas. The Early Cretaceous is characterized by deltaic and lacustrine deposits, and the Late Cretaceous by marine deposits, namely chalk, sandstone, and siliceous rocks. The extensive Cretaceous shallow submergence favored the accumulation of fossil fuels, e.g., coal, oil, and gas. Other important products derived from Cretaceous strata are bauxite (Europe) and gold, silver, copper, lead, and zinc (circum-Pacific island areas).

The Nevadan orogeny took place in the Sierra Nevada and Klamath Mountains of western North America in Late Jurassic and Early Cretaceous times. The Laramide Revolution shaped the Rocky Mountains in Late Cretaceous and Early

Tertiary times. Mountain building of the South American Andean System climaxed in the middle of the Late Cretaceous. Volcanic activity is indicated by volcanic seamounts of Cretaceous age in the central and western Pacific. Climatic zones, as indicated by floral and faunal distribution, are generally divided between the tropical to subtropical equatorial regions and the cooler, higher latitude regions. Warmer ocean temperatures are indicated by oxygen-isotope determination in belemnites. Except for mountain glaciers, glaciation was absent during the Cretaceous.

Ammonites were predominant during the Cretaceous (some reached a diameter of 2 m), but became extinct by the end of the period. Belemnites also disappeared at this time. Bivalves and gastropods are variously distributed and serve as zone fossils when ammonites are absent. Echinoderms, crinoids, sponges, brachiopods, and bryozoans occupied numerous marine habitats. Dinosaurs, the dominant vertebrates of the Cretaceous, were found on every continent, but became extinct by the end of this period. Other vertebrates include pterosaurs and two diving birds, Ichthyornis and Hesperornis. Mammals were still indistinct. Early Cretaceous land plants (conifers, gingkoes, ferns) differed little from those of the Jurassic; angiosperms became important in mid-period. Late Cretaceous flora included magnolias, figs, poplars, willows, and plane trees.

crevasse   1. *Glacial crevasse*. 2. A deep opening in the earth that appears after an earthquake. 3. A break or breach in a river bank, such as a natural *levee*.

crevice   Any narrow opening, cleft, or rift in a structure or formation. Often used in the sense of *crevasse*.

crinoid   An echinoderm of the class Crinoidea, commonly called "sea lily." Range, Lower Cambrian to present. Fossils show that some were free swimming, but most were anchored to the sea bottom by a stem or stalk made of variformed discs, surmounted by a cup or calyx bearing radially directed tentacles. These living forms are attached to the sea bottom, but most are unstalked. Crinoids are important Paleozoic index fossils.

crinoidal limestone Sedimentary rock composed almost totally of fossil crinoidal parts, esp. the stem plates.

cristobalite A silicate mineral, SiO_2, of two polymorphs: a high-temperature polymorph that is stable above 1470° C, and a low-temperature form that is stable below 268° C. It occurs in cavities in felsic igneous rocks.

critical angle The minimum *angle of incidence* at which there is total reflection, or when an optic, acoustic, or electromagnetic wave travels from one medium to a second that is less refractive.

critical point For a given substance, the *critical point* represents the set of temperature and pressure conditions at which two phases become physically indistinguishable. The temperature and pressure at the critical point are the *critical temperature* and *critical pressure*. At temperatures above the critical temperature the substance can exist only in the gaseous state, no matter how great the pressure exerted. At this temperature, both liquid and gaseous phases merge and become identical in all properties. The *critical pressure* is the minimum pressure necessary to liquefy a gas at its critical temperature.

critical pressure See *critical point.*

critical temperature See *critical point.*

critical velocity Although hydrological engineers have several other meanings for the term, the most common definition of *critical velocity* is that velocity at which the nature of the flow for a given fluid changes from *laminar* to *turbulent* or vice versa.

crocidolite A silicate of sodium and iron, an asbestiform variety of riebeckite. Syn: *blue asbestos.*

Croixan (croy-an) The name for the North American Upper Cambrian, named for the exposures about St. Croix Falls, Minnesota.

cross-bedding See *bedding.*

cross-correlation A measure of the similarity of or a method for the comparison of sequences of data; e.g., a measure of the similarity of two seismic wave forms. Cf: *correlation.*

crosscut In mining, a tunnel or level of access across the county rock so as to intersect a vein or ore-bearing structure.

crossed Nicols In a polarizing microscope, two Nicol prisms arranged so that their respective planes for transmitting polarized light are at right angles to each other, in which position light emerging from one prism is disrupted by the other.

crossopterygian A lobe-finned bony fish of the order Crossoptergii, believed to be the predecessor of the modern lungfishes. The order appeared in Middle Devonian time and includes the *coelacanths.*

cross ripple mark *Interference ripple mark.*

cross-section A *cross-section,* also called *a transverse section,* is made by a plane that cuts the longest axis of an object at right angles. The term is also used for the pictorial representation of the features exhibited by such a plane cut; e.g., a cross-cut of an orebody or fossil. A *longitudinal cross-section* is a diagram drawn on a vertical plane and parallel to the longer axis of a given feature; e.g., a section drawn parallel to the axis of a fossil. Despite the difference between the two terms, *cross-section* is commonly used for any cut that shows transected geologic features.

crude oil *Petroleum* as it emerges in its natural state, before distillation or cracking.

crush breccia A breccia formed basically in place by fragmentation, as a result of faulting or folding. Cf: *crush conglomerate; autoclastic.* Syn: *cataclasite; tectonic breccia.*

crush conglomerate A rock similar to *crush breccia*, but the fragments of which are more rounded. It closely resembles a sedimentary conglomerate. Syn: *cataclasite; tectonic conglomerate.*

crushing strength The compressive stress required to cause a given solid to fail by fracture.

crust A term often imprecisely used to mean the lithosphere, or used as a convenient lay expression for the near-surface part of the earth. In geology, however, it is strictly defined as the outermost layer of the earth, the lower boundary of which is the *Mohorovičić discontinuity* (Moho). Part of the crust is equivalent to the *sial* (silica and alumina), and part to the *sima* (silica and magnesia). The entire crust represents less than 0.1% of the earth's total volume, and is non-homogeneous in composition and distribution. The *continental crust* is distinct from the *oceanic crust* in age, as well as in physical and chemical characteristics.

Continental Crust: The most ancient parts of the continental crust are about 4 b.y. old. It is constantly modified by tectonics, volcanism, sedimentation, and erosion, and is composed of two sublayers that together have a thickness of 25 to 60 km. The thicker crust, though often located beneath mountains, does not necessarily bear a relation to altitude. The continental crust is an assemblage of igneous, sedimentary, and metamorphic rocks rich in elements such as silicon and potassium, and holds much more uranium than does the oceanic crust. Granite and gneiss are dominant in the shields. The density of the upper layer is ~ 2.7 gm/cm^3; the velocity of crustal

seismic P waves varies greatly: › 6.2 km/sec to ‹ 6.5 km/sec. The *Conrad discontinuity* is believed to separate the upper from the lower continental crust. Along certain points of this postulated boundary, the abrupt increase of seismic wave velocities indicates a density of 3.0 gm/cm^3 for the lower continental crust.

Oceanic Crust: This part of the crust is much younger than the continental crust, the oldest region of the ocean basin being less than 200 m.y. in age. The thickness of the ocean crust (5 to 10 km) is much less than that of the continental crust; its density is in the order of 3.0 gm/cm^3. Oceanic crust is constantly being generated along mid-ocean ridges, and destroyed as it moves to a subduction zone. (See *sea-floor spreading.*) *Gabbros* and other rocks make up the lower layer of the oceanic crust, and *basalts* are the main material of the upper. There are three suboceanic layers as defined in terms of the velocities of P waves that travel through them. However, the velocity pattern in the crust is complex, and the idea that the velocity in these layers is constant has been seriously questioned. Therefore, instead of using absolute wave velocity as an indicator, the velocity gradient, i.e., the change in wave velocity with depth, is used. See also: *core; mantle.*

crustacean Any arthropod of the subphylum Crustacea, characterized chiefly by two pairs of antenna-like appendages in front of the mouth and three pairs behind it. Present groups are represented by shrimp, crabs, lobsters, copepods, and isopods; most forms are marine. They are commonly found in various strata, but are not geologically important. Range, Cambrian to present.

crustal plate A *lithospheric plate.*

cryoconites Tubular depressions that are melted into glacier ice at locations where areas of dark particles absorb the sun's rays.

cryolite A white or colorless

(transparent to translucent) mineral, Na_3AlF_6, usually massive; crystals are rare. The only important deposit is at Ivigut, Greenland, where it is found in granite. Formerly used as a source of aluminum, it is now used in the manufacture of certain glass and porcelain, and as a metal-cleaning flux.

cryomorphology The subspecialty of geomorphology pertaining to the various processes of cold climates.

cryopedology The study of processes associated with intensive frost action and the occurrence of frozen ground, including techniques used to surmount or minimize problems connected with it.

cryoplanation Degradation of a land surface by processes associated with intensive frost action that is augmented by the action of moving ice, running water, and other agents. Cf: *altiplanation; equiplanation.*

cryosphere The part of the earth's surface that is frozen throughout the year.

cryptocrystalline Said of a crystalline aggregate that is so finely divided that individual crystals cannot be recognized or distinguished under an ordinary microscope. Cf: *microcrystalline.*

cryptoexplosion structure A large, roughly circular impact structure, the origin of which is unknown or only supposed. Many such structures are believed to derive from crater-generating meteorites, and others from uncertain volcanic activity. The term *cryptoexplosion structure* includes or largely explains the term *cryptovolcanic structure.*

cryptoperthite An extremely fine-grained, intercombined sodic and potassic feldspar on which lamellae are discernible only by means of X-rays or with the use of an electron microscope. Cf: *perthite.*

cryptovolcanic structure See *cryptoexplosion structure.*

Cryptozoic Term meaning "hidden life." It is that part of geologic time represented by rocks in which evidence of life is slight and only of primitive forms. Cf: *Phanerozoic.*

cryptozoon Presumed fossil formations, such as the cryptozoon "reefs" of the Cambrian. These have been largely determined to be *concretions* or *stromatolites.*

cryoturbation *Congeliturbation.*

crystal A homogeneous solid body in which the atoms or molecules are arranged in a regular, repeating pattern that may be outwardly manifested by plane faces.

crystal axes Three imaginary lines in a crystal (four in the hexagonal system) that pass through its center; the structure and symmetry of a particular crystal are described by the relative lengths of these lines and their mutual angular orientation. The ratio of unit distances along the axes to the angle between them defines the *axial elements.* The *axial ratio* is obtained by comparing the length of a crystal axis with one of the lateral axes taken as unity. Crystal axes may coincide wholly or in part with axes of symmetry. Syn: *crystallographic axes.*

crystal class One of the 32 possible combinations of symmetry in which crystals can form; such combinations are the only possible arrangement of symmetry axes and planes in which these elements all intersect at a common point. Syn: *point group.* See also: *crystal system.*

crystal flotation Floating of lightweight crystals in magma. Cf: *crystal settling.*

crystal fractionation See *magmatic differentiation.*

crystal gliding *Translation gliding.*

crystal habit The characteristic or general form of crystals; e.g., cubic, prismatic. Some minerals are always and everywhere bounded by faces of the same crystal form, whereas in other instances, a mineral at one locality may have a different appearance or habit from the same mineral at another locality.

crystal lattice The three-dimensional arrangement of the atoms and ions that form a crystal. Auguste Bravais, a French crystallographer, was the first to show that there can only be 14 possible basic kinds of crystal lattices. Thus, they are commonly called *Bravais lattices*.

crystalline Related to the nature of a crystal, i.e., having a regular atomic or molecular structure. Ant: *amorphous*.

crystalline rock A rock consisting of minerals in a clearly crystalline state. Also, a rather general term for igneous and metamorphic rocks as distinct from sedimentary rocks.

crystallinity The degree to which a rock shows crystal development, or the extent of that development. See *holocrystalline; hypocrystalline; macrocrystalline; microcrystalline; cryptocrystalline*.

crystallization The gradual formation of crystals from a solution or a melt.

crystallization differentiation See *magmatic differentiation*.

crystallization interval In a cooling magma, the interval of temperature between the formation of the first crystal and the disappearance of the last drop of liquid from the magma. For a given mineral, it is the temperature range over which that particular phase is in equilibrium with liquid. Syn: *freezing interval*.

crystalloblast A crystal of a mineral formed completely by metamorphic processes. One type of crystalloblast, an *idioblast*, is a mineral constituent of a metamorphic rock formed by recrystallization and bounded by its own crystal faces. A *xenoblast* has formed in a rock during metamorphism without developing its characteristic crystal faces, and a *hypidioblast* is a subhedral crystal formed during metamorphism.

crystalloblastic Said of a crystalline texture formed by metamorphic conditions of high viscosity and regional pressure, as distinct from igneous rock textures that are the result of a sequential crystallization of minerals under conditions of low viscosity and near-uniform pressure.

crystallographic axes *Crystal axes*.

crystal mush Magma that is partially crystallized.

crystal settling In a magma, the sinking of crystals of greater density, resulting in crystal accumulation. Syn: *crystal sedimentation*. Cf: *crystal flotation*.

crystal system One of six groupings or classifications of crystals according to common symmetry characteristics that permit them to be referred to the same crystallographic axis. *Isometric* crystals are characterized by three mutually perpendicular axes of equal length. Syn: *cubic system*. *Tetragonal crystals* have three mutually perpendicular axes, the vertical one of which is shorter or longer than the two horizontal axes, which are of equal length. Crystals of the *orthorhombic system* are characterized by three mutually perpendicular symmetry axes, all of different relative length. *Monoclinic crystals* show three unequal crystallographic axes, two of which intersect at an oblique angle, and the third of which is perpendicular to the plane formed by the other two. *Triclinic crystals* lack symmetry other than a possible center; they are characterized by three unequal axes that are mutually oblique. Crystals of the hexagonal system have three lateral axes of equal length intersecting at a 60° angle to each other, and a vertical axis of different length perpendicular to the other three.

cubic cleavage   See *cleavage.*

cubic packing   See *packing.*

cubic system   *Isometric system.*

cuesta   Etymol: Sp., "sloping ground." A low, asymmetrical ridge with a gentle slope on one side, conforming with the dip of the underlying strata, and a steep face on the other. Cuestas are the result of erosion on gently sloping sedimentary rocks. The top, which is formed on a resistant layer, remains as the cliff face is

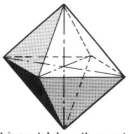

(A) Isometric crystals have three mutually perpendicular axes of equal length.

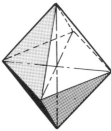

(D) Monoclinic crystals have three axes, all of different length, only two mutually perpendicular.

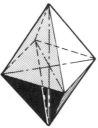

(B) Tetragonal crystals have three perpendicular axes, two of equal length.

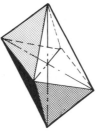

(E) Triclinic crystals have three axes, all of different length and none perpendicular.

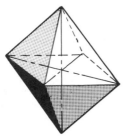

(C) Orthorhombic crystals have three perpendicular axes, all of different length.

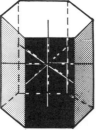

(F) Hexagonal crystals have four axes, three of equal length at a 60° angle to each other, the fourth a different length and perpendicular to the others.

Types of crystals

eroded back. Cuestas are found primarily on recently uplifted coastal plains and on geologically old folds or domes. Syn: *wold*. See also: *hogback*.

culm A local term, esp. in eastern Pennsylvania, for fine particles of anthracite coal. In certain regions, tremendous culm piles accumulate and eventually are overgrown with vegetation. The term is also applied to certain rocks consisting of sandstones and shales with some fine bands of crushed coal.

culmination The highest point of some geological structural feature; e.g., a dome or anticline. Several culminations may be found along a mountain range. The highest topographic point of an eroded-fold mountain may or may not be coincident with the culmination.

cumulo-volcano *Volcanic dome.*

cupriferous *Copper-bearing.*

cuprite A secondary copper mineral, Cu_2O, found in the oxidized zone of copper deposits associated with limonite and other copper minerals. Its common occurrence as ruby-colored isometric crystals has given it the name of "ruby copper."

cuprous Of or relating to copper.

Curie point The temperature at which ferromagnetic material becomes paramagnetic. The *upper Curie point* is the temperature above which ferroelectric materials become depolarized. Below the temperature of the *lower Curie point*, some ferroelectric materials are depolarized.

current bedding See *bedding.*

current ripple mark See *ripple mark.*

curvature correction An adjustment made to an observation or calculation to compensate for earth's curvature. In geodetic leveling, allowances are made for the joint effect of curvature and atmospheric refraction.

cusp One of a sequence of sharp, seaward-directed projections of beach material, separated at more or less regular intervals by shallow lunate troughs. The cusp-to-cusp distance depends upon the general shore contour, beach material, and wave and tidal patterns; thus, a certain cusp-and-trough design is often characteristic of a particular stretch of beach.

cutoff 1. *Chute cutoff.* 2. *Neck cutoff.* 3. For map making and cross-sections, a cutoff is a boundary, normal to the bedding, indicating the areal borders of a particular stratigraphic unit that is not defined by natural features. 4. An impermeable barrier placed within or beneath a dam to block or deter seepage.

cutoff grade The lowest grade of mineralized material, i.e., material of lowest assay value considered as ore in a particular deposit.

cutoff limit *Assay limit.*

cutoff spur See *meander.*

cutout Mass of siltstone, sandstone, or shale occupying the space of an erosional channel cut into a coal seam.

cwm Welsh word for *corrie.*

cycle 1. A recurring period of time, esp. one in which certain events or phenomena repeat themselves. 2. A series of occurrences repeated at regular intervals in the same order and terminating under conditions that are the same as at the beginning; e.g., a *cycle of sedimentation*. 3. A series of events in which, upon completion, the last phase is unlike the initial phase.

cycle of denudation *Cycle of erosion.*

cycle of erosion Also called

geomorphic or *geographic cycle*, this is a theory of the evolution of landforms. First set forth by W.M. Davis, it assumes progressive development of landforms. The initial uplift that elevates the land is followed by dissection and denudation of the region by streams. Further erosion reduces it to an old-age surface, or a *peneplain*. The cycle could be interrupted at any time by further uplifts and returned to the "youthful" stage (rejuvenation). The theory also assumes that the history of a landscape proceeds directly from a known stage according to a predetermined framework but, at present, it is generally thought that the initial conditions of a region do not necessarily determine the final form. Instead, there appears to be an eventual isostatic readjustment, i.e., a dynamic equilibrium between crustal elements of the earth and the processes that act upon them.

cycle of sedimentation   1. An ordered, repeated sequence of associated processes and conditions recorded in a sedimentary deposit. 2. The deposition of sediments in a basin between the onset of any two successive marine transgressions. 3. A *cyclothem*. Syn: *sedimentary cycle*. See also: *rhythmite*.

cyclic evolution   The concept that in lineages of many life forms, evolution took a strong and rapid initial course, which was followed by a long phase of slow and moderate change. Overspecialized or inadaptive forms became extinct during a final brief episode.

cyclosilicate   One of the group of *silicates* in which the SiO_4 tetrahedron forms rings by linking together. Regardless of the number of such links in a ring (3, 4, or 6), all the rings have a 1:3 ratio of silicon to oxygen. Beryl, $Be_3 Al_2 Si_6 O_{18}$, is a cyclosilicate. Syn: *ring silicate*.

cyclothem   A recurring series of beds deposited during a sedimentary cycle of the type that pervaded during the Pennsylvanian Period. Such cycles consist of recurring series of sandstones, clays, coals, and limestones, indicating repeated oscillations of the sea level, with intervening periods during which terrestrial conditions necessary for the growth of coal plants obtained. Marine sediments typically occur in the upper half of a cyclothem and non-marine sediments in the lower half. Because most cyclothems are incomplete, a theoretical cyclothem called the *ideal cyclothem* is used to represent, within specified stratographic intervals and regions, the optimum succession of deposits during a full sedimentary cycle. Both experimental data and theoretical assumptions are used to construct such a cyclothem.

cylindrical fold   A fold described geometrically as a surface generated by a line moving through space in a position parallel to itself. The axis in such folds is perpendicular to the plane represented by the girdle, which is the projection of the plane containing all perpendiculars to the bedding. Cf: *conical fold*.

cylindrical projection   See *map projections*.

Cyprus-type deposit   A copper deposit of the type found in Cyprus. Such deposits are believed to form on the sea floor. Water in sea-floor rocks, heated by magma, reacts with the sea-floor basalt and takes into solution many minerals present in the rock. As rising hot water is cooled and diluted by cold sea water, elements such as copper, nickel, and cadmium, and others as sulfides, are precipitated in rock fractures.

cystoid   A member of an extinct group of primitive echinoderms. Range, Cambrian-Ordovician to Late Devonian. The cystoid, saclike body of these animals, had a stalk attached to the sea floor; the body may be less symmetrical than that of other echinoderms. Some cystoids are important index fossils.

D

dacite A light-colored igneous rock that is the approximate extrusive equivalent of *quartz diorite*. Its principal minerals are plagioclase and quartz, which occur as phenocrysts in a glassy to microcrystalline groundmass. Mafic phenocrysts of any composition may also be present. Dacite is difficult to distinguish from *rhyolite* without microscopic examination.

dalles A term sometimes used for rapids in a deep, narrow river or stream flowing through a gorge or canyon, for example, the Dalles of the Columbia River. In Wisconsin, the geologic feature called the *Dells* is a channel system cut into Cambrian sandstone, through which the Wisconsin River flows. The ravine complex was originally called the *Dalles*, very likely from the French, *dalle*, meaning "flagstone," in reference to the clearly fissile sandstone that borders the channel ways.

darcy A unit expressing the permeability coefficient of a rock; e.g., in calculating the flow of water, gas, or oil. The customary unit is the *millidarcy* (md), equivalent to 0.001 darcy.

Darcy's law A law stating that the flow rate of a fluid through a porous material varies directly with the products of the pressure gradient causing the flow and the permeability of the material. The formula that is derived from Darcy's law assumes laminar flow and negligible inertia. It is used in studies of water, gas, and oil from underground structures.

dark mineral Any one of a group of rock-forming minerals showing a dark

color in thin section. Examples are biotite, augite, hornblende, and olivine.

Darwinism The doctrine proposing that organic evolution derives from variation and selection of preferred individuals through *natural selection*. Syn: *Darwinian Theory*.

dating methods Various methods used to analyze substances, and to determine when they were formed. The particular method chosen depends upon the approximate magnitude of age. The *radiometric* dating technique involves the ratio of radioactive parent atoms to their daughter atoms, and comparison of this ratio to the known *half-life* of the radioactive element. Radiometric tests can be used to determine the age of crystallization by metamorphism, the age of fold mountains, and the age of sedimentary rocks in which new materials formed during deposition. The precision of the calculations depends on how accurately the half-life of the radioactive element is known, whether any of the parent elements and/or their daughter elements have been removed from or added to the rock after formation, and the accuracy of the assumption that the time interval for rock formation is short, compared with the total age of the rock.

Listed below are the four principal elements used to determine age in methods based on radioactive properties. Of all the radiometric methods used in geochronology, the *uranium-lead method* is the most useful, especially for rocks more than 100 million years old. Uranium-lead determinations are usually made on zircon and sphene, since they contain uranium and are associated with many igneous and metamorphic rocks. The *potassium-argon method* can be used to date rocks ranging in age from the very beginning of geologic time to as recently as 30,000 years ago. This method, too, is less precise for younger rocks, because of the relatively small amount of daughter product that has formed in them. Minerals such as hornblende, nepheline, biotite, and

Parent Isotope	Stable Daughter Element	Half-life in Years
uranium (U-238)	lead (Pb-206)	4.5 billion
uranium (U-235)	lead (Pb-207)	0.7 billion
potassium (K-40)	argon(Ar-40)	1.3 billion
rubidium (Rb-87)	strontium (Sr-87)	47.0 billion

muscovite are used in K-Ar dating, inasmuch as one of these occurs in most igneous and metamorphic rocks. Any rocks older than 20 million years can be dated by the *rubidium-strontium* method; younger rocks can also be dated by this method if they contain enough Rb-87. Rubidium occurs in such minerals as muscovite, biotite, and all potassium feldspars.

Carbon-14 (radiocarbon) is often used to date materials and events that are relatively recent; it is particularly valuable for use on materials less than 50,000 years old, as this time span is too brief for the more slowly disintegrating elements to produce measurable amounts of daughter products. The radiocarbon method has been widely used in dating archeological finds, and provides the best dating for the Holocene and Pleistocene. The method is based on the assumption that the isotope ratio of carbon in the cells of living things is identical with that in air because of the balance between photosynthesis and respiration. Although several sources of error are introduced into this method, such as the inconstant supply of C-14, both in time and latitude, it has proved to be enormously useful. See *Carbon-14*.

Varves, which are pairs of thin sedimentary layers deposited within a 1-year period, have been used to correlate lake deposits for many thousands of years, back into the Pleistocene and Holocene. Each varve pair consists of a fine winter layer and a coarse summer layer. The ages of deposits are determined by counting the pairs in a sample sediment core. See also: *rhythmites*.

Dendrochronology, or *tree-ring dating*, is a technique of dating and interpreting past events, especially paleoclimates, based on tree-ring analysis. A core, extending from bark to center, is taken from a tree. After the rings are counted and their widths measured, the sequence of rings is correlated with "deciphered" sequences from other cores. Tree-ring counts are used to calibrate dates that have been determined by carbon-14 dating. In so doing, it becomes necessary to modify the assumption of a constant ratio of radioactive to nonradioactive carbon at the time when the organic material to be dated ceased to exchange with the atmosphere.

datum   A fixed or assumed point, line, or surface used as a base or reference for the measurement of other values. It is also that part of a bed of rock on which structure contours are represented. A *datum level* is any level surface (usually mean sea level) from which elevations are reckoned. The term *datum plane* applies to some permanently established horizontal surface that serves as a reference for water depth, ground elevation, and tidal data. In seismology, the datum plane is used to eliminate or minimize local topographic effects to which seismic velocity calculations are referred. Syn: *reference plane*.

daughter element   An element formed from another through radioactive decay; e.g., radium is the daughter element of thorium.

Davisian   Relating to the concept and writings of W.M. Davis (1850-1934), an American geologist who was responsible for many new geomorphologic concepts, such as the geomorphic cycle.

dead glacier   One that does not have

an *accumulation area*. Ant: *active glacier*. See also: *glacier*.

dead line The level in a batholith dividing the metalliferous region above from the economically barren region below.

dead sea A body of water in which there are no normal aquatic organisms, and from which evaporites are being or have been precipitated. The Dead Sea in the Near East is the type locality.

dead valley *Dry valley*.

death assemblage *Thanato-coenosis*.

debouch In physical geography, to emerge from a comparatively narrow valley out upon an open plain or area. *Debouchement*, also *debouchure*, is a mouth or outlet of a river or pass.

debris 1. Any surface accumulation of material (rock fragments and soil) detached from rock masses by disintegration. Syn: *rock waste*. 2. Rock and soil material dumped, dropped, or pushed by a glacier, or found on or within it. 3. Interplanetary material, including cosmic dust, meteorites, comets, and asteroids.

debris avalanche A slide of rock debris in narrow channels or tracks down a steep slope, often initiated by heavy rains. Common in humid regions.

debris cone 1. Fine debris, piled in a conical mound, lying at the angle of repose on certain landslide-transported boulders. 2. A steep-sloped alluvial fan, usually consisting of coarse material. Syn: *alluvial cone*. 3. A mound of snow or ice on a glacier, topped with a debris layer thick enough to prevent ablation of the material underneath. Syn: *dirt cone*.

debris fall The nearly free falling of weathered mineral and rock material from a vertical or overhanging face. Debris falls are very common along the undercut banks of streams.

debris flow Moving mass of mud, soil, and rock fragments of which more than half the particles are larger than sand size. Such flows range in velocity from less than 1 m/yr to 160 km/hr. Cf: *mudflow*.

debris line *Swash mark*.

debris slide The rapid sliding or rolling of predominantly unconsolidated earth debris without backward rotation in the movement.

decke German equivalent of *nappe*, but occasionally used in English language literature.

declination The horizontal angle between true north and magnetic north; varies according to geographic location. An instrument that measures magnetic declination is called a *declinometer*.

declivity 1. A slope that descends from some reference point. Ant: *acclivity*. 2. A surface gradient.

décollement Etymol: Fr., "unsticking." A detachment structure of strata, associated with overthrusting. It results in disharmonic folding, i.e., independent patterns of deformation in the rocks above and below the décollement. In attempting to explain the extreme disharmonic folding in the Jura Mountains, many proposed theories have assumed the presence of a décollement.

decomposition *Chemical weathering*.

deconvolution A technique designed to restore a wave configuration to the shape it is assumed to have taken prior to filtering. It is applied to seismic reflection and other data for the purpose of improving the visibility and clarity of seismic phenomena.

decrepitation To heat or roast a mineral until it crackles or until crackling ceases.

dedolomitization A process resulting from contact metamorphism, wherein dolomite is broken down into its two components, $CaCO_3$ and $MgCO_3$. The former crystallizes into a coarse calcite, whereas the latter further breaks down into magnesium oxide and carbon dioxide. The magnesium oxide will usually occur in the rock as brucite, whereas in the presence of silica, magnesium silicates, e.g., forsterite, are formed. See also: *dolomitization*.

deep A readily discernible depression on the ocean floor, generally implying depths exceeding 5500 m (3000 fathoms). Syn: *abyss*.

deep-focus earthquake An earthquake whose focus is at a depth of 300 to 700 km. Syn: *deep earthquake*. Cf: *shallow-focus earthquake; intermediate-focus earthquake*.

deep-scattering layers Sound-reflecting oceanic layers that frequently appear on the echograms of depth recorders. The layers generally rise to the surface at dusk and descend to their daytime depths at sunrise. This phenomenon has been attributed to the vertical migration of marine fauna.

deep-sea cones See *submarine fans*.

Deep Sea Drilling Project A program sponsored by the U.S. Government with the support of the National Science Foundation to investigate the history of earth's ocean basins by means of drilling in deep water from the research vessel *Glomar Challenger*. Abbrev: D.S.D.P.

deep-seated Said of geological features and processes originating or located at depths of 1 km or more below earth's surface; *plutonic*.

deep-well disposal The disposal of liquid waste material by injection into specially constructed wells that penetrate deep, porous, and permeable formations; the formations hold mineralized groundwater, and are vertically confined by more or less impermeable beds. Such a method is used for the disposal of many industry-related liquid wastes. Syn: *deep-well injection*.

defect lattice A crystal lattice in which the usual lattice perfection is interrupted by an omission or by inclusion of an extraneous item. The technique of creating lattice imperfections is basic to semiconductor technology.

deflation The removal of material from a desert, beach, or other land surface by the action of wind. In many cases, particularly in desert regions, deflation lowers the surface, thus producing *deflation basins*.

deflecting force See *Coriolis force*.

deflection of the vertical The angle at a given point on the earth between the vertical, as defined by gravity, and the direction of the normal to the reference ellipsoid, where the normal is drawn through the given point. Sometimes referred to as *deviation* or deflection of the plumb line.

deformation A general term for processes of rock deformation resulting from various earth forces; they include faulting, folding, shearing, compression, and extension.

deformation ellipsoid *Strain ellipsoid*.

deformation fabric The spatial orientation of the components of a rock as imposed by external stress. During such stress conditions, minerals are translated or rotated, or new minerals develop in common orientation. Syn: *tectonic fabric*.

deformation plane The plane that is parallel to the direction of movement and normal to a flow surface; the a-c plane of structural petrology.

deformation twinning Twinning produced by deformation and gliding. A portion of the crystal structure is sheared in such a way as to produce a mirror image of the original crystal. Deformation twins may be distributed throughout the crystal, or can take over most or all of a crystal. Syn: *mechanical twinning*. See *twinning*.

degradation A gradational process that produces a general leveling of land by removal of material through erosive processes. Ant: *aggradation*. Cf: *denudation*.

degrees of freedom The number of independent variables, e.g., temperature, pressure, and concentration, in the different phases that must be specified in order to describe a system completely. See also: *phase rule*.

delayed runoff Water from precipitation that penetrates the ground and later discharges into streams via springs and seeps. The term is also used for water that is temporarily stored as snow or ice.

delay time In *refraction shooting*, the delay times are two factors, D_1 and D_2, the sum of which is equal to the intercept time; D_1 is that component associated with the shot end of the trajectory, and D_2 with the detector end. Because the depths computed from intercept times represent the sum of the respective depths below the shot and below the detector, most refraction results are intrinsically ambiguous. The concept of delay time is a technique for separating the depths at the two ends of the trajectory. The depths at each end of the trajectory can be determined if the intercept time can be separated into its component delay times.

If the interface between the media is horizontal, each of the two delay times is half the intercept time.

$$D_1 = z \frac{\sqrt{V_1^2 - V_0^2}}{V_1 O_2}$$

and $D_2 = z \frac{\sqrt{V_1^2 - V_0^2}}{V_1 V_0}$.

See also: *intercept time*.

deliquescent Capable of liquefying through the absorption of water from the air.

delta A sedimentary deposit of gravel, sand, silt, or clay, formed where a river enters a body of water. Sediments pile up to sea level as streams course over the *delta plain*, a nearly level surface that becomes the landward part of the delta. Lakes or marshes may be formed along the margin or within a part of the sea enclosed by the accumulation of deltaic deposits. Deltas are crossed by many distributaries that flow through to open water, depositing bars in channels and building up levees along the banks. The form and size of a delta depend upon the preformation coastline, the hydrologic regime of the river, and the local waves and tides. Most are arcuate or shaped like the Greek letter *delta* (Δ). Some, e.g., that of the Mississippi, exhibit a digitate or "*bird-foot*" plan that is formed by numerous distributaries flanked by low levees.

Three sets of beds are often observable in a delta. *Bottomset beds* consist of finer sediments deposited as horizontal or gently inclined layers on the floor of the area where the delta is forming. *Foreset beds* have a steeper dip and consist of coarser sediments; these represent the advancing head of the delta and the greater part of its total bulk. *Topset beds*, lying above the foreset beds, are actually an extension of the alluvial valley of which the delta is the terminal section.

Differences among the three beds are not always clear. They are most obvious in deltas formed by small streams flowing into protected or confined areas.

deltageosyncline An *exogeosyncline.*

demersal Refers to species that are included in the *nekton* because of their swimming capability, but whose food source is mainly benthic animals. Skates, cuttlefish, nautiloids, and some benthic and reef-inhabiting fish are examples of demersal species.

dendrite A branching pattern of deposited materials, often on limestone surfaces, that resembles a tree or fern. Dendrites are often mistaken for fossil flora, but are really mineral incrustations produced on or in a rock by such minerals as pyrolusite, MnO_2, or pyrite. *Moss agate* is an example of a mineral with a dendrite inclusion; native copper can also form dendrites. See *pseudofossil.*

dendritic drainage pattern See *drainage patterns.*

dendrochronology The study and correlation of *tree rings* for the purpose of dating events in the recent past. See *dating methods.*

dense 1. Said of any system whose parts or components are closely crowded together. 2. Said of fine-grained aphanitic rocks in which grain size usually averages less than 0.05 to 0.1 mm. 3. Said of a mineral or rock characterized by a comparatively high specific gravity.

density 1. The mass or quantity of a given substance per unit volume of that substance, usually expressed in grams per cubic centimeter (gm/cm^3). 2. The quantity of any entity per unit volume or per unit area; e.g., faunal population in a region. 3. The character of being dense, compact, or close.

density currents See *turbidites; turbidity currents.*

density stratification The stratification of water in a lake as a result of density differences. It is usually caused by temperature changes, but also may be caused by differences in the quantity of dissolved material, e.g., where a less dense freshwater surface layer overlies denser salt water. See also: *thermal stratification.*

dentate Having teeth or toothlike projections.

denudation Despite several different interpretations of this term, the basic concept is that denudation involves the exposing of deeper rock structures whose land surface has been worn down by *erosion. Denudation* and *erosion* are often used synonymously, but an eroded landscape is not necessarily a denuded one, e.g., *karst.* See also: *degradation.*

deoxidation sphere *Bleach spot.*

depletion The act of seriously diminishing or exhausting a quantity of something, e.g., natural resources.

deposit The term refers to any sort of earth material that has been accumulated through the action of wind, water, ice, or other agents. Cf: *sediment.* It may also mean a *mineral deposit.* As a verb, it means to lay or put down, or to precipitate.

depositional remanent magnetization See *natural remanent magnetism.*

depth of compensation See *isostasy.*

depth zone of emplacement One of the regions within earth's crust where the relations of plutons and their country rocks give rise to different metamorphic phenomena. The uppermost zone, the *epizone,* extends no deeper than 6.5 km, and the temperature of the country rock does not exceed 300° C. It is the typical environment of ring dikes, cone sheets, and laccoliths. Sedimentary and volcanic country rock show weak contact metamorphism, and local extreme mineralization and hydrothermal alteration.

In the *mesozone*, at a depth of 6.5 to 15 km, the country rock temperature ranges from 300° to 500° C. Plutons in this region are commonly stocks and batholiths built by multiple intrusions. Low-grade metamorphic rocks are metamorphosed again and deformed by pluton emplacement. Metamorphism is a combination of contact and regional types. The country rock of the lowermost zone, the *catazone*, at depths of 11 to 20 km and temperatures of 450° to 600° C, is medium to high-grade metamorphic. Sharp contacts are absent. Plutons may consist of gneissic pods or lenses, and metamorphic effects are widespread. As indicated by chemical analyses, the theoretically calculated mineral composition of a metamorphic rock at the epizone, mesozone, or catazone are called the *epinorm, mesonorm*, and *catanorm*, respectively.

deranged drainage pattern   See *drainage patterns*.

desalination   The removal of dissolved salts from seawater to derive fresh water. Distillation is the most common method employed; others include ion exchange and freezing of the seawater.

desert   In general, a region where precipitation is less than potential evaporation. Although they may be hot or cold, deserts are always characterized by a scarcity of vegetation. Hot deserts, found mainly in the southwestern U.S., Africa, and Saudi Arabia, are typified by sand *dunes, loess, desert pavement*, and *ventifacts*, all wind-produced formations. Other hot-desert features include *evaporites, salinas*, and *salt lakes*. Hot deserts occur in areas removed or cut off from moisture. They can occur in dry-air regions and on coasts along which there are cold-water currents and onshore winds. Cold deserts, found in Greenland and Antarctica, have moisture-poor atmospheres because of their very low temperatures. These deserts are

characterized by *patterned ground, solifluction deposits*, and *felsenmeer*.

desert armor   Syn: *desert pavement*.

desert crust   1. A hard layer containing binding matter (e.g., calcium carbonate, gypsum), exposed at the surface in a desert area. 2. *Desert varnish*. 3. *Desert pavement*.

desert mosaic   Syn: *desert pavement*.

desert pavement   A residual layer of wind-polished, closely concentrated rock fragments covering a desert surface due to the removal by wind of loose silt and sand. Syn: *desert crust; desert armor; desert mosaic*. See also: *lag gravel; boulder pavement*.

desert polish   A slick, polished surface on rocks of desert regions, produced by wind and sand abrasion. Sometimes used as a synonym of *desert varnish*.

desert rose   Any of various flowerlike groups of crystals, e.g., *barite roses*, found in loose sand and sagebrush areas of deserts.

desert varnish   A dark, enamel-like film of oxides of iron and manganese, formed on the surface of rocks in stony desert regions. It is thought that desert varnish is a surface deposit of mineral solutions exuded from within. Syn: *desert patina; desert lacquer; desert crust*. Cf: *desert polish*.

desiccation breccia   A breccia composed of fragmented, mud-cracked polygons together with other sediments. Syn: *mud breccia*.

desiccation crack   A crack in fine sediment produced by water loss or drying; esp. a *mud crack*.

desiccation polygon   See *nonsorted polygon*.

desilication   The extraction of silica from a rock by the chemical dissociation of silicates. Silica may also be removed by the chemical reaction of a body of magma with the surrounding rock. In warm climates, the silica is commonly removed from soils when large quantities of rainwater percolate through the soil.

Desmoinesian   Upper Middle Pennsylvanian of North America.

detached core   The inner part of a fold that has been detached or pinched from the parent structure by some process, usually severe compression and folding. The term core is used even though a fold may be composed entirely of one kind of rock.

detector   Any device that "notices" appropriate or required information; e.g., radioactivity detector; seismic detector.

detrital   Pertaining to *detritus*, esp. of mineral grains transported and deposited as sediments that were derived from pre-existing rocks either within or outside the area of deposition. Cf: *clastic; allogenic.*

detrital ratio   *Clastic ratio.*

detrital remnant magnetism   See *natural remnant magnetism.*

detritus   1. Any loose rock or mineral material resulting from mechanical processes such as disintegration or abrasion, and removed from its place of origin. Cf: *debris.* 2. Any fine debris of organic origin; e.g., plant detritus in coal.

deuteric   A term applied to those reactions between primary magmatic minerals and water-saturated solutions that separate from the same magma body at a late phase of its cooling profile. Syn: *epimagmatic.* See also: *autometamorphism.*

development well   A well drilled within the area of an oil field, where the area is either known (as by investigative deduction) or technologically proved to be productive. Cf: *exploratory well.*

devitrification   The conversion of glass to a crystalline state. Since all volcanic glasses are unstable under atmospheric conditions, they devitrify in time to a stable crystalline arrangement; aphanitic and spherulitic textures are a result of devitrification. Because of their inherent metastability, most natural glasses are not older than the *Cenozoic.* See also: *glass; glassy.*

devolatilization   The reduction of volatile components during *coalification*, which results in a proportional increase in carbon content. The higher the level of devolatilization, the higher the *rank* of coal.

Devonian   A time interval of the Paleozoic Era, during which rocks of the Devonian System were formed (345 to 395 million y.b.p.). The period was named for the type area of Devon, England. Because this area lacked sufficient fossils for correlation, the stages of the Devonian were established in the fossiliferous marine deposits of Ardennes, Belgium. This system shows widespread continental and desert conditions, but marine deposits are typically more widespread. In Europe, the continental deposits are called *Old Red Sandstone.* Rocks of Devonian age are found in all continents, and are sources of iron ore, lodes of tin, zinc, and copper, and of oil and evaporites.

Certain areas of Europe have become the world standard for the many subdivisions of the Devonian. Devonian equivalents in New York State form the North American standard. Ammonoid cephalopods, conodonts, brachiopods, and corals are used to stratigraphically define marine deposits; in non-marine deposits, freshwater fish and spores are used. By international agreement, the zone of the graptolite, *Monograptus uniformia*, is taken as the base of the Devonian, and the zone of the ammonoid cephalopod, *Gottendorfia*, as the top. Particular types of Devonian sediments characterize various areas of continental and marine environment, and give distinctive rock types. Invertebrates of the Devonian Period are essentially of types established by the time of the Ordovician: brachiopods, corals, stromatoporoids, crinoids, trilobites, and gastropods. Molluscan groups are well represented; freshwater clams first appear. The origin of the ammonoids during the Devonian is of great significance. Giant eurypterids (arthropods) are found in Old Red Sandstone facies. The first spiders, millipedes, and insects appear in the Devonian. Conodonts, which are useful for the correlation, show their greatest diversity during this period. The Devonian is called "The Age of Fishes," since freshwater and marine varieties proliferate. The fishes include several kinds of armored fish (placoderms), jawless fish (ostracoderms), sharks (Middle Devonian), and the first bony fish. Early Devonian (sometimes placed in Late Silurian) denotes the emergence of fish with true jaws and the appearance of the first tetrapods. Lobe-fins, primitive airbreathing fishes, are dominant during the period; lungfish make up one tribe of lobe-fins, and crossopterygians the other. Crossopterygians are acknowledged as the link between fishes and lower tetrapods, primitive amphibians. All modern algae are represented in the Devonian. Bryophytes emerge, and freshwater algae and fungi occur in particular places. Primitive land plants found in the early Devonian become diversified by the mid- to late parts of the period. Two kinds of psilopsids (simple rootless plants) have been found, as well as horsetail rushes (herbaceous sphenopsids), ferns, and "tree ferns." Primitive gymnosperms (e.g., callixylon) are known in the Devonian, and scale trees (lycopsids) are plentiful by late in the period.

Evidence from the Devonian supports the theory that the present continents were formerly united in some way. In the Northern Hemisphere the linking of North America, Greenland, and Europe allows the concept of a single Old Red Sandstone land mass. The union of the southern continents to Gondwanaland at

this time is imperative in explaining the later distribution of glacial deposits. The *Caledonian* orogeny, which climaxed in the Middle Devonian, was experienced in northwestern Europe. The *Antler* and *Acadian* orogenies of North America occurred during the Devonian. All the disturbances were accompanied by volcanism. Wide distribution of evaporite basins in the Northern Hemisphere, coals in Arctic Canada, and carbonate *reefs* suggest a warm climate over large areas during the Devonian. Growth lines of Devonian corals indicate that the Devonian year was 400 days and the lunar cycle about 30 1/2 days. Paleomagnetic evidence shows that an equator passed from California to Labrador and from Scotland to the Black Sea. See also: *continental drift; Paleozoic.*

dextral Inclined, pertaining to, or spiraled to the right; e.g., the clockwise direction of coiling of gastropod shells. Ant: *sinistral.*

dextral fault *Right-lateral fault.*

diabase An intrusive rock, ophitic in texture, consisting chiefly of pyroxene and plagioclase feldspar, most commonly labradorite. In mineral composition, diabase is a *gabbro*; its grain size is intermediate between *basalt* and normal gabbro. In Great Britain, this rock is called *dolerite.*

diachronous Said of formations where boundaries cut across time planes or biozones. Diachronism may arise in various ways. It will occur if the base of a formation is transgressive, or if the upper part has been affected locally by erosion in areas of shallow water, or by nondeposition. As deposition proceeds, diachronism may also be caused by the lateral shifting of facies boundaries. Syn: *time-transgressive.* Cf: *synchronous.*

diagenesis The sum of all changes, physical, chemical, and biological, to which a sediment is subjected after deposition, but before weathering or metamorphism. The formation of unconsolidated sediments in place, i.e., early diagenesis, is sometimes called *syngenesis.* Some diagenetic changes, called *halmyrolysis,* occur at water-sediment interfaces; however, most diagenetic activity occurs after burial. Examples of hamyrolytic activity include the modification of clay minerals and the alteration of biotite flakes into glauconite. During deep burial, compaction and lithification are the primary diagenetic processes.

diagnostic mineral A mineral whose presence in a rock indicates whether the rock is *oversaturated* or *undersaturated.*

diagonal fault *Oblique fault.*

diagonal joint A joint whose strike is in a direction that lies between the strike and direction of dip of the associated rocks. Syn: *oblique joint.*

diagonal slip fault *Oblique-slip fault.*

dialysis A technique of separating compounds in solution or suspension, based on their dissimilar rates of diffusion through a semipermeable membrane. For example, some colloidal particles will diffuse readily, others slowly, and some will not move through at all. Cf: *osmosis.*

diamagnetic Having a magnetic permeability slightly less than unity, and a small, negative magnetic susceptibility. The permeability of bismuth, the most diamagnetic substance, is equal to 0.99998. In a magnetic field, the induced magnetism of a diamagnetic substance is in a direction opposite to that of iron. Cf: *paramagnetic; ferromagnetic.*

diamond Mineral of the isometric system, a crystalline form of carbon, dimorphous with graphite. Diamond is the hardest mineral known, and has perfect cleavage. Excellent crystals occur in the kimberlite of South Africa. Colorless and colored varieties (yellow, brown, gray, green, and black) are valued as gemstones. See also: *diatreme; industrial diamond.*

diamond bit A diamond-studded rotary-drilling bit, used for coring and drilling in very hard rock. This variety of rotary drilling, called *diamond drilling,* is a well-known method of prospecting for mineral deposits. See also: *industrial diamond.*

diamond pipe See *diatreme.*

diapir Etymol: Greek, "to pierce." A vertical columnar plug of less dense rock or magma that is forced through a more dense rock. The process, called *diapirism,* may be produced by gravitational forces (heavy rocks causing lighter rocks to rise), tectonic forces, or a combination of the two. Diapirism forms both igneous and non-igneous intrusions. These may be in the shape of waves, domes, mushrooms, or *dikes,* and are often associated with salt domes (a non-igneous intrusion) or salt anticlines.

diaspore An orthorhombic mineral, $AlO(OH)$, associated with boehmite and gibbsite as a constituent of some bauxites. Usually found in foliated or stalactitic pale-pink, gray, or greenish aggregates.

diastem A theoretical, minor, undetected break in a sedimentary sequence due to non-deposition or slight erosion. The concept is based on inferences drawn from the discrepancies between calculated sedimentation rates and those resulting from observational analysis.

diastrophism A term for all movement of the earth's crust resulting from tectonic processes. This includes the formation of continents, ocean basins, plateaus, and mountain ranges. Diastrophic processes are usually classified as *orogenic* (mountain building with deformation), and *epeirogenic* (regional uplift without significant deformation). Syn: *tectonism.*

diatom Any of numerous microscopic unicellular marine or freshwater algae having siliceous cell walls. They resemble delicate glass structures, and are a major food source for copepods, larvae, etc. Diatom ooze is an ocean-bottom deposit made up of the shells of those algae. See *pelagic deposits*.

diatomaceous earth An ultra-fine-grained siliceous earth composed mainly of *diatom* cell walls. It is used as an abrasive and in filtration. *Kieselguhr* is a synonym.

diatomite See *diatomaceous earth*.

diatom ooze See *pelagic deposits*.

diatreme A general term for a pipelike structure formed deep within the earth's crust by the explosive energy of magmatic gases, or by the explosion of heated groundwater (phreatic explosion). In strong explosions, such pipes are blown out without production of magmatic material. With weaker explosions, the penetrated rocks are shattered but there is no large-scale ejection of material; breccia-field diatremes are formed in this way. Diatremes that are fully blown out may be later filled with minerals deposited by hydrothermal solutions. The diamond-bearing kimberlite pipes of South Africa are well-known diatremes. Diamonds are formed at a depth of nearly 200 km at the temperature of molten rock. A rare type of volcanic eruption transports them to earth's surface so rapidly that the diamonds do not revert to more normal surface forms of carbon, such as graphite, even with the reduction in pressure and temperature.

dichroism See *pleochroism*.

dichroscope See *pleochroism*.

dickite A crystallized clay mineral of the kaolin group, having the same composition, $Al_2Si_2O_5(OH)_4$, as kaolinite, differing only in details of atomic structure and certain physical properties.

differential erosion See *erosion*.

differential thermal analysis A technique used to observe change of state, and to determine the temperature at which a change occurs. Two samples having an identical heat capacity are heated, and the temperature differences between them are noted. The difference between them becomes especially conspicuous when one of the samples changes state. The temperature at which this occurs is called the transition temperature.

differential weathering Weathering that occurs at varying rates because of the intrinsic differences in rock composition and resistance, or differences in the intensity and degree of erosion. Differential weathering helps to develop and modify many upstanding forms such as columns, pillars, and rock pedestals.

differentiation 1. *Magmatic differentiation*. 2. Sedimentary differentiation.

diffraction 1. The phemonenon exhibited by wave fronts that, upon passing the edge of an opaque body, are modulated, thus effecting a redistribution of energy within the front; in light waves this is detectable by the presence of dark and light bands at the edge of a shadow. 2. The bending of waves, esp. sound and electromagnetic waves, around obstacles in their path. 3. Generation and transmission of seismic wave energy in accordance with *Huygen's principle*.

diffraction pattern The interference pattern of lines formed when rays such as light rays or X-rays are passed through a diffraction grating. Each substance has a characteristic diffraction pattern.

digital Referring to the representation of measured quantities in discrete units. A digital system is one in which information is stored and operated as a system of digits in a code, in contrast to an *analog* system.

digitation   A subsidiary recumbent anticline attached to a larger recumbent fold, somewhat resembling fingers extending from a hand.

dike   1. A tabular body of intrusive igneous rock that cuts across the layering or structural fabric of the host rock. (Cf: *sill.*) It may be a composite or multiple intrusion. Dikes may be fine, medium, or coarse-grained, depending on their composition and the combination of their size and length of their cooling period. A group of linear or parallel dikes is called a *dike set.* When in radial, parallel, or *en échelon* arrangement, a group of dikes is called a *dike swarm*; in such a configuration, their relationship with the parent body may not be directly observable. See also: *ring dikes and cone sheets.* 2. *Clastic dike.* 3. An embankment or wall built to prevent flooding of a lowland area. See *intrusion.*

dike ridge   *Dike wall.*

dike wall   A topographic feature formed when a dike is more erosion-resistant than the rock on either side. A dike ridge several kilometers long and about 15 m high radiates from Ship Rock (a *volcanic neck*) in New Mexico. Even though they may resemble *hogbacks,* there is a clear distinction: dike walls have developed from igneous intrusions, whereas hogbacks are features of stratified rocks. See *hogback.*

dilatancy   An inelastic increase in the bulk volume of a rock during deformation by squeezing, owing to the opening and extension of small cracks. Dilatancy begins when the stress reaches about half the breaking strength of the rock. It has been demonstrated by laboratory experiments that measurable physical changes accompany dilatancy; such effects include changes in the electrical resistivity of the rock and in the velocity of elastic waves traveling through the rock. These findings were verified in different locations. The assumption of the growth of strain-associated microcracks in crustal

rocks before an earthquake, thus causing dilatancy, serves as the basic model of precursory seismic effects. See also: *earthquake prediction.*

dilatational wave   P wave.

dilation   Deformation by change in volume but not configuration. Also spelled *dilatation.*

dilation vein   A mineral deposit occurring in vein space created by the bulging of a crack, rather than in veins formed by the replacement of wall rocks.

dimension stone   Stone that is cut or quarried in accordance with specific dimensions.

dimorphism   1. Crystallization of an element or compound into two distinct crystal forms; e.g., diamond and graphite, pyrite and marcasite. 2. The characteristic of having two distinct forms in the same species, e.g., male and female; also said of animals that produce two different kinds of offspring.

dinoflagellate   A biflagellated, single-celled microorganism, chiefly marine, with features belonging to both the plant and animal kingdoms. Range, Triassic to present.

dinosaur   Etymol: "terrible lizard." The collective term *dinosaurs* is given to that particular group of reptiles that dominated or were prominent among Mesozoic life forms. They ranged in size from 30 cm to 26 m. They were carnivorous, herbivorous, bipedal, or quadrupedal. Most were terrestrial, but there were also aquatic and semi-aquatic representatives. Range, Triassic to Cretaceous.

diopside   A monoclinic inosilicate of the pyroxene group, $CaMg(SiO_3)_2$. Its color may be pale-green, blue, whitish, or brown. Violane is a purple, manganese-bearing variety, and dark-green varieties contain chromium. Diopside occurs in contact metamorphic rocks, esp. in certain dolomitic marbles.

diorite A dark-colored, coarse-grained plutonic rock composed essentially of plagioclase feldspar (oligoclase-andesine), hornblende, pyroxene, and little or no quartz; it is the approximate intrusive equivalent of *andesite*. With an increase in the alkali feldspar content, diorite grades into *monzonite*.

dioxide An oxide containing two atoms of oxygen per molecule of compound; e.g., MnO_2.

dip 1. The angle in degrees between a horizontal plane and an inclined earth feature such as a rock stratum, fault, or dike. Measurement is made perpendicular to the strike and in the vertical plane; it thus determines the maximum angle of inclination. See also: *true dip; apparent dip.* 2. The angle between a reflecting or refracting seismic wave front and the horizontal, or the angle between an interface associated with a particular seismic event and the horizontal. 3. *Magnetic inclination.*

dip calculation Dip is computed from the difference in reflection times of a seismic pulse that travels along two different paths after impinging on a reflecting surface; e.g., a reflecting interface. In such a technique, a series of geophones is located equidistant from each other on either side of a *shot point*. A shot is fired and a reflector, at some point R, will transmit the pulse back to the geophones along different routes. The difference in reflection times between the most easterly and most westerly geophones is a measure of the amount of dip of the reflector, i.e., the reflecting surface. See also: *moveout.*

dip fault A fault that strikes essentially parallel to the dip of the adjacent strata; i.e., its strike is perpendicular to the strike of the adjacent strata. Cf: *strike fault; oblique fault.*

diphotic zone That level in bodies of water at which sunlight is faint and little photosynthesis is possible. Cf: *aphotic zone; euphotic zone.*

dip joint A joint that strikes parallel or nearly parallel to the direction in which the bedding schistosity or gneissic structure dips. Cf: *strike joint.*

dipmeter A finely defined *resistivity log.* The dipmeter sonde consists of four microresistivity pads mounted at 90° angles from each other; these are applied to the walls of a borehole. The sequence of readings at each pad will deviate slightly from the others, depending on the dip of the strata. Such measurements indicate the inclination and direction of the borehole as well as the *dip* and *strike* of the strata. This information is valuable in the plotting of oil-field structures. See *well logging.*

dip separation The distance between formerly adjacent beds on either side of a fault surface, measured directly along the dip of the fault. Cf: *dip slip; strike separation.*

dip shift The relative displacement of rock units parallel to the fault dip, but beyond the fault zone itself. Cf: *dip slip.*

dip shooting A system of seismic surveying in which the chief concern is determination of the dip, or dip and depth, of a geological marker that will serve as a reflecting interface. Although one can correlate seismic reflections along all shooting profiles, and map the depths to the markers, if the reflection from a particular formation is distinctive, it is recognizable over a fairly wide area. See also: *dip calculation.*

dip slip The component of the net slip that is measured parallel to the dip of the fault plane. Cf: *strike slip; dip shift.*

dip-slip fault A fault along which rock displacement has been predominantly parallel to the dip. Cf: *strike-slip fault*. See also: *fault*.

dip slope   The stope of the land surface that more or less conforms with the inclination of the underlying rocks.

directional drilling   The drilling of a well during which azimuths and deviations from the vertical are controlled. One of its purposes is to establish multiple wells from a single location; e.g., an offshore platform. Syn: *slant drilling*.

directional log   See *well logging*.

direct runoff   See *runoff*.

dirt cone   A glacial *debris cone*.

disappearing stream   A surface stream that disappears underground into a *sink*.

discharge   The amount of water flowing down a stream channel; it is measured in cubic meters per second, and data for such measurement is always taken at a specified location along a given stream. Discharge may also refer to *sediment discharge*.

disconformable   1. Relating to a disconformity. 2. Said of the contact of an igneous intrusion where the contact is essentially non-parallel with the internal structure of the intrusion. Cf: *discordant*.

disconformity   See *unconformity*.

discontinuity   1. Any interruption in sedimentation that may represent a time interval in which sedimentation has ceased and/or erosion has occurred. 2. In structural geology, abrupt changes in rock type caused by tectonic activity. 3. A zone within the earth in which the velocity of seismic waves shows a sharp change; e.g., *Conrad Discontinuity; Mohorovičić Discontinuity*.

discontinuous deformation   Deformation by fracture rather than flow, as distinct from *continuous deformation*.

discontinuous reaction series   See *reaction series*.

discordant   1. Said of intrusive igneous bodies, such as dikes and batholiths, whose margins cut through the bedding or foliation of the country rock. Ant: *concordant*. Cf: *disconformable*. 2. Structurally *unconformable*; i.e., said of strata in which parallelism of bedding or structure is absent. 3. Radiometric ages, determined by more than one method for the same specimen, are said to be discordant if they disagree beyond experimental error. Ant: *concordant*. 4. Said of topographic features whose elevations are not the same or similar. Ant: *accordant*.

discovery well   The first well to reach gas or oil at a formerly unproductive depth, or in a yet unproven area; it is variously referred to as a successful *wildcat, outpost well, deeper-pool test*, or *shallower-pool test*.

disharmonic fold   A fold that is not uniform throughout the stratigraphic column. In profile, it shows considerable variation in the layers through which it passes.

disintegration   1. A syn. of *mechanical weathering*. 2. *Radioactive decay*.

dislocation   1. A lattice imperfection in a crystal structure. 2. *Displacement*.

dispersed elements   Trace elements that occur in different minerals in a rock. Because their relative concentrations can differ significantly, the dispersed element composition of a magma can therefore impose constraints on the nature of its source rock.

dispersion   1. The type of geographic distribution of the individuals in a plant or animal population; e.g., random, uniform, clumped. 2. Variation of the refractive index of a substance with the wavelength of light. It is because of this property that a prism is able to generate a spectrum. 3. A continual spreading out of a disturbance into trains of waves, given that the initial disturbance is confined to a finite range of x values, and that the medium is unlimited.

displacement The term generally used for the relative movement of the two sides of a fault, as measured in any specified direction. The term also designates the specific distance of such movement. Syn: *dislocation.*

dissection The cutting of ravines, gullies, or valleys by erosion, esp. by streams.

dissociation The reversible decomposition of a complex substance into simpler components, as produced by variation in physical conditions; e.g., the decomposition of water into hydrogen and oxygen, after the water is heated to a high temperature, and the recombination of the liberated elements to again form water, when the temperature is lowered. The temperature at which a compound breaks up in such a reversible manner is the *dissociation point.* A temperature point at which a given dissociation is presumed to occur is called the *dissociation temperature*; it is, in fact, a range of temperatures, owing to variations in pressure or composition and, simply indicated, the temperature at which the rate of a given dissociation becomes significant under given conditions.

dissolution *Solution.*

dissolved 1. *Dissolved load.* 2. The total amount of dissolved organic and inorganic material contained in a water sample.

dissolved load That fraction of the total stream load that is carried in solution or suspension. Syn: *dissolved solids; solution load.* Cf: *bed load.*

distal 1. Referring to an ore deposit formed at a significant distance from the source of its constituents. 2. Said of a sedimentary clastic deposit that was laid down far from its source location. Ant: *proximal.*

distillation 1. The volatilization or evaporation and subsequent condensation of liquid; e.g., when water is boiled in a retort and the steam collected in a cool collecting vessel. 2. The purification or concentration of a substance, or the separation of one substance from another through the process described in 1. 3. A *fossilization* process, in which the loss of volatile components from an organic substance results in preservation of the substance as a carbonaceous residue.

distributary A system or set of independent channels frequently formed by rivers or streams, such as where they enter their deltas. The pattern is well defined by the Mississippi. Ant: *tributary.*

distributive province The environment that encompasses all rocks contributing to the formation of a contemporaneous sedimentary deposit, and those agents or factors responsible for their distribution. Cf: *provenance.*

disturbance A term used in geology literature for a lesser or minor orogeny such as the *Palisades disturbance* at the close of the Triassic. Cf: *event.*

divergence 1. (Meteorol.) A net flow of air outward from a given volume of the atmosphere; i.e., a decrease of mass within that given volume. Mathematically, it is the expansion of the atmospheric velocity vector field. High-pressure areas are generally divergent near the earth's surface. Divergence results in a decrease of cyclonic vorticity. Cf: *convergence.* 2. The separation of ocean currents by horizontal flow, usually the result of upwelling. Cf: *convergence.* 3. (Paleontol.) Adaptive radiation.

divergent plate boundary *Accreting plate boundary.*

diverted stream   See *stream capture.*

diviner   *Dowser.*

divining rod   See *dowsing.*

D layer   See *seismic regions.*

dogtooth spar   A variety of calcite that occurs as pointed crystals somewhat resembling a dog's teeth.

dolerite   In the U.S., *dolerite* is a syn. of *diabase.* In British usage, the term is used instead of *diabase,* since among British petrologists *diabase* applies to altered basalts and dolerites. The adj. *doleritic* is a preferred syn. of *ophitic* in European usage.

doline   Etymol: Serbo-Croatian, *dolina,* "vale." A term originally used in the U.S. to mean a *solution sink,* i.e., a *sinkhole* that developed downward by solution beneath a soil mantle, without physical disturbance of the ambient rock. Sinkholes produced by the collapse of rock above an underground void were called *collapse sinks.* At present, at least in the U.S., *doline* is a syn. of *sinkhole.* See also: *karst.*

dolomite   The term applied to both the sedimentary rock and the mineral $CaMg(CO_3)_2$. The name *dolostone* was proposed over 40 years ago for the rock to distinguish it from the mineral; although such a term would be useful, it has not been adopted. Dolomite forms by the replacement of limestone (magnesium for calcite). Such penecontemporaneous replacement can either retain or obliterate textures and features in the limestone. It is common for certain features to be preserved and others destroyed, but there does not yet seem to be a generally accepted explanation for such selectivity. Occurrences of penecontemporaneous dolomitization, ancient and modern, are grouped into those related to evaporite deposits and those that are not. The former dolomites are produced as the end stages of evaporite deposits, preceded by calcite and gypsum; a conclusive theory for the origin of extensive dolomites not associated with evaporites is still awaited.

dolomitic limestone   See *dolomitization.*

dolomitization   The process by which limestone ($CaCO_3$) becomes dolomite by the substitution of magnesium for the original calcite, either wholly or in part. Replacement by dolomite is common in organisms whose original hard parts were composed of calcite or aragonite (corals, brachiopods, echinoderms). In some rocks all fossil remains may be dolomitized and the matrix unaltered. In some limestones only the smallest grains are converted to dolomite, the larger crystals remaining unchanged. Dolomitization occurs most commonly very soon after the limestone has formed. See also: *dolomite; limestone; dedolomitization; replacement.*

dolostone   See *dolomite.*

dome   1. Any rounded landform or rock mass resembling the dome of a building. 2. An uplifted structure, circular or elliptical in outline; doming may be gentle or steep; e.g., salt domes. 3. *Lava dome.* 4. *Volcanic dome.*

dome mountain   Geomorphology recognizes two distinctive types of dome mountain: *igneous* and *tectonic.* The igneous type includes the stripped *laccolith domes* (Henry Mountains of Utah), the complex laccolith (or lopolith) intrusion in which igneous material has penetrated the surrounding sediments (La Plata Mountains, Colorado), and the stripped pluton, a dome-shaped igneous intrusion found in smaller batholiths (Stone Mountain, Georgia). The second type of dome is the result of tectonic uplift over a wide radius, leading to dips that radiate in all directions. Examples of tectonic dome mountains in North America include the Black Hills dome of South Dakota.

dormant volcano A volcano that is not actively erupting but that has erupted within historic time and is considered likely to do so again in the future. As noted from the imprecision of the time limits, there is no clear distinction between a dormant and an active volcano.

dot chart A graphical technique used to compare the gravity effects of assumed subsurface structures with observed gravity values, when the masses are irregular. The technique generally involves the use of a *graticule*, a transparent template, which is superimposed over a cross-section of the formation whose gravity effect is to be determined. The template is usually a fan-shaped pattern of lines forming compartments that increase in area as the distance from the vertex increases. The vertex is placed over the gravity station or the section, and the gravity effect of any body included in the section can be determined by counting the compartments it covers on the template. Each compartment represents a vertical gravity contribution at the observing station (at the vertex), and the number of dots within the compartment indicates the amount of this effect.

double refraction *Birefringence.*

downbuckle A downfolding of earth's crust due to compression and associated with the formation of oceanic trenches. Cf: *tectogene.*

downthrown Referring to that side of a fault that appears to have moved downward relative to the other side. The amount of downward vertical displacement is called the *downthrow*. Rocks on the downthrown side of a fault are referred to as the *downdip block*, and are described as *downfaulted.* Cf: *upthrown; heave.*

downward enrichment *Supergene enrichment.*

downwasting 1. *Mass wasting.* 2. A part of the ablation process, during which glacial thinning occurs; this may eventually lead to stagnation or lack of forward motion.

dowsing The practice of locating groundwater, mineral deposits, or other concealed objects with the aid of a forked stick called a *dowsing rod* or *divining rod.* Prior to the nineteenth century, groundwater was thought to flow in definite rivers like surface water, and that a lucky well would therefore reach one of these streams and produce abundantly. Because of the uncertainty of where to dig a well, help was sought from *dowsers*, also called *diviners* or *water witches*; such persons supposedly were gifted with supernatural powers that enabled them to find the underground streams. The belief still persists in many regions. As a dowser walks about with a dowsing rod, it is supposed to dip sharply when he is over an underground stream. Although there is no scientific basis for it, dowsing has had a fairly high degree of success.

drag 1. Bending or distortion of strata on either side of a fault, caused by the friction of the moving blocks along the surface of the fault. These distortions generally record some components of relative movement, but are not reliable, as some faults may reverse their movement and some retain drag from earlier displacements. 2. Rock and ore fragments separated from an ore body, and found in and along the fault zone. Syn: *drag ore.*

drag fold See *fold.*

drag mark A long groove or striation produced by an object as it is dragged by a current across a soft surface. The cast or impression of such a mark, such as might appear on the underside of the overlying bed, is also called a drag mark. Cf: *groove cast.*

drag ore *Drag*, def. 2.

drainage The removal of excess meteoric water by rivers and streams. A river network is related to the geological structure of a region. In the case of young drainage systems, it is determined by the existing rocks and superficial deposits. In mature systems, the courses may have been determined by strata that were subsequently removed. See also: *drainage patterns.*

drainage basin Region drained by a particular stream or river system, i.e., an area that contributes water to a particular river or channel system. The amount of water reaching the reservoir depends on the size of the basin, total precipitation, evaporation, and losses due to absorption. The term "catchment area" is often used instead by hydrologists. Syn: *watershed.*

drainage density A term that describes the relative spacing of a stream network, expressed as the ratio of the total length of all streams within a drainage basin to the area of the basin. Thus it is a measure of the dissection, or of the *topographic texture*, of the area. The drainage density depends on the climate, geology, and physical characteristics of the basin.

drainage patterns A drainage pattern is the particular arrangement or configuration that is collectively formed by the individual stream courses in an area. Drainage patterns reflect the influence of factors such as initial slopes, inequalities in rock hardness and structural controls, and the geological and geomorphic history of the drainage basin. The most frequently encountered drainage patterns are described here. *Dendritic patterns* are the most common. They are characterized by irregular branching of tributary streams in many directions at almost any angle, but usually less than 90°. They develop on rocks of uniform resistance to erosion, and are most likely to be found on nearly horizontal sedimentary rocks or in areas of massive igneous rocks. A special dendritic pattern, the *pinnate*, is one in which tributaries to the main stream are

subparallel and join it at acute angles, thus resembling a feather. This is believed to indicate the effect of very steep slopes on which the tributaries formed. *Trellis patterns* show an arrangement of subparallel streams, usually aligned along the strike of the rock formations or between parallel topographic features deposited by wind or ice. However, they are most prominent in regions of tilted or folded sedimentary rocks. A *parallel pattern* is one in which a series of streams and their tributaries flow parallel or almost parallel to one another, and are regularly spaced. It is indicative of an area of pronounced and uniform slope. *Barbed drainage patterns* are formed by tributaries that join the mainstream in sharp bends oriented upstream. These patterns can occur at or near the headwater portions of systems, and are usually the result of stream piracy that has caused a reversal of the flow direction of the mainstream. In a *rectangular drainage pattern*, both the mainstream and its tributaries exhibit right-angled bends. This pattern is indicative of a stream course that is determined by a prominent fault or joint system. *Radial drainage*, a pattern in which streams diverge from an elevated central area, develops on volcanic cones and other types of isolated conical hills. *Centripetal patterns* occur where drainage channels converge into a central depression. They are characteristic of sinkholes and craters, and large structural basins. *Annular drainage*, a roughly ringlike concentric pattern, is found around domes that are encircled by alternating belts of strong and weak rock. The annular pattern occurs when the stream erodes the weaker rocks around the dome. A *deranged pattern* is marked by irregular stream courses that flow into and out of lakes and have only a few short tributaries. It occurs in areas of more recent glaciation, where the preglacial drainage has been obliterated and the new drainage has not had sufficient time to develop any coordinated pattern.

drainage system See *drainage.*

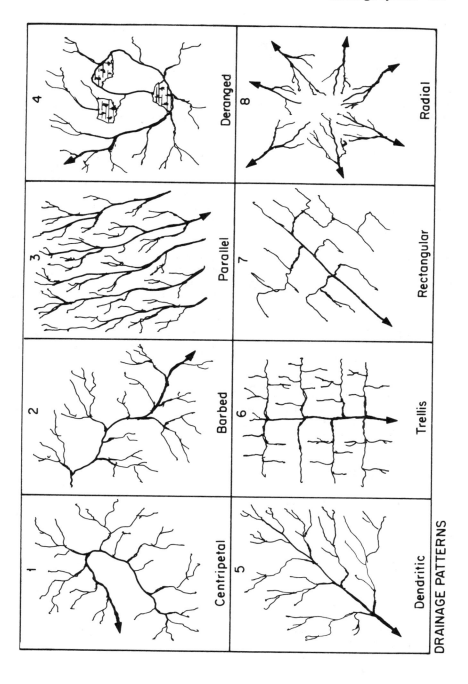

DRAINAGE PATTERNS

draw A small, natural drainage way with a shallow bed; the dry bed of a stream. The term is also used for a slight depression that leads from a valley to a breach between two hills.

drawdown The lowering of the water level of a well as a result of the withdrawal of water. See also: *cone of depression.*

dreikanter See *ventifact.*

driblet See *spatter.*

driblet cone See *hornito.*

driblet spire See *hornito.*

drift A general term for all rock material that has been carried by glaciers and deposited directly from the ice or through the agency of meltwater. See also: *glacial drift.*

drift curve A graph of a series of gravity readings taken at a single field station, plotted against intervals during the day. The curves show a continual variation of the gravity readings with time, known as *drift.* This is caused by imperfect elasticity of the gravimeter springs, which are thus subject to a slow creep over long time intervals.

drift ice 1. Floating ice, e.g., icebergs or floe fragments. 2. *Pack ice.*

drift theory One of the theories of coal origin, according to which the plant material from which coal develops has been transported from its place of growth to another locality, where it accumulates and where coalification occurs. Ant: *in-situ theory.*

drilling mud A suspension of clay (usually bentonite) in water or oil used in *rotary drilling* to lubricate and cool the drilling bit and carry rock cuttings to the surface. Syn: *drilling fluid.* See *rotary drilling.*

dripstone General term for calcite or other mineral deposits formed in caves by dripping water. This includes *stalactites, stalagmites,* and similar deposits formed by flowing water. See also: *cave onyx; flowstone; travertine.*

drive pipe A pipe driven into a bored hole to prevent caving or to exclude water.

drowned valley A valley that is partly flooded by a sea, lake, or reservoir. It may result from a rise of sea level, subsidence of the land, or a combination of both. The lower ends of the Delaware and Susquehanna river valleys were drowned.

drumlin Elongated elliptical hill consisting of unconsolidated material and commonly occurring in swarms. Drumlins are formed under the margin of glaciers or from the erosion of older *moraine* by reglaciation. The long axes are approximately parallel to the direction of ice movement and are typically steepest and highest at the end that faced the advancing ice, and slope gently in the direction of movement. Cf: *roches moutonnées.* Most drumlins are composed of clayey *till,* but some have bedrock cores and many contain sand and gravel sediments. Drumlin swarms or fields may contain as many as 10,000 drumlins, as in central New York State, Wisconsin, Canada, and Northern Ireland. These may have been formed by the molding action of a glacial ice sheet on newly deposited moraine; however, there is still no generally accepted explanation of their origin. See also: *glacier.*

druse Etymol: Czech. *druza,* "bit of crystallized ore." 1. A crust of small crystals that develop along the walls of a cavity or mineral vein. The crystals are euhedral and usually of the same minerals as those of the enclosing rock. 2. A cavity of this sort. 3. Aggregates of subhedral crystals with distinct faces, often encrusted with small crystals; e.g., quartz, chert, oölites. 4. A hole or bubble in glacial ice filled with an air-ice crystalline mixture. See also: *vug; amygdule; geode.*

drusy 1. Describes a crystalline aggregate whose surface is covered with a layer of small crystals; e.g., *drusy quartz.* 2. Describes rocks containing numerous druses (in the sense of cavity). Essentially synonymous with *miarolitic.*

dry delta 1. *Alluvial fan.* 2. *Alluvial cone.*

dry lake 1. A basin formerly containing a lake. 2. A *playa.*

dry-snow avalanche An avalanche consisting of dry, loose snow; it is the fastest-moving of the snow avalanches, capable of a speed of 450 km/hr. See: *avalanche.*

dry valley A valley in which streams once flowed but now has no or little running water. It may be the result of a fall in the water table, a climatic change, or of stream capture. Dry valleys are common in areas underlain by limestone and chalk. Syn: *dead valley.* See also: *windgap.*

D.S.D.P. *Deep Sea Drilling Project.*

DTA *Differential thermal analysis.*

ductility As used in geology, the ability of a solid material to undergo more or less plastic deformation before it ruptures. Cf: *brittleness.* In the geological sense, a rock that is able to sustain, under a given set of conditions, 5 to 10% deformation before rupture or faulting is said to be *ductile.* Cf: *brittle.*

dumpy level A leveling instrument that has a telescope rigidly connected to the vertical spindle, and which can therefore rotate only in a horizontal plane.

dune A mound, ridge, or hill of windblown sand. Dunes occur where there is a source of sand, sufficient wind to carry it, and a suitable land surface on which to deposit it and where it can accumulate. Two types of dune are related to topography. *Climbing dunes* form where the wind rises over an abrupt topographic break; e.g., the dunes of Panamint Valley, Calif. *Falling dunes* form where sand is blown across a cliff and falls into a sheltered depression where it can accumulate. Dune forms are greatly influenced by even sparse vegetation. Abundant sand and forceful winds form *transverse dunes* in both barren and shrubby deserts. A transverse dune is at right angles to the wind and may reach heights of 4 to 5 m before breaking up. Less sand and sparser vegetation favor the formation of *longitudinal dunes.* These are long, narrow dunes, generally symmetrical, trending parallel with the direction of the prevailing winds. Moderate winds may form *parabolic dunes* where the vegetation grows quickly enough to hold the parts of the dune that advance most slowly. Parabolic dunes often have an elongate or "hairpin" shape, and their points face upwind (the reverse of *barchans*). Many form by blowouts of formerly stabilized sand.

dune complex A group or aggregate of mobile and fixed sand dunes whose existence does not necessarily depend on an obstruction or topographic break. A complex usually attains its maximum development on relatively flat terrain.

dunite A variety of peridotite composed almost completely of olivine, but which can contain such accessory minerals as pyroxene, plagioclase, and chromite. It may represent upper-mantle material.

durain See *lithotype.*

duricrust A general term for the hard crust on the surface or within the upper horizons of a soil in a semi-arid climate. Among the several methods of its formation, one is by accumulated deposits from mineral-bearing waters that evaporate during the dry season. See also: *calcrete; silcrete; caliche.* Cf: *hardpan.*

DUNES

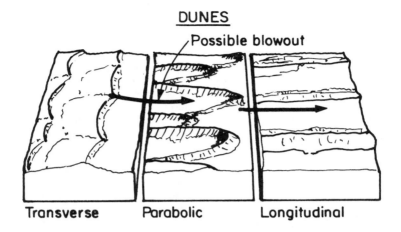

Possible blowout

Transverse   Parabolic   Longitudinal

duripan A soil layer characterized by cementation by silica. Duripans occur mainly in volcanic areas having arid or semi-arid climates. See also: *hardpan*.

dyke Br., *dike*.

dynamic metamorphism The sum total of the processes and effects of orogenic movements and stress variations in changing old rocks to produce new ones that exhibit structural and mineralogical changes due to shearing and crushing at low temperatures and significant recrystallization at higher temperatures. Also called *dynamo-metamorphism*. See *metamorphism*.

dynamothermal metamorphism See *metamorphism*.

dystrophic lake A lake in which nutrient matter and oxygen are sparse. It is rich in humic matter, which consists mainly of undecomposed plant fragments. Dystrophic conditions are often associated with acidic peat bogs. Cf: *eutrophic lake; oligotrophic lake*.

E

earth 1. That planet of our solar system that is fifth in size of the nine major planets and, as one of the inner planets, is third in order of distance from the sun. Earth's equatorial radius is 6378 km; polar radius, 6357 km; equatorial circumference, 40,075 km. Its path about the sun is nearly circular. See also: *crust; mantle; core*. 2. In civil engineering, a general term for matter that can be conveyed with a power shovel, scraper, or loader. 3. Any of several metallic oxides that are difficult to reduce; e.g., alumina, zirconia. Cf: *rare earths*. 4. An organic deposit that has remained generally unconsolidated; e.g., *diatomaceous earth*. 5. *Fuller's earth*.

earth current Direct or alternating electric current flowing through the ground and which, by electromagnetic induction, causes irregular currents to flow and to interfere with the reception of transmitted signals. Syn: *ground current; telluric current*.

earthflow A term for a gravity process and landform on a hillside that shows downslope movement of rock and soil within defined lateral boundaries. As fluidity increases, earthflows sometimes grade into mudflows. Cf: *mudflow.*

earth hummock A low, rounded *frost mound*, generally 10 to 20 cm in height, consisting of a central core of fine material surmounted with vegetation. Earth hummocks occur in arctic and alpine regions, where they form in groups to produce a nonsorted *patterned ground.*

earthquake A rumbling or trembling of the earth produced by the sudden breaking of rocks in response to geological forces within the earth. According to the *elastic rebound theory,* the elastic property of rocks permits energy to be stored during deformation by tectonic forces. When the strain exceeds the strength of a weak part of the earth's crust, e.g., along a geological fault, opposite sides of the fault slip, generating elastic waves that travel through the earth. An idealized model of an earthquake source would show the rupture of the fault as originating at a point on the fault surface called the *seismic focus* or *hypocenter.* The rupture surface progresses along the fault plane at a certain velocity and stops after it has spread across a certain area. The length and width of this area, and the location of the hypocenter and *epicenter,* are the main parameters for describing the essential characteristics of an earthquake source.

In recent years, the properties and origins of both earthquakes and volcanoes have been re-examined in the context of plate tectonics. According to the concept, earthquakes and volcanoes are manifestations of the motion of crustal plates. Shallow earthquakes occur where plates are colliding or separating, and deeper earthquakes occur where the lithosphere plunges into the mantle (subduction zone). Earthquakes are regarded as the most direct evidence for subduction slabs; this is deduced from the

fact that the focal mechanisms of moderate and deep-focus earthquakes indicate stresses that are compatible with a bending, brittle plate under gravitational force. Many shallow earthquakes are caused by movement of one crustal block with respect to an adjacent block along a fault plane. In a *normal fault* the blocks move as though being pulled apart. In a *thrust fault* the overlying block is forced up the dip of the fault plane. (See *fault.*) In a *strike-slip fault* the two blocks slide past each other. (See *strike-slip fault.*) It is possible to tell what type of fault motion has occurred from an analysis of the seismic waves generated by an earthquake. See also: *earthquake measurement; earthquake prediction; seismograph; seismic waves; plate tectonics.*

earthquake engineering The study of the behavior of foundations and structures, and of the materials used in their construction; also, techniques that would lessen the effect of earthquakes on such structures. Syn: *engineering seismology.*

earthquake intensity See *earthquake measurement.*

earthquake magnitude See *earthquake measurement.*

earthquake measurement The determination of the relative energy generated by an earthquake involves the measurement of both its *intensity* and *magnitude.* Earthquake *intensity,* the effects of an earthquake at a particular location, is measured by some standard of relative measurements such as damage. Some of the many factors that contribute to earthquake intensity at a certain point on the earth's surface include the location of the focal region, the "triggering" mechanism of the earthquake inside this region, the quantity of energy released, and local geological structure.

Various intensity scales have been formulated. For many years the *Rossi-*

Forel scale was the most widely used. The *Mercalli scale*, which was formulated later, was, like the Rossi-Forel scale, an arbitrary scale of earthquake intensity. Numbers from I to XII (I to X⁺ is the Rossi-Forel system) were assigned to conditions ranging from "detectable only by instrument" to "almost total destruction." The scale was modified and updated by American researchers to adapt it to North American conditions. It replaced the Rossi-Forel scale and is known as the *modified Mercalli scale.*

The *magnitude* of an earthquake is based on the amount of strain energy released, as recorded by seismographs. It is measured by the *Richter scale*, a numerical scale that describes an earthquake independently of its effect on people, objects, or buildings. It has a range of values from -3 to 9, and there is only one number for each earthquake. Negative and small numbers represent earthquakes of very low energy; higher values indicate earthquakes capable of great destruction. The scale of magnitude actually has no upper limit but, in practice, no earthquake greater than 8.9 has ever been recorded. That constraint, however, is intrinsic to the earth, not to the scale.

The Richter scale was based on Richter's observation that the seismograms for any two earthquakes, regardless of the distance from the recording station (hence regardless of the distance from the focus), show a constant ratio between the maximum amplitude of their surface waves. Because of the enormous variation in amplitude that is possible between earthquakes, the scale range is compressed by taking the logarithm (base 10) of the ratio. Thus, an increase of one in the rating of an earthquake would indicate a tenfold increase in the seismic-wave amplitude. Universal application of the Richter system was realized after formulating the mathematics required to standardize seismograms from different machines in various areas of the world. See *seismic moment.*

earthquake prediction　Methods and techniques of forecasting the occurrence of an earthquake, its magnitude, location, and time, over either a short or long term. Some of the most promising earthquake precursors at present include: regional crustal strain; uplift and tilting of the ground surface, as gauged by geodetic methods; decreases in the ratio of compressional-wave velocity to shear-wave velocity (V_P/V_S); anomalous changes in water level; and radon gas concentration in wells. Foreshocks sometimes precede the main shock. The duration of such anomalies appears to be correlated with the magnitude of the predicted earthquake; e.g., a magnitude 5 earthquake seems to be preceded by premonitory changes for about 3 months, whereas an earthquake of magnitude 7 is preceded by anomalous patterns for about 10 yr. The interpretation of earthquake precursors follows a basic physical model grown out of laboratory experiments. The model assumes the formation of microcracks in the rocks as strain builds up just prior to an earthquake. The opening cracks increase the volume of the strained rocks; this alters the seismic-wave velocities through the rocks, and is perhaps responsible for changes in electrical resistivity of the rocks, uplift and tilting, and increased quantities of radon in well water. See *dilatancy.*

The U.S. program for earthquake prediction is still not sufficiently supported to realize prediction within the next decade. Lack of funds has seriously deterred the research efforts of universities and industries. Russian field experiments have yielded many, if not most, of the acknowledged anomalous precursors of earthquakes. The U.S.S.R. strategy in prediction study is to monitor several experimental sites, as compared with the greater emphasis in the U.S. on monitoring specific areas in California. The Japanese are currently stressing surveys every few years, over ranges of more than 20,000 km, and systematically observing many precursory phenomena.

The Chinese approach to prediction has certain unique features; it uses the findings of large numbers of students and peasants in remote areas. Pertinent data, based on all accepted precursory anomalies, are being obtained at 5000 points. In addition, the Chinese are giving serious consideration to pre-earthquake animal behavior. See also: *earthquake; earthquake measurement.*

earthquake swarm The occurrence of a large number of lesser earthquakes in a particular region over an interval of time, perhaps several months, without the event of a major earthquake.

earthquake wave *Seismic wave.*

earthquake zone An area or region in which fault movements occur. Such an area sometimes coincides with a location of volcanic activity.

earth science A broad term for all earth-related sciences. Although it is often used as a syn. for geology or geological science, it includes, in its wider scope, meteorology and physical oceanography.

earth tide Rising and falling of the earth's solid surface in response to the gravitational pull of the sun and the moon, i.e., the same forces that produce the tides of the sea. Earth tides, like ocean tides, show the greatest rise when the moon and sun are aligned (*spring tide*), and least rise when the moon and sun are perpendicular to each other (*neap tide*). However, earth tides do not show as large a displacement as ocean tides; the maximum displacement of earth tides is less than 1 m.

earthy 1. Consisting of, or of the nature of, soil. 2. Describing a type of fracture akin to that of hard clay.

echinoderm Any of the entirely marine invertebrates belonging to the phylum Echinodermata, having an endoskeleton composed of numerous calcite plates. Living forms have both pentamerous and bilateral symmetry. The fossil history of echinoderms dates to the Lower Cambrian, but their evolution is not clear because of incomplete fossil records. Asteroids, crinoids, and echinoids are members of this phylum.

echinoid A class of echinoderms to which sea urchins and sand dollars belong. Range, Ordovician to Recent. Echinoids have a hollow test of limy plates that is characteristically spiny, spherical, and radially symmetrical, but may become elongated and bilaterally symmetrical in more advanced forms. Simple classification of echinoids divides them according to these two forms of symmetry. Many extinct species are used as Paleozoic and Mesozoic *index fossils.*

echogram Graph of the sea-floor contour made by echo sounding. Echograms are used to establish contour maps and physiographic profiles.

echo sounding Since water is an excellent conductor of sound energy, a generated sound pulse will bounce off a reflecting surface and return to its source as an echo. With a known speed of sound in the water, the time lapse between the initial pulse and return of the echo can be converted into distance or depth. See *echogram; sounding.*

eclogite Granular metamorphic rock composed essentially of garnet (almandine-pyrope) and pyroxene (jadeite-diopside), but with the bulk of its chemical composition very similar to that of basalt. The relatively high density (3.4 to 3.5 gm/cm^3) suggests a high pressure of crystallization.

eclogite facies A set of metamorphic mineral assemblages of basaltic composition represented by omphacitic pyroxene and almandine-pyrope garnet. Minerals of this assemblage were once thought to have formed within a limited temperature-pressure range; later work has indicated that the range may be much wider.

ecology The study of interrelations of living organisms and their environment, including both biological and physical factors. adj. *Ecologic; ecological.* Syn: *bionomics.* See also: *paleoecology.*

economic geology The study and analysis of formations and materials that can be useful or profitable to man: fuels, water, minerals, metals.

ecosphere Those portions of the universe suitable for the existence of living organisms, in particular, the *biosphere.*

ecotope The habitat of a particular organism. See also: *biotope.*

écoulement *Gravity sliding.*

eddy A swirling or circular movement of water that develops wherever there is a flow discontinuity in a stream. Some of its causes may be velocity changes, flow separation, or changes in water density.

edrioasteroid A class of extinct echinoderms found as fossils in Cambrian to Mississippian (Lower Carboniferous) marine rocks. They were attached or fixed organisms, and include the earliest echinoderm representation. Edrioasteroids had bodies formed of irregular, flexible, polygonal plates. Their body shape varied from discoid to cylindrical.

effective permeability The ability of earth materials to transmit one fluid, such as gas, in the presence of other fluids, such as water or oil. See also: *absolute permeability; relative permeability.*

effective porosity The percent of the total volume of a given rock or earth material that consists of interconnecting voids. Cf: *porosity.*

efflorescence 1. A white powdery or mealy substance produced on the surface

of a rock or mineral in an arid or semi-arid region by loss of water or crystallization upon exposure to air. 2. The process, either through evaporation or chemical change, by which a rock or mineral becomes encrusted with crystals of such a salt.

effluent adj. Flowing out or forth. n. 1. A stream flowing out of a larger stream or out of a lake. Ant: *influent.* Cf: *effluent stream.* 2. An outflow or an effluence; e.g., liquid discharges as waste.

effluent stream A stream whose channel lies below the water table and into which water from the zone of saturation flows.

effusion The flow of a fluid onto the earth's surface, or the rock formed by such a flow. Cf: *extrusion.*

effusive *Extrusive.*

einkanter See *ventifact.*

ejecta 1. Material flung from a volcanic vent. Synonym for *pyroclastic material.* 2. Material such as glass and rock fragments thrown out of an impact crater during its formation.

elastic Said of a body capable of changing its length, volume, or shape in direct response to applied stress, and of instantly recovering its original form upon removal of the stress. Cf: *plastic.*

elastic aftereffect *Creep recovery.*

elastic bitumen *Elaterite.*

elastic constant One of various coefficients, expressed in units of stress, that defines the elastic properties of matter.

elastic deformation A deformation that disappears upon the release of stress. The term is frequently used for deformation in which the relation between stress and strain is a linear function.

elastic discontinuity A boundary between rock strata of differing density, or having different elastic properties. Seismic waves passing through rocks that are not uniform in all directions, or from one kind of rock into another (e.g., limestone to shale), are refracted or reflected. Where either a P or an S wave meets an elastic discontinuity, it generates two new P waves and two new S waves. See also: *seismic waves.*

elastic limit The maximum stress that a body or material can sustain and beyond which it cannot return to its original shape or dimensions. Syn: *yield point.*

elasticoviscosity A process involving both elastic and viscous behavior. A material is said to be elasticoviscous if it shows evidence of early deformation under elastic conditions, and if this is followed by a continuously generated, permanent strain for as long as the stress is present. The latter component is sometimes called *equivalent viscosity*, because true viscosity applies only to fluids in which the rate of flow is proportional to force; i.e., Newtonian flow.

elastic rebound theory A theory of earthquake genesis first propounded by H.F. Reid in 1911. In substance, it states that movement along a geological fault is the sudden release of a progressively increasing elastic strain between the rock masses on either side of the fault. The strain is the reaction to deformation by tectonic forces. Movement returns the rocks to a condition of little or no strain. Some of the first persuasive evidence for the elastic rebound theory came from observations of slip along the San Andreas Fault in the 1906 San Francisco earthquake.

elaterite A dark-brown asphaltic bitumen derived from the metamorphism of petroleum; used in building.

E layer See *seismic regions.*

electrical log See *well logging.*

electrical resistivity See *resistivity, electrical.*

electrolysis A method of dissociating a chemical compound by passing an electric current through it, with subsequent migration of the resulting positive and negative ions to the negative and positive electrodes, respectively.

electromagnetic prospecting A method of *geophysical prospecting* in which electromagnetic waves are generated at the surface. When such waves encounter a conducting formation or ore body, currents are induced in the conductors, and these currents generate new waves that are radiated from the conductors and detected by surface instruments. Inhomogeneities in the electromagnetic field, as observed on the surface, would indicate variations in the conductivity beneath the surface, and imply the presence of anomalous masses or substances.

electron diffraction pattern The interference pattern observed on a screen when an electron beam is directed through a substance, each substance having a unique pattern. Electron diffraction patterns supply basic crystallographic information.

electron microscope An instrument in which a beam of electrons from a cathode is focused by magnetic fields so as to form an enlarged image of an object on a fluorescent screen. The electron beam functions in a manner similar to that in which a beam of light functions in an optical microscope. Because of the shorter wavelength of electrons, the electron microscope has a very high resolving power.

electrum A natural alloy of gold and silver.

element A substance that can only be broken into simpler parts by radioactive decay, and not by ordinary chemical means. An element consists of atoms that all have the same atomic number.

elevation   1. Vertical distance from mean sea level to some point on earth's surface; i.e., *height* above sea level. In present-day surveying terminology, "elevation" refers to heights on the earth, whereas altitude refers to heights of spatial points above earth's surface. 2. General term for any topographically elevated feature.

elevation correction   In seismic surveys, an adjustment of observed reflection or refraction time values with respect to an arbitrary reference datum. Adjusted time values may be higher or lower than the true values, depending on whether the data were lower or higher than the seismic shot. See also: *gravity anomaly.*

elutriation   A process for separating finely divided particles into sized fractions in accordance with the rate at which they rise or sink in a slowly rising current of water or air of known and controlled velocity. Elutriation is used in the mechanical analysis of a sediment, and in the removal of a substance from a compound.

eluvial   1. Pertaining to eluvium. 2. The eluvial phase of a dune cycle is marked by dune degradation, rather than growth. Soil-forming processes, creep, and slope wash predominate. This phase is initiated when vegetation becomes heavy enough to halt deflation. Cf: *eolian.* 3. Said of a deposit formed by rock disintegration at the place of origin; no stream transport is involved.

eluviation   The downward movement of materials in suspension or solution that are being carried through the soil by descending soil water. Cf: *illuviation; leaching.*

eluvium   An accumulation of disintegrated rock found at the site where the rock originated. Cf: *alluvium.*

embayed   Characterized by the presence of a bay or bays; e.g., an *embayed coast.*

embayment   1. The process by which a bay is formed; also, the bay itself. 2. The incursion of a downwarped region of stratified rocks into other rocks. 3. The penetration of a crystal by another. 4. The *corrosion* of a crystal or xenolith by the magma in which it has formed.

emerald   A deep green variety of beryl, highly valued as a gemstone. It crystallizes in hexagonal prismatic forms, occurring chiefly in mica-schists. Its color is attributed to the presence of chromium.

emergence   An exposure of land areas that were formerly under water; the exposure may be a result of uplift of the land or decline of the water level. Evidence of land emergence is demonstrated by sands containing marine shells and fish remains at many levels in the stratigraphic column.

emery   A finely granular, dark-gray to black, impure variety of corundum, mixed with magnetite or hematite. Useful as a polishing and grinding material.

emplacement   The *intrusion* of igneous rock into a particular place or position; also, the development of ore deposits in a particular location. See also: *forcible intrusion; stoping; migmatite.*

emulsion   A colloidal dispersion of one liquid in another.

enantiomorphism   In crystallography, the existence of two chemically identical crystal forms that are mirror images of each other. adj. *enantiomorphous.*

enantiotropy   A condition or relation between polymorphs in which one may convert to the other at a critical temperature and pressure.

enargite   A gray, black, or grayish-black mineral, Cu_2AsS_4, occurring as metallic orthorhombic crystals. It is an ore of copper, closely resembling stibnite.

endellite   A name used in the U.S. for a

clay mineral, $Al_2Si_2O_5(OH)_4\cdot4H_2O$. Synonymous with *halloysite* of European literature.

end member One of the two extremes of a series; e.g., types of sedimentary or igneous rocks, minerals, or fossils.

end moraine See *moraine*.

endogenetic A term applied to processes that originate below the earth's surface, including extrusive and intrusive igneous activity, crustal warping, faulting, and folding. Such processes are responsible for crustal features (batholiths, anticlines), geological features (volcanoes, mountain ranges), and ore deposits. Syn: *endogenic; endogenous*. Cf: *exogenetic*.

endothermic Said of a chemical reaction that is accompanied by the absorption of heat. Cf: *exothermic*.

end product Any stable *daughter element* of radioactive decay. The term is also used to mean the *end products of weathering*. In general, these are quartz, sand, hard clay, and salt.

en échelon Etymol: Fr., *en échelon*, "in step." A term describing geological features, such as faults or folds, that are in staggered or overlapping arrangement. *En échelon faults* are parallel, relatively short faults that overlap each other; their strikes are the same but the dips differ. *En échelon folds* are those in which anticlines support anticlinal mountains that rise above a flat plain. The axial trace of the fold is then *en échelon* relative to the folds themselves.

engineering geology Application of the geological sciences to engineering procedures so that geological factors pertinent to the location and construction of engineering projects will be recognized and provided for. Syn: *geological engineering*.

englacial Said of something that is contained, embedded, or carried within a glacier or ice sheet; e.g., till, or drift. Syn: *intraglacial*.

enrichment The action of natural agencies that increases the relative amount of one constituent mineral or element contained in a rock. This may be due to the selective removal of other constituents, or to the introduction of increased amounts of other constituents from external sources. The process may be: 1. *Mechanical*, e.g., the transport of light material, such as quartz, away from heavier material, such as gold. 2. *Chemical*, e.g., downward-filtering, copper-containing solutions converting chalcopyrite, $CuFeS_2$ (34.5% copper) to covellite, CuS (66.4% copper). In this process the copper ions in solution replace the original iron. Such a process is often call *secondary* or *supergene enrichment*.

enstatite A rock-forming mineral of the orthopyroxene group, $MgSiO_3$; an important primary constituent of basic igneous rocks. Also present in many meteorites, both metallic and stony. With the substitution of iron for magnesium, enstatite grades into bronzites and, with the addition of more iron, into hypersthene.

enterolithic Said of sedimentary structures of small intestine-like or ropy folds, one of the primary sedimentary structures described from evaporite units. Such folds represent crumpling in an evaporite, caused by the swelling of anhydrite during the hydration to gypsum. See also: *chickenwire anhydrite*.

Entisol In U.S. Department of Agriculture taxonomy, a soil order described by dominance of mineral soil materials and an absence of defined horizons.

entrainment The process of picking up and transferring or carrying along, such as the collection and movement of sediment by water currents. Also, to trap or incorporate bubbles in a liquid.

entrenched meander An *incised*

meander showing little if any contrast between the slopes of the two valley sides of the meander curve. Cf: *ingrown meander.*

entropy A property, related to the Second Law of Thermodynamics, that can be equated with the disorder in a system. Given a system of different particles, the more even the distribution, the greater the particle disorder and the greater the entropy. Thus, in a stratigraphic unit of different kinds of rocks, as the composition approaches that of a single component, the entropy approaches zero. In terms of energy, the more evenly distributed the concentration of some form of energy in a system, the greater the entropy, and the less the opportunity to perform work. For example, hot magma intruded into cooler rocks causes a flow of heat into the wall rocks, where work is performed in volumetric expansion. Without the uneven concentration of thermal energy, which creates an energy gradient, no work could be performed.

envelope A term often used for the outer part of a fold. Cf: *core.*

environmental geology The application of geological concepts to problems created by man, and their effects on the physical environment.

Eocene The epoch of the Tertiary period between the Paleocene and Oligocene Epochs. See *Cenozoic Era.*

eolian Pertaining to or caused by wind. Loess and dune sands are eolian deposits, and ripple marks in sand are formed by wind. *Ventifacts, yardangs,* and *zeugen* are products of eolian erosion.

eon A division of geologic time, the longest time unit next in order above *era.* The Phanerozoic Eon includes the Paleozoic, Mesozoic, and Cenozoic Eras. The term is also used for an interval of 1 billion (10^9) years.

Eötvös effect The vertical component of Coriolis acceleration, observed when taking gravity measurements while in motion. The velocity over the surface of the meter that is recording the gravity data adds vectorially to the velocity caused by the earth's rotation; this changes the centrifugal acceleration and thus the apparent gravitational attraction. The Eötvös correction in mgal (*milligal*) for a meter with a velocity of K knots at an azimuth angle α and latitude $\emptyset$ is $E = 7.503 \, K \cos \emptyset \sin \alpha + 0.004154 K^2$. The Eötvös uncertainty dE, in terms of direction uncertainty d α, and velocity uncertainty dK, is $dE = (7.503 \, K \cos \emptyset \cos \alpha) \, d\alpha + (7.503 \cos \emptyset \sin \alpha + 0.008308 \, K) \, dK$.

Eötvös torsion balance A type of instrument used in geophysics for measuring curvature and gradients in a gravitational field. This differs from the Cavendish torsion balance, which measured curvature alone; many present-day instruments measure gravity differences directly by determining the gravitational acceleration. The Eötvös balance consists of two equal weights at different heights, connected by a rigid frame. The system is suspended by a tension wire in such a way that it can rotate freely in a horizontal plane about the wire. The gravity gradient is indicated by the amount of torque causing rotation; the higher the torque of the wire, the greater the gradient. See also: *geophysical exploration.*

Eötvös unit A unit of gravitational gradient or curvature, 10^6 mgal/cm.

epeiric sea *Epicontinental sea.*

epeirogenesis *Epeirogeny.*

epeirogeny The uplift and subsidence of large portions of the earth's crust that have produced broader topographic features of continents and oceans, such as basins and plateaus, in contrast to *orogeny,* which is responsible for mountain ranges. Syn: *epeirogenesis.* See also: diastrophism.

ephemeral stream A stream that flows (carries water) only during periods of precipitation and briefly thereafter. Its channel is dry most of the year, and may be referred to as an arroyo, or gully. Such streams occur in arid and semi-arid regions, but may be found in any area where the channel bed is at all times above the water table. Cf: *intermittent stream*.

epibole *Acme-zone.*

epicenter An area on the surface of the earth directly above an earthquake focus.

epiclastic Said of sediments consisting of weathered or eroded fragments of pre-existing rocks, where the fragments have been transported and deposited at the surface. Also said of the rocks formed of such sediments; e.g., sandstone and conglomerate. Cf: *autoclastic*.

epicontinental Located on the continental shelf or on the continental interior, e.g., an *epicontinental sea* is a sea within a continent, or on the continental shelf. Syn: *inland sea; eperic sea.* Cf: *mediterranean*.

epidote A green monoclinic mineral, $Ca_2(Al,Fe)_3Si_3O_{12}(OH)$. Common in regional and contact metamorphic rocks, esp. in green schist facies.

epieugeosyncline A deeply subsiding trough lying above an intruded and deformed *eugeosyncline*. It is associated with narrow uplifts and limited volcanic activity. Syn: *backdeep*. See also: *geosyncline*.

epigene adj. Formed at or originating at or near the earth's surface, as opposed to *hypogene*.

epigenesis 1. Changes, exclusive of weathering and metamorphism, that affect sedimentary rocks after they are compacted. 2. Alteration in the mineral profile of a rock because of external effects taking place at or near the earth's surface.

epigenetic 1. Formed or originating on the earth's surface. Cf: *hypogene*. 2. Describes minerals introduced into pre-existing rocks. Such deposits may occur as tabular or sheetlike lodes, or in various other forms.

epinorm See *depth zone of emplacement*.

epithermal deposit See *hydrothermal deposit*.

epizone See *depth zone of emplacement*.

epoch 1. An interval of geologic time longer than an *age* and shorter than a *period*, during which the rocks of a particular *series* were formed. 2. An inexact, short space of geologic time, such as a *glacial epoch*.

equal-area projection A type of map projection, including the Albers projection and the Mollweide projection, in which regions on the earth's surface that are of equal area are represented by equal areas. Syn: *homolographic projection*. See also: *map projections*.

equant 1. Term applied to a crystal having the same or nearly the same dimensions in all directions. Syn: *equidimensional; isometric.* Cf: *tabular; prismatic*. 2. A sedimentary particle is said to be equant if its length is less than 1.5 times its width. 3. Said of a rock in which most of the grains are equant.

equatorial projection A map projection, such as the Mercator projection, in which center points of the projection are at the equator, and the polar axes are vertical. See also: *map projections*.

equiplanation Development of terrace-like surfaces by the reduction of land without the loss or gain of material and without reference to a base level. Ex:

marine lacustrine or fluvial terraces. Cf: *altiplanation, cryoplanation.*

equipotential surface   A surface along which the gravity is constant at all points, and for which the gravity vector is normal to all points on the surface. The *geoid* is an equipotential surface. See also: *geopotential.*

equivalent adj. Agreeing or corresponding in geological age or position within the stratigraphic profile. Strata in different regions that have been formed at the same time, or contain the same fossil types, are said to be equivalent.

era   A geologic time unit during which rocks of the corresponding *erathem* were formed. It ranks one order of magnitude below *eon;* e.g., Paleozoic, Mesozoic, Cenozoic.

erathem   The largest formal *chronostratigraphic unit* generally recognized, followed by *system;* it consists of the rocks formed during an *era* of geologic time, such as the Paleozoic erathem composed of the systems from the Cambrian through the Permian.

erg   An undulating plain occupied by complex sand dunes produced by wind deposition, such as is found in parts of the Sahara. Called *koum* in Turkestan. Cf: *hammada; reg.*

Erian   Middle Devonian of North America.

erosion   The wearing away of any part of the earth's surface by natural agencies. These include mass wasting and the action of waves, wind, streams, and glaciers. Fundamental to the process of erosion is that material must be picked up and carried away by such agents. Evidence for erosion is widespread; the retreat of marine cliffs, deposition of fluvial material, and the cutting of great canyons; e.g., the Grand Canyon. *Differential erosion* is a difference in the erosion rate on individual features as well as over large

areas. Rate variation is due to the differential resistance of certain outcrops; e.g., *rock pedestals.* Large-scale differential erosion can be seen in the formation of *hogbacks.* Cf: *denudation.*

erratics   Glacially transported stones and boulders. Erratics may be embedded in *till* or occur on the ground surface. They range in size from pebbles to massive boulders weighing thousands of tons. Their transport distances range from less than 1 km to more than 800 km (500 miles). Erratics composed of distinctive rock types can be traced to their point of origin and serve as indicators of glacial flow direction. Certain erratics found in sediments of a type that would preclude glacial transport may have been deposited by icebergs.

eruption   See *volcanic eruption.*

eruption cloud   A rolling mass of atomized lava, partly condensed water vapor, dust, and ash emitted from a volcano during an eruption, and rising to great heights. Syn: *ash cloud; dust cloud; volcano cloud.*

escarpment   A high, more or less continuous cliff or long, steep slope situated between a lower, more gently inclined surface and a higher surface. Escarpments, which may be formed by faulting or erosion, occur as plateau margins, or one may form the steep face of a cuesta. Syn: *scarp.*

esker   Etymol: Celtic, "ridge of mountains." A long, narrow, sinuous or straight ridge formed of stratified glacial meltwater deposits, usually including large amounts of sand and gravel. Most eskers are probably deposited in channels beneath or within slow-moving or stagnant ice. Their general orientation runs at right angles to the ice edge. Some eskers originate on the ice and have ice cores. Typical eskers show a fairly constant relationship between their individual dimensions. Long eskers never exceed 400 to 700 m in width and 40 to 50 m in height; smaller eskers of 200 to 300 m may be 40 to 50 m wide and 10 to 20 m high. Well-known esker areas are found in Maine, Canada, Sweden, and Ireland. Cf: *kame.* See also: *drumlin.*

essental mineral A mineral component that is basic to or required in the classification of a rock. Cf: *accessory mineral.*

essexite An alkaline igneous rock essentially composed of plagioclase, orthoclase, pyroxene, and biotite, with accessories of olivine, apatite, and magnetite. It occurs as small plutons or as differentiated marginal portions of gabbroic plutons; sometimes found as xenoliths in lava flows.

estuary A widened mouth of a river valley where fresh water intermixes with seawater, and where tidal effects occur. An elongated portion of a sea that is affected by fresh water is also called an estuary. Estuaries are formed where a deeply cut river mouth is drowned following a land subsidence or a rise in sea level.

etch figure A technique, used for revealing crystal symmetry, in which solvent is applied to a crystal surface. The shapes of etch markings that appear on the crystal faces vary with the solvent and with the mineral species; however, for a given solvent, the etching effect is the same for all faces of a crystal form.

etching In geomorphology, the gradation of the earth's surface by differential weathering, mass wasting, and sheetwash (broad, continuous sheets of running water). Etching is particularly significant in areas of diverse rocks that exhibit extreme contrast in their weathering rate. Interstream degradation may play the most important role in the etching concept of erosion.

ethane A colorless, odorless, flammable gas, C_2H_6, of the methane series, present in natural gas and crude petroleum.

eugeocline *Eugeosyncline.*

eugeosyncline See *geosyncline.*

euhedral Said of mineral crystals that develop fully and show well-formed polyhedral facial planes. Syn: *automorphic; idiomorphic.* Cf: *anhedral; subhedral.*

euphotic zone An ocean zone in which light penetration is sufficient for photosynthesis. It commonly extends from 60 to 70 m, but may be greater. The zone is shallowest over the continental shelves because light is blocked by the great quantities of suspended sediment there. Cf: *aphotic zone.*

Eurasian-Melanesian belt A belt of tectonic activity extending from the Mediterranean to the Celebes, where it joins the *circum-Pacific belt.*

eurypterid Extinct group of brackish-water *arthropods,* probably predaceous, closely related to horseshoe crabs. Their flattened, segmented bodies were covered with chitin, and some grew to lengths of almost 3 m. The cephalothorax had two pairs of eyes, four pairs of walking legs, one oarlike pair of legs, and one pair of pincers. Eurypterid fossils are not used for zonation. Range, Ordovician to the end of the Permian.

eustatic Pertaining to worldwide changes in sea level, as distinct from local changes.

eutaxitic A term used to describe a fabric typical of welded ash-flow tuffs, in which lapilli, pumice, and glass shards are flattened into a disc-like form more or less parallel to the plane of deposition. Eutaxitic fabric is created by the simultaneous welding and compaction of glassy particles that are part of the lava flow when ejected. The most intense welding and compaction occurs at the lower portion of the flow, but not right at the base.

eutectic   Pertaining to a system of two or more solid phases and a liquid, where the components are in such proportion that the melting point of the system is the lowest possible with these components.

eutectic point   The lowest temperature at which a mixture can be maintained in a liquid phase or state; the lowest melting or freezing point of an alloy.

eutrophication   The process whereby an aging aquatic ecosystem, such as a lake, supports great concentrations of algae and other aquatic plants due to a marked increase in concentration of phosphorus, nitrogen, and other plant nutrients. Algal blooms on the surface prevent the light penetration and oxygen absorption necessary for underwater life.

eutrophic lake   A lake characterized by large quantities of dissolved plant nutrients (phosphorus and nitrogen) and a seasonal deficiency of oxygen in the bottom layers owing to deposits containing significant amounts of rapidly decaying organic material. Cf: *oligotrophic; dystrophic.*

euxinic   Term used to describe an environment characterized by large masses of stagnant, deoxygenated water, which thus fosters reducing processes. The Black Sea is the standard example of a land-locked euxinic sea, and is referred to as the *Euxinic* or *Euxine Sea.*

evaporite   A sediment deposited from a saline solution as a result of extensive or total evaporation of the water. Anhydrite is the first evaporite mineral to precipitate from seawater. Although they are very soluble, evaporite minerals have been found in rocks 3.5 b.y. in age. The variety and abundance of minerals so produced depend on the initial composition of the body of water. Nearly 70 evaporite minerals are known: 27 are sulfates, 27 are borates, and 13 are halides. Seawater is the originating solution of most evaporite deposits. Non-marine, saline lake and playa evaporite waters generally have initial compositions quite different from seawater and, as a result, nonmarine evaporite deposits can contain minerals rarely formed from seawater; e.g., mirabilite, glauberite, borax.

event   1. *Seismic event.* 2. A more or less catch-all term for any occurrence in which probable tectonic importance is implied by some geologic evidence.

exfoliation   The separation of successive thin, onionlike shells (*spalls*) from bare surfaces of massive rock, such as granite or basalt. It is common in regions of moderate rainfall. Geologists have attributed some small-scale exfoliation to diurnal temperature variations. Large-scale exfoliation or sheeting can be caused by a decrease that occurs in compressional forces when previously covered rocks are exposed. The slow development at the exposed rock surface of clay minerals, which involves an increase in volume, can also produce separation. The resulting differential moisture content between the surface and inner layer causes flaking. *Spheroidal weathering* is a small-scale form of exfoliation restricted to boulder-size rock material, and may occur at depth.

exogenetic   Said of geological processes that act at or near the earth's surface, such as weathering and denudation. The term is also applied to rocks, landforms, and ores that are derived through such processes. Syn: *exogenic; exogenous.* Cf: *endogenetic.*

exogeosyncline   See *geosyncline.*

exothermic   Pertaining to a chemical reaction that is accompanied by the liberation of heat. Ant: *endothermic.*

exploratory well   A well drilled in an unproved area or to an untried depth, either to seek a new pool of gas or oil, or for the possibility of broadening the area of a known field. Syn: *test well; wildcat well.*

explorer's alidade   *Gale alidade.*

explosive index   The percentage of pyroclastics found among all the products of a particular volcanic eruption.

exsolution   The separation of a homogeneous solid solution into distinct crystal phases. For example, at 600° C an initially homogeneous feldspar exsolves into a potassium-rich feldspar solid solution and an albite-rich feldspar solid solution. The two feldspars usually segregate within the original crystal as thin lamellae, forming the intergrowth *perthite.*

extinction angle   If a thin section of a birefringent mineral is placed under a petrographic microscope between crossed Nicol prisms, the angle through which it must be rotated from a given plane until darkness is obtained is the extinction angle. This angle is often used as a means of mineral identification.

extrusion   A term for the emission of fairly viscous lava onto the earth's surface, or for the rock formed by this. Cf: *effusion.*

extrusive rocks   A term applied to rocks that result from cooling, solidification, and the formation of igneous materials above the surface of the earth. Such rocks are differentiated from *plutonic* or *intrusive* rocks, which have solidified from magma below the earth's surface. As magma moves upward from the deeper layers, the resultant heat and pressure may cause the rocks above to break or move, permitting the magma to effuse through such breaks onto the surface. Because the material cools rather rapidly, an extrusive rock is generally finely textured. Extrusive rocks are synonymous with *volcanic* rocks. Cf: *intrusive rock.* See *magma.*

F

fabric   1. An inclusive term for both the *texture* and *structure* of a rock body. The fabric supplies information about the geological processes that produced a particular rock. For example, the spatial orientation of the particles and crystals of which a sedimentary rock is composed implies certain conditions of deposition and consolidation. 2. In igneous rock, a particular orientation of columnar amphiboles in a granite connotes differential flow in the magma containing the amphiboles. 3. The physical description of a soil according to the spatial arrangement of its particles and voids.

fabric axis   In structural petrology, one of the three orthogonal axes used as references in the orientation of fabric elements, and in the description of folding and symmetry of movement in deformed rocks. Syn: *tectonic axis.* Cf: *a axis; b axis; c axis.*

fabric diagram   An equal-area or stereographic projection of components of a rock fabric (fabric elements), used in structural petrology. Syn: *petrofabric diagram.*

face   1. The principal or most evident side or surface of a landform, such as a cliff face. 2. One of the planar bounding surfaces of a crystal; *a rational face.* 3. The surface on which mining operations are in progress, or the place that was last worked.

facet   1. One of the small, polished plane surfaces on a cut gemstone. 2. A nearly planar surface produced on a rock by the action of water or sand. 3. Any plane surface resulting from faulting or erosion, and cutting across the general slope of the land.

facies An assemblage or association of mineral, rock, or fossil features reflecting the environment and conditions of origin of the rock. It refers to the appearance and peculiarities that distinguish a rock unit from associated or adjacent units. Cf: *lithofacies*. The term *facies* is used in so many contexts that the reader of current geological literature should be certain of the kind of facies that is meant. See also: *biofacies; petrographic facies; sedimentary facies*.

facies evolution Gradual lateral or vertical changes in the type of rock or fossils found in contemporaneous sedimentary deposits. Such variations reflect a change in the depositional environment.

facies fauna A group of animals characteristic of specific stratigraphic facies or adapted to life in a particular environment.

facies fossils Assemblages that represent a particular paleoenvironment and, therefore, a correspondingly restricted distribution.

facies map A type of map on which is shown the distribution of sedimentary facies that occur within a given geologic unit; e.g., a *lithofacies map*. See also: *isolith map*.

facies tract A system of interconnected sedimentary facies of the same age, including the areas of erosion from which the sediments of the facies originate.

fairy stone *Staurolite*.

falling dune See *dune*.

family 1. Basic unit of the *clan* in

igneous rocks. 2. In biological classification, a category between order and genus. 3. An ecologic community consisting of only one kind of organism. 4. A category of soils, as used by the U.S. Department of Agriculture.

fan 1. *Alluvial fan*. 2. Fan-shaped mass of solidified lava.

fan fold See *fold*.

fanglomerate A *conglomerate* composed of an alluvial fan deposit that has become cemented.

fan valleys See *submarine fans*.

fathometer A name for a type of *echo sounder*. The graphic record produced by a fathometer is called a *fathogram*; it is a type of *echogram*.

fault A fracture in earth materials, along which the opposite sides have been relatively displaced parallel to the plane of movement. Fault lengths may range from a few centimeters to hundreds of kilometers, and displacement may be of comparable magnitude along the fault plane (fracture zone). Some areas of the earth's surface are cut by thousands of faults of varying size, whereas in other areas faults are almost totally lacking. Faulting may mark the walls of a fault plane with *slickensides*; other evidence of faulting is the presence of *fault breccia, fault gouge*, or *mylonite*. In sedimentary strata, *drag* may be evidenced in fault zones, a result of frictional resistance to slippage. There may be no surface indication of faulting in an area where the soil cover is deep. Block displacement on the opposite sides of a fault plane is usually recorded in relation to sedimentary strata, or to dikes and veins. The apparent movement of a fault may be very different from the actual movement if erosion has obliterated the evidence. One basic rule in ascertaining the movement of a fault is that a fault must always be younger than the youngest rock it cuts, and older than the oldest lava or undisturbed sediment across it. Movement along faults may occur as continuous creep or as a series of abrupt jumps of a few meters over a few seconds. During quiet intervals that separate these jumps, the stress increases until frictional forces along the fault plane are overcome. Most earthquakes are caused by such abrupt movement along faults.

Faults occur in all possible orientations from horizontal to vertical, and relative movement is possible in any direction on each fault. Although such variety makes their classification difficult and arbitrary, faults are differentiated according to their angle of inclination and their relative and apparent motion. 1. *Normal* (or *gravity*) *faults* are produced by vertical compression. The *hanging wall* appears to have slipped downward relative to the *footwall*. The angle of dip is generally 45 to 90°. Such faults are found throughout the world. See also: *oblique-slip fault; hinge fault.* 2. *Thrust faults* are produced by horizontal compressional forces. The hanging wall appears to have moved up and over the footwall at a dip of 45° or less. Large thrust faults are found in the Ridge and Valley Province of Virginia and Tennessee. 3. A *reverse fault* is a fairly steep-dipping fault along which the hanging wall appears to have moved upward with respect to the footwall. 4. A *dip-slip fault* is a normal or reverse fault on which the only component of movement lies in a vertical plane normal to the fault strike. 5. A *strike-slip fault* is one along which movement has been predominantly parallel to the strike. See also: *right-lateral fault; left-lateral fault; oblique-slip fault; transcurrent fault; transform fault.*

fault basin Depression separated from the surrounding area by faults.

fault block A unit of earth's crust either completely or partly bounded by faults. During faulting and tectonic activity it functions as a unit. See also: *fault-block mountain.*

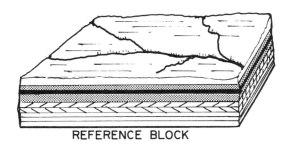

A reference block prior to faulting; drainage left to right.

REFERENCE BLOCK

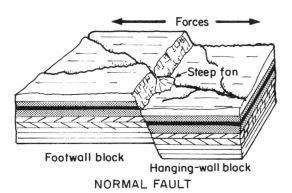

A steeply inclined fault along which the hanging wall block has moved downward.

NORMAL FAULT

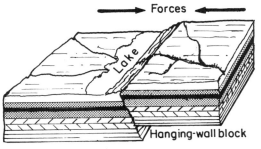

Forces → ←

A steeply inclined fault along which the hanging wall block has moved upward.

Lake

Hanging-wall block

Footwall block
REVERSE FAULT

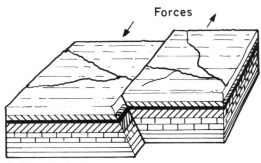

Forces

A dip-slip fault is a normal or reverse fault on which the only component of movement lies in a vertical plane normal to the strike of the fault surface.

Strike-slip fault

fault-block mountain   A mountain formed by *block faulting*; i.e., isolated by faulting, and categorized as structural or tectonic. They are typically developed in the Basin and Range Province of Nevada, Utah, and Arizona, and parts of New Mexico and California. The geomorphic characteristics of fault-block mountains include more or less rectilinear borders, a general lack of continuity of formation, fresh fault scarps, and occasional seismic revival. The uplifted blocks may have been stripped of younger formations that covered them, and are thus relict landforms. When worn down to the old stage, fault-block mountains can no longer be recognized as such on geomorphic evidence. Fault-block mountains are formed from crustal segments showing wide variation from one mountain to another in rock type and history. This feature distinguishes them from *fold mountains*, which, for long dis-

tances along a strike, have undergone similar sedimentary and tectonic histories. The *normal fault* boundaries of block-fault mountains are evidence of crustal extension, whereas fold mountains are products of crustal compression. Syn: *block mountain*. See also: *graben; horst*.

fault breccia   Angular fragments resulting from the shattering or crushing of rocks during movement along a fault or in a fault zone. See also: *fault*.

fault escarpment   *Fault scarp*.

fault gouge   A fine-grained, claylike substance formed by the grinding of rock material as a fault develops; it may also be formed by decomposition caused by circulating solutions.

faulting   The action or process of fracturing and displacement that produces a *fault*. Evidence of faulting may be demonstrated by such criteria as: 1. discontinuity of structures; 2. repetition or omission of strata; 3. The presence of features characteristic of fault planes; 4. sudden changes in sedimentary facies; 5. physiographic features, such as springs, aligned along the base of a mountain range.

fault line Intersection of a fault with the ground surface. It usually varies in form from straight to moderately sinuous. Sometimes referred to as *fault trace*. See also: *fault scarp; fault-line scarp; fault*.

fault-line scarp A cliff, basically parallel to the fault trace, that has retreated because of extensive erosion, leaving the actual fault line buried beneath sediment. The fundamental difference between *fault scarp* and fault-line scarp is that the former is produced by movement along the fault, whereas the latter is produced by differential erosion along the fault line. An *obsequent fault-line scarp* faces in the opposite direction from the original fault scarp, and occurs at a lower stratigraphic level. In this case, the original fault scarp has been completely eroded away. Erosion of what was originally the uplifted block continues until an escarpment is formed, whereby the uplifted block becomes the lowermost level. A *resequent fault-line scarp* faces in the same direction as the original fault scarp. The original topographic difference of the two sides of the fault has been completely removed, and erosion continues at successively lower stratigraphic levels that were not part of the original terrain.

fault plane The surface along which rock masses on opposite sides of a fault rub against each other. In some instances, it may be a warped and irregular surface. Also referred to as a *fault zone* or *fault surface*. See also: *fault*.

fault scarp A cliff formed by an upthrown fault block from 1 to 10 m high and trending along the *fault line*. It is the evidence of the latest movement along the fault. Cf: *fault-line scarp*. See also: *fault*.

fault set A group of parallel or near-parallel faults associated with a particular deformational event. Two or more interconnecting fault sets constitute a *fault system*.

fault surface The surface along which the dislocation on a fault has occurred. Cf: *fault plane*.

fault system See *fault set*.

fault zone A fault evidenced by a zone consisting of a network of many small faults, or by a zone of fault breccia or mylonite. Fault zones may be hundreds of meters wide.

faunal break An abrupt change in a stratigraphic sequence from one faunal fossil assemblage to another.

faunal province A region characterized by an assemblage of specific fossil fauna widely distributed within it. The so-called Atlantic Province includes outcrops in Britain, Scandinavia, Poland, Germany, southeastern Newfoundland, Nova Scotia, and the environs of Boston, Mass. It is characterized by a sequence of trilobite fauna by which correlations can be made within the province.

faunal succession The chronologic sequence of life forms through geologic time. Fossil fauna and flora succeed one another in a definite, recognizable order, as expressed in the Law of Faunal Assemblages: *Like assemblages of fossils indicate like geological ages for the strata containing them.*

faunule 1. A collection of fossil fauna gathered from a bed extending over a very limited geographic area. Syn: *local fauna*. 2. An association of fossil fauna found in a single stratum or a few contiguous layers, in which members of a single community predominate.

fayalite A silicate mineral, Fe_2SiO_4, the iron end-member of the olivine group. It crystallizes in the orthorhombic system, and is found mainly in igneous rocks.

feeder 1. The channel through which magma passes from the magma chamber to some intrusion. 2. An opening in a rock through which mineral-bearing solutions can travel. 3. *Tributary.*

felspar *Feldspar.*

feldspar A group of rock-forming minerals that make up about 60% of the earth's crust. Their general formula is $MAl(Al,Si)_3O_8$, where M can be K, Na, Ca, Ba, Rb, Sr, or Fe. Feldspars occur in all types of rock, but are essential constituents of most igneous rocks; the kind and amount of feldspar present are used as a means of classifying these rocks. Feldspars are divided into potassium feldspars, some of which are triclinic, and sodium-calcium feldspars (plagioclase), all of which are triclinic. Potassium feldspar includes the monoclinic form (orthoclase) and the triclinic form (microcline). Sanidine is an orthoclase in which sodium replaces some potassium. The plagioclase feldspars form a complete series from the pure sodium member, albite, to the calcium end-member, anorthite. Members with compositions intermediate between the two extremes are listed in order of increasing calcium content: albite, oligoclase, andesine, labradorite, bytownite, and anorthite. Most feldspar is colorless, white, or light gray, but may also be brown, yellow, red, green, or black. Moonstone, a gem variety, is milky opalescent; other gem varieties are aventurine and amazonite, a deep-green microcline. Adj: *feldspathic.*

feldspathic Said of a rock or mineral aggregate containing feldspar. The percentage of feldspar present must usually be within certain limits. For example, a *feldspathic sandstone* contains from 10 to 25% feldspar and is intermediate in composition between a quartz sandstone and an *arkosic sandstone.* Approx. syn: *subarkose.*

feldspathoid A rock-forming mineral similar to the feldspars but containing less silica. In those igneous rocks that are *silica-undersaturated*, feldspathoids take the place of feldspars; i.e., in those igneous rocks that crystallize from magmas containing too little silica to form feldspars. Feldspathoids never occur in the same rock with quartz. Nepheline and leucite are the most abundant feldspathoids.

felsenmeer Etymol: Ger., "sea of rock." Another name for *block field.*

felsic Acronymic word derived from *feldspar* and *silica* and applied to light-colored silicate minerals such as quartz, feldspar, and feldspathoids. Also describes the rocks that contain an abundance of one or all these minerals. Cf: *mafic.*

felsite Any light-colored aphanitic igneous rock composed chiefly of quartz and feldspar. When *phenocrysts* (usually of quartz and feldspar) are present, the rock is called a felsite *porphyry.* The term is also used for a rock showing *felsitic* texture.

felsitic A textural term ordinarily used to describe dense, light-colored igneous rocks that contain microcrystalline or cryptocrystalline aggregates. Originally used to convey the idea of a mineral substance now identified as a mixture of feldspar and quartz. Cf: *felsic.*

felted See *felty.*

felty (felted) A textural term for dense, holocrystalline igneous rocks, or the dense, holocrystalline groundmass of porphyritic igneous rocks consisting of tightly and flatly pressed microlites. These are generally feldspar, interwoven in an irregular fashion. Many andesites and trachytes contain crowded feldspar microlites that show subparallel arrangement as a result of flow, with the microlite interstices occupied by microcryptocrystalline material. This texture is also called *pilotaxitic* or *trachytic*.

felty *Pilotaxitic.*

femic An acronymic term derived from iron (Fe) and magnesium (Mg) and applied to the group of *standard normative* dark-colored *minerals* of which these elements are basic components. These minerals include the olivine and pyroxene molecules and most normative accessory minerals. See *CIPW classification* (normative composition). The corresponding term for ferromagnesian minerals actually present in a rock is *mafic.* Cf: *salic.*

fen *See bog.*

fenster *Window.*

Ferrel's law A statement of the effect of the *Coriolis force* in the earth's wind systems. It is the dynamic concept that currents of air or water in the Northern Hemisphere are deflected to the right and those in the Southern Hemisphere to the left. This deflection, the result of the Coriolis force, is nearly balanced by the pressure gradient but, because of friction, is not completely balanced. Cf: *geostrophic motion.* See *Coriolis force.*

ferricrete 1. An amalgam of surface sand and gravel cemented into a mass by iron oxide. 2. A ferruginous *duricrust.* Cf: *calcrete; silcrete.*

ferriferous Iron-bearing; said particularly of a mineral that contains iron, or of a sedimentary rock that contains more iron than is usually the case. Cf: *ferruginous.*

ferroalloy A metal whose principal use is to be alloyed with iron to produce steel of a special type or quality. Included among the ferroalloys are manganese, nickel, tungsten, chromium, cobalt, molybdenum, and titanium.

ferromagnesian adj. Containing iron and magnesium. In petrology, it is applied to certain dark silicate minerals, particularly amphibole, olivine, and pyroxene, and to igneous rocks containing these minerals as dominant components. Cf: *femic; mafic.*

ferromagnetic Having a magnetic permeability very much greater than unity. Certain elements (iron, nickel, cobalt), and alloys with other elements (titanium, aluminum) exhibit relative permeabilities up to $10°$ C. A ferromagnetic substance, e.g., iron, at a temperature below the Curie point, can show a magnetization in the absence of an external magnetic field. The magnetic movements of its atoms have the same orientaton. Cf: *diamagnetic; paramagnetic.*

ferruginous Pertaining to iron or containing it; e.g., a sandstone cemented with iron oxide, or with a cement containing iron oxide. Also said of a red or rust-colored rock whose color is due to the presence of ferric oxide.

fetch 1. The distance that prevailing winds travel over a body of water, thus generating a wave system. 2. Area of generation of a given wave system. Coastlines, frontal systems, or isobar curvature may be used to define the boundaries of an area of fetch.

fibroblastic A type of fabric in metamorphic rock bodies, characterized by mineral of even grain size and fibrous habit. The fabric is due to solid-state crystallization during metamorphism.

fibrous texture An array of circular or rodlike crystals found in certain mineral deposits, such as asbestos.

fiducial time A time noted or marked on a seismic record to correspond to some arbitrary time. The marks may help in synchronizing different records, or may indicate some reference base.

field classification   Examination or preliminary analysis of fossils or hand specimens of rock or minerals, commonly with the aid of a hand lens.

field geology   That part of geology that is practiced by direct observation in the field.

field well   A well drilled for gas or oil within the vicinity of a pool already proved for production.

filiform   *Capillary.*

fill   1. Man-made deposits of soil, rock, or various debris materials used for building embankments, filling in soggy ground, extending a shoreline into a lake, or filling unused mine workings. 2. Sediment deposited by any agent, in such a manner that it fills or partially fills a valley.

filter   A device used in a seismograph circuit to control the frequency characteristics of the recording system. Low frequencies are usually excluded in order to prevent *ground roll* and noise of other types from interfering with reflection of seismic waves. High frequencies are attenuated in order to remove wind noise and other extraneous effects.

filter pressing   A theoretically possible process of *magmatic differentiation* wherein the melt in a mushlike, crystal-rich magma separates from the crystals, either by draining or by being pressed out. It might occur on the floor of mafic intrusions, where the weight of the accumulating crystals forces some of the melt out of the underlying crystal mush.

fine aggregate   The portion of an aggregate composed of particles smaller than approximately 4.76 mm in diameter. Cf: *coarse aggregate.*

fine-grained   1. Said of the texture of a sedimentary rock in which the average diameter of the particles is less than one-sixteenth mm (silt size and smaller). 2. Said of the texture of an igneous rock whose particles have an average diameter of less than 1 mm. Syn: *aphanitic.*

fines   Very small particles in a mixture of particles of various sizes. The term is also used for finely crushed ore, mineral, or coal.

fine sand   Sand whose particles are from 0.125 to 0.25 mm in size.

finger lake   A long narrow lake of the type that comprises the Finger Lakes of western New York State, a group of eleven elongated lakes in digitate arrangement. They occupy very deep, steep-walled troughs that drain to the north. It is thought that their great depths may be the result of glacial erosion concentrated in preglacial valleys. The development of the Finger Lakes paralleled and was contemporaneous with that of the Great Lakes, but on a smaller scale.

fiord   *Fjord.*

fireclay   A clay that can withstand high temperatures without disintegrating or turning pasty. *Flint clays*, nonplastic fireclays that are extremely hard, are microcrystalline clay rocks composed predominantly of kaolin. Much fireclay is derived from underclays beneath coal beds, but not all underclays are fireclays. Fireclays form where surface conditions permit most minerals, except kaolinite and illite, to be leached out. Fireclay is rich in hydrous aluminum silicate, and is used widely to manufacture clay crucibles and firebrick, and as a binder in molding sands. Syn: *refractory clay.*

firedamp   A combustible and highly explosive gas consisting chiefly of methane; formed esp. in coal mines.

fire fountain   See *lava fountain.*

fire opal   See *opal.*

firn   Etymol: Ger., "of last year." Form of snow intermediate between freshly fallen snow and ice. A *firn field*, more accurately called névé, is a wide expanse of glacier surface over which snow accumulates and becomes firn. After lasting through one summer melt season, snow becomes firn by alternate melting and freezing, condensation, and compaction. The point at which firn becomes glacial ice is not defined by universal standards. The transformation is effected when densities of 0.4 to 0.55 are reached, although higher densities are often recorded. The density at which the densifying material sometimes becomes impervious to water is 0.55. Although some literature uses névé and firn synonymously, many specialists do not equate them. *Firn* is the term used for the process and the material, whereas *névé* is used as a locational term for the position of the snow accumulation area. See also: *glacier.*

firn line The lower limit of firn in summer; above it is the zone of net accumulation, below it the zone of net loss. The firn line is the line at which the average velocity of glacial flow is generally greatest. Syn: *firn limit.*

first arrival The first energy to reach a seismograph from a source of seismic activity. First arrivals on reflection records are used to obtain information about a *weathered layer*, they are often used as a basis for refraction studies.

First Law of Thermodynamics Another name for the *Law of Conservation of Energy*: The total amount of energy in any system is conserved in any process or, as a more general statement, the total amount of mass and energy in a system is conserved in any process.

fissility 1. The property of splitting or dividing readily along closely spaced, parallel planes; e.g., bedding in shale or cleavage in schist. adj. *Fissile.* Any sedimentary rocks, esp. fine-grained varieties, tend to part parallel to the stratification, thus exhibiting *bedding fissility.* This is very probably caused by long, platy mineral grains aligned more or less parallel to the stratification.

fission The spontaneous or induced division of the nucleus of an atom into nuclei of lighter atoms, accompanied by the release of great amounts of energy. Elements whose atomic nuclei can be so divided are *fissionable.* The property is peculiar to the heavy elements; e.g., uranium, plutonium. Cf: *fusion.*

fissure A rather general term meaning an extensive cleft, break, or fracture in a rock formation, volcano, or the earth's surface. It is also used in the sense of *crevasse.*

fissure eruption Volcanic eruption from a fissure or series of fissures rather than from a central vent. In distinction to *central vent eruptions,* which occur in many tectonic settings, fissure eruptions favor *mid-ocean ridges.* Tensional forces in such areas may cause breaks in the earth's crust, with the ensuing formation of long, narrow vertical fissures. The intrusion of magma into deeper parts of these cracks forms *dikes,* and sufficiently great magmatic pressure forces the magma to the surface, resulting in basalt lava spills at several points along the dike. It should be noted that the majority of dikes are not formed in this geographic setting, and that most do not reach the surface. See also: *flood basalt; volcanic eruption, sites of.*

fissure polygon See *nonsorted polygon.*

fissure vein A mineral deposit of veinlike shape, characterized by clearly defined walls, as opposed to extensive country-rock replacement.

fixed carbon A solid, finely divided black elemental carbon whose presence distinguishes *coal* from *peat.* The higher the proportion of fixed carbon, and the lower the content of volatiles, the higher the *rank* of coal.

fjord   A long, narrow, steep-sided inlet of mountainous, glaciated coasts. It results where the sea invades a deeply excavated glacial trough after the glacier has melted. The side walls are characterized by *hanging valleys* and high waterfalls. The threshold at the seaward end may be bedrock or the site of the terminal *moraine*. If bedrock, it marks the place where the glacier's erosive strength decreased. Fjords are found in Norway, Greenland, Alaska, British Columbia, Patagonia, Antarctica, and New Zealand.

flagstone   A rock, such as micaceous sandstone or shale, that can be split along bedding planes into slabs suitable for flagging.

flame structure   A sedimentary structure in which flame-shaped pods of mud have been compressed and forced upward into an overlying layer.

flame test   A method used to detect the presence of certain elements in a substance. A sample of the given material is heated over a Bunsen burner and the color of the flame observed. Particular elements impart a characteristic color to the flame: e.g., sodium, yellow; barium, green; strontium, crimson.

flank   *Limb*.

flank eruption   See *volcanic eruption, sites of*.

flaser gabbro   A coarse-grained cataclastic gabbro in which flakes of chlorite or mica swirl around lenses (augen) of quartz and feldspar; recrystallization and formation of new minerals is apparent. Cf: *augen structure*.

flaser structure   A fabric structure of dynamically metamorphosed rocks in which large, ovoidal grains that have survived deformation are surrounded by highly sheared and crushed material (mylonite). The general appearance is that of a flow texture. Cf: *augen structure*.

F layer   See *seismic regions*.

flexible   A mineral that can be bent without breaking, but which will not return to its initial form, is said to be flexible.

flexible sandstone   *Itacolumite*.

flexure   *Hinge*.

flexure-flow fold   A fold in which the mechanism of folding is a compressive force acting parallel to the bedding. Usually the rocks are sufficiently plastic so that instead of rupturing, there is some flow within the layers. This results in a shortening and thickening of hinge areas, and lengthening and thinning of limbs. Cf: *flexure-slip fold*.

flexure-slip fold   A fold in which the beds making up bedding planes slide or slip past one another. Of two adjacent beds, the upper one moves away from the syncline axis relative to the lower bed. The thicknesses of individual strata do not change, and the resultant folds are parallel. This concept is important in the interpretation of certain types of drag folds. Cf: *flexure-flow fold*.

flint   A dark-gray or black quartz mineral that is similar in composition to *chert*; each term is frequently used as a synonym for the other.

flint clay   See *fireclay*.

float   A term for loose rock fragments that are often found in the soil on a slope, even in the absence of clean outcrops. Insofar as the fragments cannot creep uphill, their source must be upslope from the location where they are found. Thus, floats are useful in fixing the upper contact of their source rock. *Float ore* is a type of float composed of fragments of vein material, usually found downbed from the outcrop containing the vein.

flocculation   The process of forming aggregates or compound masses of particles; e.g., the settling of clay particles in water. In mining technology, ore particles are coagulated by reagents that promote the formation of aggregates, after which excess water is removed.

floe A sheet of floating ice, smaller than an ice field, found esp. on the surface of the sea.

floe till *Berg till.*

flood basalt Extremely fluid basaltic lava that erupts as a series of horizontal flows in rapid succession (geologically), covering vast areas; generally believed to be the product of fissure eruption. Flood basalts are extruded along mid-oceanic ridges during crustal extension. In Iceland, which straddles the Mid-Atlantic Ridge, the eruptions from the Laki fissure in 1783, covering an area of almost 600 square km, were a surface manifestation of flood basalts that flowed from that ridge. *Lava plains* and *lava plateaus* indicate that flood basalts have also occurred in inland areas. A *lava plain* is a broad stretch of level or nearly level land underlain by a relatively thin succession of such flows. An extensive elevated tableland underlain by a thick succession of basaltic flows is termed a *lava plateau*. The Deccan Plateau in India, covering 250,000 square km, was created 100 m.y.a. by basaltic flood eruptions. Similar lava floods created the Columbia River Plateau of the northwestern U.S. Because they have occurred in areas distant from plate margins, the inland flood basalts do not seem to relate to the present plate tectonic picture. There is some thought that they represent areas where mid-ocean ridges once formed below the continents in the late stages of continental rifting. Syn: *plateau basalt.* Cf: *shield basalt.*

flood basin A term applied to a tract of land that is inundated during the highest of floods. It also applies to a flat area between a sloping plain and a natural levee where swampy vegetation often grows.

floodplain That portion of a river valley, adjacent to the river, that is built up of alluvium deposited during the present disposition of the stream flow. It is covered with water when the river overflows during flood periods. Meandering streams are typical features of floodplains. During the process of lateral erosion, the form of a meandering stream is altered by reduction, trimming, and cutting through, until all that remains is a crescentic mark, a *meander scar*, indicating the former position of a river meander on a floodplain. The beginnings of a floodplain are represented by lunate or sinuous strips of coarse alluvium along the inner bank of a stream meander. These are called *floodplain scrolls.*

floodplain meander scar See *floodplain.*

floodplain scroll See *floodplain.*

floor 1. A rock surface on which sedimentary strata have been formed. 2. *Valley floor.* 3. The *footwall* of a horizontal orebody. 4. The bed of any body of water; e.g., *ocean floor.*

flow 1. The movement or rate of movement of water, or the moving water itself. 2. Any rock deformation that is not immediately removable without permanent loss of cohesion. 3. The mass movement of unconsolidated material in the semifluid or still plastic state; e.g., a mudflow. 4. *Lava flow.* 5. *Glacier flow.*

flowage fold A fold that occurs in relatively plastic rock in which the thickness of an individual bed is not constant. The rocks have flowed toward the syncline trough and the deformation shows no apparent surfaces of a slip. Syn: *flow fold.*

flow breccia A lava flow containing fragments of solidified lava that have become joined or cemented together by the still fluid parts of the same flow.

flow cast A ridge or other raised feature formed on the underside of a sand bed by sand that flowed into a groove or channel made in soft plastic sediment. Although the underlying rock preserves none of the structure, a flow cast may be used to determine the top and bottom of a bed. See also: *flute cast*.

flow cleavage Used as a syn. of *slaty cleavage*, because of the assumption that recrystallization of the slaty minerals during metamorphism is accompanied by rock flowage.

flow fold *Flowage fold*.

flowing artesian well An artesian well in which the hydrostatic pressure is sufficient to raise the water above the land surface. Cf: *nonflowing artesian well; flowing well*.

flowing well A well from which water or oil issues without the aid of pumping.

flow layering A feature of an igneous rock, expressed by layers of contrasting color, texture, or sometimes mineralogical composition, formed by magma or lava flow. Flow layers vary in thickness from less than 1 mm up to many centimeters. Suspended crystals existing at the time of solidification may be segregated into crystal-rich and crystal-poor zones. These, as well as mineral streaks or inclusions, indicate the direction of flow before consolidation, and are called *flow lines*. Flow layers may be planar, folded into smooth open forms, or intricately contorted. Planar layers originate with laminar flow within the moving magma; folded or contorted layers evidence a transition from totally laminar to nonlaminar or turbulent flow as the moving magma responds to local changes, such as obstruction. Syn: *flow banding*. See also: *banding*.

flow plane In igneous and metamorphic rocks, the plane along which displacement occurs. In igneous rocks it is a *flow layer;* in metamorphic rocks it is usually subparallel to the *foliation* visible in hand specimens.

flowstone General term for any mineral deposit formed by water flowing on the floor or walls of a cave. See also: *dripstone; travertine; cave onyx*.

flow texture See *flow layering*.

flow unit A unit stack of beds or sheets of lava or pyroclasts formed by a single eruption from the same volcano.

fluid inclusion A minute ($<10^{-2}$ mm diameter) bubble of liquid commonly found as an inclusion in metamorphic minerals. Such inclusions may also contain a bubble of gas and one or more crystalline phases that were probably formed as the liquid cooled from high temperatures.

flume 1. A deep, narrow gorge or defile containing a stream, torrent, or series of cascades. 2. A manmade channel or trough for conducting water, such as one used for transporting logs.

fluorapatite 1. A mineral of the apatite group, $Ca_5(PO_4)_3F$. It is a common accessory mineral in igneous rocks, high-temperature hydrothermal veins, and metamorphic rocks. It also occurs as *collophane* in marine deposits. 2. An apatite mineral in which the amount of fluorine exceeds that of chlorine hydroxyl, or carbonate. See also: *apatite*.

fluorescence The property of emitting light as a result of absorbing light. Some minerals exhibit this property when subjected to light rays of a particular wavelength. Some respond to ultraviolet light, some to X-rays, and others to cathode rays. Many minerals that are drab to the naked eye become brilliant with color when placed under the appropriate wavelength of light. The property of fluorescence is utilized in mineral ore prospecting, and in detecting the presence of oil at a surface.

fluoride A compound of fluorine, such as methyl fluoride, CH_3F.

fluorine dating A method of determining the related ages of Pleistocene or Holocene fossil bones, from the same excavation, based on the gradual combination of fluorine in groundwater with the calcium phosphate of buried bone material. See also: *dating methods.*

fluorite A clear to transparent halide mineral, CaF_2, of the isometric system. Fluorite is colorless and completely transparent when pure, but is also found in blue, purple, yellow, and green. It occurs in medium- and high-temperature hydrothermal veins in association with lead and zinc, and with silver sulfides. Its most important use is in the manufacture of hydrofluoric acid.

fluorspar Commercial name for *fluorite.*

flute An assymetical, spoon- or scoop-shaped groove or depression in a sedimentary bed. It is usually seen as a *flute cast.* Such impressions are formed by non-laminar, sediment-bearing currents that can scour a soft stream bottom.

flute cast A structure found on the underside (the sole) of some sandstone or siltstone beds, esp. in environments where turbidity currents have been common. It is formed by the filling of a *flute.*

fluting 1. A differential erosion in which the surface of an exposed coarse-grained rock, such as granite, becomes ridged or corrugated with *flutes.* 2. The process of generating a *flute* by the abrasive effect of water on a soft or muddy surface. 3. The formation by a glacier of grooves or furrows on the face of a rock mass that is blocking the advance of the glacier.

fluvial Of or pertaining to rivers; living or growing in a river or stream; produced by the action of a river or stream; e.g., fluvial deposits. See also: *fluviatile.*

fluvial cycle of erosion The sculpture and reduction of a land mass to base level by running water, specif. the action of rivers and streams.

fluviatile Pertaining to, belonging to, or peculiar to rivers, esp. the physical products of river action; e.g., a *fluviatile dam,* which is a dam formed in a stream channel by tributary-deposited sediment.

fluviation The sum of activities and processes in which rivers or streams are the agents.

fluvioglacial *Glaciofluvial.*

flux (*Physics*) 1. Rate of flow of fluid, particles, or energy. 2. Quantity expressing the strength of a force field in a particular area. (*Chem.; metall.*) 1. A substance used to refine metals by combining with impurities to form an easily removable molten mixture. 2. A substance used to remove oxides from fused metal and thus to prevent further oxidation (ex: soldering). 3. In refining scrap or other metal, a salt that combines with nonmetallic impurities, causing them to coagulate or float.

flux density The thermal, magnetic, or electric flow per unit of cross-sectional area. Radiation flux density (electromagnetic radiation) involves a volumetric density rather than a flow across a surface.

fluxing ore An ore smelted chiefly because it contains fluxing agents, rather than for its metal content. Fluxing agents do not have to be added during the reduction of such ores.

fluxstone In metallurgical processes, a limestone, dolomite, or other rock used to lower the fusion temperature of the ore, or to combine with any impurities present, to form a fluid *slag.*

flying magnetometer *Airborne magnetometer.*

flysch A sedimentary deposit typically consisting of interbedded shale and sandstone. Most of the youngest strata involved in the folding and nappes of particular mountains, e.g., the Alps, are flysch deposits. Flysch is considered an indicator of the approach of tectonic uplifts. Cf: *molasse.*

focal sphere An arbitrary sphere drawn about the focus or hypocenter of an earthquake, to which body waves recorded at the earth's surface are referred.

focus See *seismic focus.*

fold A bending, warping, or buckling in bedding, cleavage, foliation, or other planar features in rock, varying in size from a few millimeters to several kilometers across. Folds are best displayed by a bedded or layered formation, e.g., sedimentary or volcanic rocks, or their metamorphosed equivalents. However, any layered or foliated rock, e.g., gabbro or granite, may show folds. Although their classification has resulted in a great variety of terms, folds are generally identified by the attitude of their axial plane, their overall appearance, and their mode of formation.

An *anticline* (or *antiform*) is a fold that is bowed upward and has older rocks at the center. In its simplest form, the two limbs of an anticline dip away from each other.

But the term also includes folds, as shown, where the two limbs dip in the same direction at different angles. A *syncline* (or *synform*) is defined as a fold that is bowed downward. In its simplest form, the two limbs dip toward each other, but many other varieties are also known. Because the younger rocks are normally in the center of the fold, the term *syncline* includes any fold where younger rocks are so located. Some other varieties of folds are shown in the figures. A *symmetrical* or *upright fold* is one in which the axial plane is essentially vertical; in an asymmetrical fold the axial plane is inclined. An *overturned fold*, or *overfold*, has its axial plane inclined, and both limbs dip in the same direction, usually at different angles. The *inverted limb* is the one rotated through more than 90° to its present position. The *normal limb* is right-side up.

In a *recumbent fold* the axial plane is essentially horizontal. The strata in the inverted limb are usually much thinner than the corresponding strata in the normal limb, and the strata in either limb are thinner than at the bend. An *isoclinal fold* is one in which the two limbs dip in the same direction at equal angles. The figure shows three attitudes of the axial planes in such folds. A *chevron fold* is characterized by sharp and angular hinges; a *box fold* has a broad, flat crest with hinges on either side. A *fan fold*, which may be anticlinal or synclinal, is one in which both

PARTS OF A FOLD

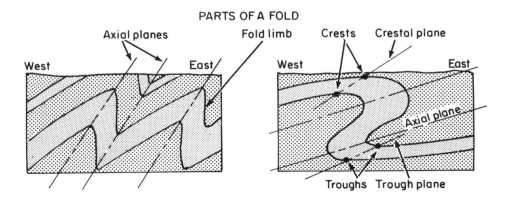

Some Varieties of Anticlines

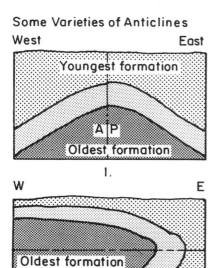

1.

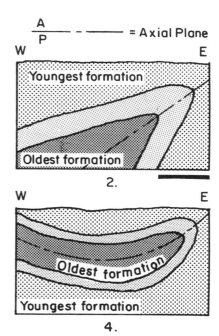

$\dfrac{A}{P}$ — — — — = Axial Plane

2.

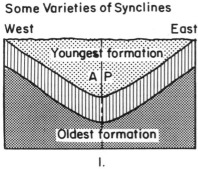

3.

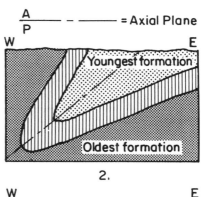

4.

Some Varieties of Synclines

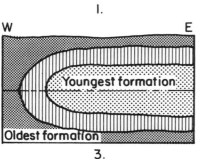

1.

$\dfrac{A}{P}$ — — — — = Axial Plane

2.

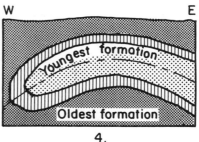

3.

4.

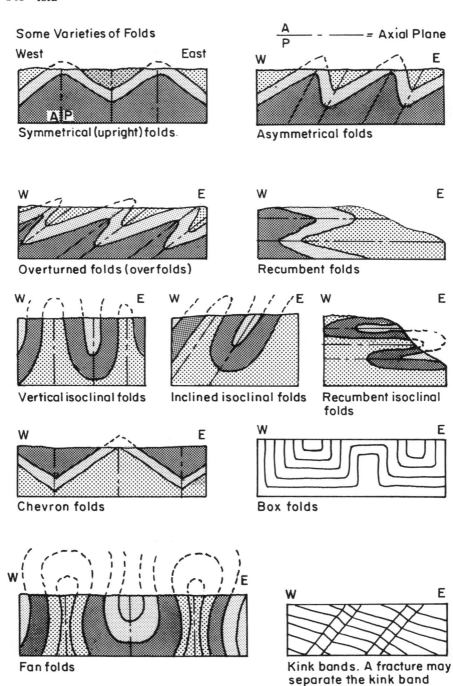

Some Varieties of Folds

$\frac{A}{P}$ - - ———— = Axial Plane

West   East

Symmetrical (upright) folds.

Asymmetrical folds

Overturned folds (overfolds)

Recumbent folds

Vertical isoclinal folds

Inclined isoclinal folds

Recumbent isoclinal folds

Chevron folds

Box folds

Fan folds

Kink bands. A fracture may separate the kink band from the rest of the beds

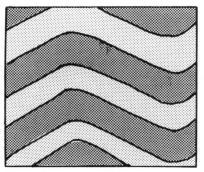

Open folds

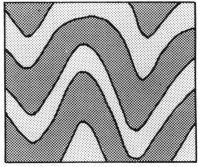

Closed folds

limbs are overturned. *Kink bands* are narrow bands in which the beds assume a dip that is gentler or steeper than the adjacent beds. The deformation producing a *closed fold* has been sufficiently intense to cause flowage of the more mobile beds. An *open fold* is one in which such flowage has not occurred. *Drag folds* are formed when a competent ("strong") bed shears past an incompetent ("weak") bed. Axial planes are inclined at an angle to the bedding of the competent bed. Folding in which the thickness of the bedded strata is greater near the hinges than on the limbs is called *similar folding*. In *parallel (concentric) folding*, the beds are of uniform thickness.

fold belt   Orogenic belt.

fold mountains   Mountains that have been formed by large-scale and profound folding. Studies of typical fold mountains, such as the Alps, indicate fold involvement of the deeper crust and upper mantle, as well as of the shallow upper crust.

fold system   A set or group of congruent folds that have been produced by the same tectonic episode.

foliation   1. A planar arrangement of the textural or structural features in any variety of rock. Certain foliation develops at the time of rock formation. Igneous

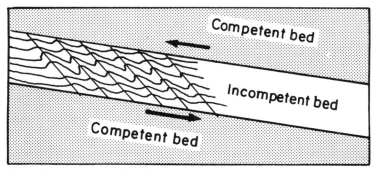

Drag folds resulting from shearing of beds past each other

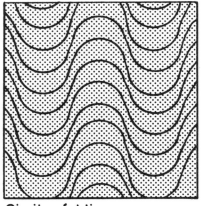

Similar folding.

Parallel folding.

rocks become foliated while in the molten state, as crystals orient themselves in parallel planes directed by the laminar flow of the remaining liquid. Minerals whose crystals form plates, e.g., mica, are most likely to form foliate patterns. Many metamorphic rocks exhibit foliation; this is the form of closely spaced layering due to the parallel orientation of some of the minerals. Foliated rocks that have not been thermally metamorphosed after the foliation process split easily along the foliations. Metamorphic foliation is often mistaken for stratification, esp. in slates. 2. A layered structure produced in the ice of a glacier by plastic deformation. Syn: *banding.*

fool's gold   A popular name for any of several minerals that resemble gold and have been mistaken for it; specif. pyrite and chalcopyrite.

footwall   The mass of rock below a fault plane, ore body, or mine working; in particular, the wall rock beneath a fault or inclined vein. Cf: *hanging wall.*

foram   Abbreviated name for *foraminifer.*

Foraminifera   See *Protozoa.*

forcible intrusion   A mode of magma

emplacement in which space for intrusion is forcibly created. It is obvious in the case of many batholith intrusions that wall rocks have been shoved aside. This is shown by the deflections of the wall rock schistosity away from its regional trend toward parallelism with the walls of the pluton.

foredeep   1. An elongate crustal depression that borders an orogenic belt or island arc on the convex side. 2. *Exogeosyncline.*

foreland   An area marginal to an orogenic belt, toward which folds have been overfolded and thrusts have moved. The side from which the surface rocks have moved in the *hinterland.*

forelimb   The steeper of the two limbs of an asymmetrical anticline. Cf: *backlimb.*

fore reef   The seaward side of a reef. Parts of it may be slopes covered with reef debris; other areas may be vertical walls built by deposits of marine organisms.

foreset bed   See *delta.*

foreshock   A small tremor that precedes a larger earthquake by an interval ranging from seconds to weeks or even longer. Foreshocks of an earthquake all have approximately the same epicenter. As a rule, major earthquakes occur without detectable warning in the nature of minor foreshocks. However, there is evidence that some major earthquakes are preceded by foreshocks; e.g., the North Idu (Japan) earthquake of 1930. Studies have shown that the tendency for foreshocks to occur is limited to particular seismic zones.

foreshore That zone of a shore or beach that is regularly covered with tidal water. Syn: *beach face*. Cf: *backshore*.

foresight A sight on a new survey point, reckoned in a forward direction, for the purpose of determining the bearing and elevation of the point. Cf: *backsight*.

formal unit A stratigraphic unit that is defined and named in accordance with the rules and guidelines of an established system. Cf: *informal unit*.

formation A fundamental geologic unit used in the local classification of strata or rocks. Formations are not classified by geological time but rather by distinctive physical and chemical features of the rock. The names of formations are often taken from geographic names of places where they were originally described. These are combined with the names of the predominating rock comprising the bulk of the formation; e.g., *St. Louis Limestone*.

forsterite A whitish to yellowish mineral, Mg_2SiO_4, the magnesium end-member of the olivine mineral group.

fosse 1. A long narrow depression between a glacier and the sides of its valley. It is due to the more rapid rate of melting that occurs here because of the added effect of heat absorbed or reflected from the valley walls. 2. A ditch, canal, defilement, or other long narrow waterway.

fossil 1. adj. Term used to indicate ex-tinction or age. A "fossil fern" connotes an ancient, extinct variety of fern; fossil rain prints or hail prints are ancient markings. Certain geological forms that have become covered or changed are referred to as "fossil features." For example, a "fossil sea" is an ancient sea bed. 2. n. Derived from Latin, *fossilis*, meaning "dug up." For many years any object that was dug out of the ground was considered to be a fossil. By the late eighteenth century, the term had been reserved for prehistoric remains that were buried by natural processes and, subsequently, permanently preserved. Fossils may have the following forms: (A) *Original soft parts unaltered*. This rarely occurs, and requires very special conditions. Examples are the preservation of woolly mammoths in *permafrost*, and the encasement of insects in fossil resin, or *amber*. (B) *Original hard parts unaltered*. Most plants and animals have some hard parts that can be fossilized. Examples are the calcitic shells of echinoderms, aragonitic shells (most mollusks), phosphatic remains of brachiopods, and chitinous skins of certain animals. (C) *Original hard parts altered*. The original hard structures of many organisms may undergo alterations with time. The nature of such alterations depends on the original material of the organism's environment, and the conditions under which its remains were deposited. (i) Carbonization or distillation. The reduction of organic tissue to a carbon residue. See also: *carbonization*. (ii) Permineralization or petrifaction. The infiltration of porous bones and shells, after burial in sediment, by mineral-bearing solutions. This type of fossil is called a *petrifaction*. See also: *permineralization*. (iii) Replacement or mineralization. This occurs when the original hard parts are removed by solution and some other mineral substance is deposited in the voids created. See also: *replacement*. 3. Evidence of organisms. Although none of the original material remains, a trace provides evidence of the structure. Common evidence includes molds and casts (of bone, teeth, shell, etc.), tracks and trails, borings and burrows, ichnofossils, coprolites (fossil fecal pellets), and *gastroliths* (stomach stones). See *casts*. See also: *pseudofossils; replacement.*

fossil assemblage   *Assemblage.*

fossil community   An assemblage whose individuals inhabited the same location in which their fossils are found, and the fossil forms are present in about the same numbers as when the organisms were alive, thus indicating no postmortem transport.

fossil fuel   A general term for any hydocarbon that can be used for fuel, esp. petroleum, natural gas, and coal. The name is derived from their subsurface origin as well as their presence in fossiliferous areas.

fossil ice   1. Ice remaining from the geologic past in which it was formed. For example, fossil ice is present along the coastal plains of northern Siberia, where remaining Pleistocene ice and the preserved remains of mammoths within this ice have been found. 2. Comparatively old ground ice in a permafrost region, or underground ice in a region where the present temperatures are not low enough to have been responsible for it.

fossiliferous   Containing or bearing fossils; as rocks or strata. For example, *fossiliferous limestone.*

fossilization   See *fossil.*

fossil plants   The oldest known fossils on earth are those of plants, the earliest of

---

Non-vascular

Thallophytes: Algae; Precambrian to Recent; some value as fossils. Bacteria and fungi; Middle Devonian to Recent; unimportant as fossils.
Bryophytes: Mosses and liverworts; Mississippian to Recent; rare as fossils.

Vascular

Lycopoda: Club mosses (found fossil), *scale trees*; Silurian to Recent.
Sphenopsids: Horsetails, scouring rushes; Devonian to Recent; known mostly as fossils.
Spore Ferns (filicineans): Devonian to Recent; many found in fossil form.
Seed Ferns (pterospermae): Mississippian to Jurassic; many found in fossil form.
Cycads: Late Permian or Jurassic to Recent; important as fossils.
Early Conifers (*cordaites*): Mississippian to Permian; known only as fossils.
Conifers: Pines, junipers, firs; Pennsylvanian to Recent; many fossil species.
Angiosperms: Flowering plants and hardwoods; Triassic to Recent; many fossil species.

which date back more than 2000 m.y., to Precambrian time. Marine rocks generally contain the remains of marine plants, but terrestrial plants carried into the sea may also be preserved. Pollen and spores are also found in ancient rocks. See *palynology*. The synopsis below provides general information.

fossil soil *Paleosol.*

fossil wax *Ozocerite.*

fractional crystallizaton 1. Separation of the component minerals of a magma at successively lower temperatures. Crystals that form at high temperatures do not come into equilibrium with the parent magma; instead, there results a series of residual liquids of increasing concentration. Syn: *fractionation*. See also: *magmatic differentiation*. 2. The separation of substances of different solubilities from a solution by precipitation, as affected by varying temperatures.

fractionation *Fractional crystallization.*

fracture 1. The way in which a mineral breaks, other than along its planes of cleavage; e.g., conchoidal fracture. 2. One of the ways in which rocks yield to deforming movements in the earth's crust; e.g., cracks, joints, faults, or other breaks. 3. Any rupture in pack ice or fast ice from a few meters to several kilometers in length. 4. Deformation resulting from a momentary loss of resistance to differential stress and a release of internal elastic energy. Syn: *rupture.*

fracture zone A zone in which faulting has occurred. It runs parallel to the fault line, and may be oceanic or continental. A fracture zone can be the site of intense seismic and volcanic activity. See *fault.*

fragmental rock 1. *Clastic rock.* 2. *Pyroclastic rock.* 3. *Bioclastic rock.*

fragmental texture A term for a sedimentary rock texture characterized by broken particles, which distinguish it from a rock of crystalline texture. A *tuff* or other pyroclastic rock is said to have fragmental texture.

franklinite A mineral of the spinel group (Zn, Mn^{+2}, Fe^{+2}) $(Fe^{+3}, Mn^{+3})_2 O_4$, having black crystals with a metallic appearance. It is an ore of zinc and is found in massive aggregates in contact metasomatic conditions.

Frasch process A method of mining elemental sulfur in which superheated water is injected into the sulphur deposits. The sulfur is melted by the water and is then pumped to the surface.

frazil ice A semisoft aggregate of ice in the form of small crystal spikes and plates. It forms along the edges of turbulent or rapidly flowing streams, but is also found in turbulent seawater.

free-air anomaly See *gravity anomaly.*

free-air correction See *gravity anomaly.*

free water 1. Water in rock or soil that can move freely in response to gravity. Syn: *gravitational water*. 2. Water that is not chemically bound to a substance and which can be removed without altering the structure or composition of the substance.

freezing interval *Crystallization interval.*

fresh water Water containing less than 0.2% dissolved salts; not necessarily potable.

friable A term referring to a rock or mineral that can be disintegrated into individual grains by finger pressure.

fringing reef An *organic reef* that grows directly against coastal bedrock and actually constitutes the shoreline; commonly found along tropical coasts. A fringed reef has a rough, shelflike surface that is exposed at low tide, and its seaward side drops sharply to the sea floor. Syn: *shore reef.* Cf: *barrier reef.*

frost action   Process of repeated freezing and thawing of water, effective in the mechanical weathering and breakup of rock. Its effectiveness depends on the presence of confined spaces and the frequency of the freeze-thaw cycle. In cold regions, it is largely responsible for *talus* and rock glaciers.

frost crack   A more or less vertical crack in rock or frozen ground, having an appreciable ice content. Frost cracks are produced by thermal contraction, and commonly intersect to form a polygonal net. Syn: *contraction crack; ice crack.*

frost-crack polygon   See *nonsorted polygon.*

frost-heaved mound   *Stone ring.*

frost heaving   Upward expansion of ground caused by the freezing of water in the upper *regolith*. If the surface is horizontal, the regolith after thawing will return to its original position. If the surface slopes, the regolith upon freezing will rise perpendicular to the slope, but upon thawing will have a downslope component producing *creep*. Frost heaving is responsible for the movement of rocks to the surface, for spongy, soft ground in the spring, and for creating highway bumps.

frost line   1. The maximum depth of frozen ground in regions without permafrost. 2. The greatest depth of permafrost. 3. That altitude below which freezing does not occur; e.g., in mountainous tropical regions.

frost mound   A general term for rounded or conical forms (hummocks, knolls) in a permafrost region. It contains a core of ice and indicates a localized and seasonal upwarp of the landscape due to

frost-heaving and/or hydrostatic pressure of groundwater. Syn: *soil blister.* See also: *pingo.*

frost splitting   *Congelifraction.*

frost stirring   A syn. of *congeliturbation* in which there is no mass movement.

frost weathering   *Congelifraction.*

frost wedging   A type of congelifraction in which ice, acting as a wedge, pries apart jointed rock.

frozen ground   Ground whose temperature is below the freezing point of water and which usually holds some amount of water in the form of ice. Syn: *gelisol.*

fulgurite   Etymol: Latin, "thunderbolt." Glassy tubes of lightning-fused rock, most common on mountain tops. Although fulgurites may form from any kind of rock, the largest have been formed from unconsolidated sand.

fuller's earth   A clay having a high adsorptive capacity, composed mostly of montmorillonite. It is widely used as a bleaching agent, as a filter, and for removing grease from fabrics.

fumarole   A vent at the earth's surface that emits hot gases, usually in volcanic regions. Fumaroles are found at the surface of lava flows, in the calderas and craters of active volcanoes, and in areas where hot intrusive, igneous rock bodies occur, e.g., Yellowstone Park. The gas temperature within a fumarole may be as high as 1000° C. Steam and CO_2 are the main gases; nitrogen, carbon monoxide, argon, hydrogen, and other gases are also present, and vary greatly in their relative proportions even in the same fumarole. Deep magmas and the heating of groundwater are the sources of fumarole gases. A *solfatara* (Ital. for "sulfur mine") is a fumarole in which sulfur gases are the dominant constituents after water. Such fumaroles with temperatures much below the boiling point of water and rich in carbon dioxide are called *mofettes*.

fungi Thallophytes of the division Fungi, a group of plants that lack chlorophyll and depend on living or dead organic material for food. Fossil fungi are rare; their range in geologic time is not definitely known, but probable remains have been reported from rocks as old as the Middle Devonian.

fusain See *lithotype*.

fusibility The property of being convertible from a solid to a liquid state by means of heat; also, the degree to which a substance is fusible.

fusiform Spindle-shaped; elongated and tapering at both ends.

fusion 1. The process of liquefying a solid by the application of heat. 2. The joining or coalescence of two or more substances, objects, etc., as by partial melting. 3. The combination of two light atomic nuclei to form a heavier nucleus. The reaction is accompanied by the release of a massive quantity of energy. Fusionable elements are among those of low atomic weight; e.g., hydrogen. Cf: *fission*.

fusulinid A foraminifer of the family Fusulinidae, similar in shape to a grain of wheat. They are important guide fossils in the Pennsylvanian system. Range, Ordovician to Triassic.

future ore *Possible ore.*

G

gabbro A group of crystalline intrusive rocks composed chiefly of plagioclase and pyroxene, commonly with small amounts of other ferromagnesian minerals, esp. olivine. Magnetite, ilmenite, and apatite are the most frequent accessory minerals. If ferromagnesian minerals predominate over the plagioclase, so that the rock is dark-colored, it may generally be called *gabbro*, although the microscopic distinction from *diorite* is based on the composition of the plagioclase. With increasing alkali-feldspar content, gabbro grades into *monzonite*. It is the approximate intrusive equivalent of *basalt*. See also: *diabase*.

gabbroic layer *Basaltic layer.*

Gale alidade A lightweight, compact *alidade* with a low pillar and a reflecting prism, through which the ocular may be viewed from above. In petroleum geology, it is often used in conjunction with the *Stebinger drum*. Syn: *explorer's alidade*.

galena A gray, metallic mineral, PbS, the main ore of lead; silver is a frequent byproduct. Galena occurs in the form of compact granular masses, and is a typical hydrothermal mineral in medium-temperature deposits.

gallery A term used for a horizontal passageway in a cave; it is also used for a horizontal conduit or channel constructed to intercept water.

gamma decay　A type of radioactivity in which certain unstable atomic nuclei emit excess energy by a spontaneous electromagnetic process. Unstable nuclei that undergo gamma decay are themselves either the products of alpha and beta decay or some other nuclear process, such as neutron capture in a nuclear reactor. See also: *radiometric dating; radioactive decay*.

gamma-ray log　See *well logging*.

gamma rays　Quanta of energy (photons) emitted by nuclei of radioactive substances. They are electromagnetic waves of the same type as X-rays, but have shorter wavelengths and higher frequencies. Gamma rays have a longer range than alpha or beta rays. They may be attenuated by the use of heavy metal shields; e.g., lead.

gangue　The valueless rock or mineral aggregates associated with an ore. Cf: *ore mineral*.

garnet　A mineral group having the basic formula: $A_3B_2(SiO_4)_3$, where A = Ca, Mg, Fe^{+2}, or Mn^{+2}, and B may be Al, Fe^{+3}, Mn^{+3}, V^{+3}, or Cr^{+2}. Although all members of the garnet group conform to the same basic formula, and have the same crystal structure (euhedral, isometric), the group exhibits a great range of chemical variation. The following names are given to the end-members of the group: *almandine* (Fe,Al), *andradite* (Ca,Fe), *grossular* (Ca,Al), *pyrope* (Mg,Al), *spessartine* (Mn,Al), *uvarovite* (Ca,Cr), *goldmanite* (Ca,V). Garnet has a vitreous luster and is found in all colors but blue. The best known form is the dark red one as in pyrope, which is widely used as a gemstone. Red-to-violet almandine is called *carbuncle*. Some garnet shows angular fracture that, together with a high hardness, makes it a good abrasive. Garnet is common in metamorphic rocks (gneisses and schists), and is often an accessory mineral in igneous rocks.

garnierite　A general name for hydrous nickel silicates. Bright-green aggregates are found as crusts or earthy masses.

gas　*Natural gas*.

gas cap　Free gas acculmulated above an oil reservoir. It forms whenever the available gas exceeds the amount that will dissolve in the oil under the existing conditions of temperature and pressure.

gas coal　Bituminous coal containing a high percentage of volatile, gas-yielding constituents, thus making it suitable for the manufacture of flammable gas.

gaseous transfer　The separation of a gaseous phase from a magma, which releases dissolved substances, generally in the upper portions of the magma, where the pressure confining it to the magma is less.

gas field　A term applying either to an individual *gas pool* or to some multiple of gas pools on a single geologic feature, or to closely associated features, such as particular types of strata.

gasification　The process of production of fuel gas from coal.

gas-oil ratio　The volume quantity of gas produced with the oil from a particular oil well, expressed in terms of cubic feet of gas per barrel of oil. Abbrev: GOR.

gas pool　An economically viable subsurface accumulation of natural gas. Cf: *gas field*.

gas sand　A sand or sandstone that contains a large or significant quantity of natural gas.

gas streaming　The formation of a gas phase at a late stage of crystallization during the process of magmatic differentiation. Escaping gas bubbles cause the partial expulsion of residual liquid from the crystal mass.

gastrolith　Also called "stomach

stones," gastroliths are highly polished, rounded stones believed to have been a part of the digestive system of certain extinct reptiles, such as plesiosaurs. The gastroliths aided in grinding the animals' stomach contents. See *fossil*.

gastropod Any mollusk of the class Gastropoda; includes terrestrial, marine, and freshwater forms of snails and slugs. Most gastropods have a single calcareous shell, closed at the apex. It is usually coiled (either dextral or sinistral) and not chambered. The paleontological classification of gastropods follows shell shape and symmetry, whereas the zoological classification is based on the character of the soft parts. Gastropods range from Cambrian to present, and are most abundant in the Tertiary, where they are important as zonal indices.

gas well A well capable of producing natural gas, or one that produces chiefly natural gas. In some areas, the term is defined on the basis of *gas-oil ratio*.

gauss The c.g.s. unit of magnetic induction. A *gaussmeter* is a magnetometer for measuring the intensity of a magnetic field.

Gauss normal epoch See *constant polarity epoch*.

geanticline An uplift of regional extent, comparable in size to a geosyncline; it may be either inside or outside a geosyncline.

Geiger-Mueller counter An instrument for detecting radiation. It consists of a tube (the Geiger tube) filled with gas at sufficiently high pressure to prevent the spontaneous flow of electrons. When a high-energy particle enters the tube and ionizes the gas, a current flows. Each current pulse, which is recorded by attached electronic equipment, represents the passage of a high-energy particle through the Geiger tube.

gel A jellylike colloidal dispersion of a solid within a liquid.

gelation 1. The process of gelling. 2. Solidification by cold; freezing.

gelifluction See *solifluction*.

gelisol Frozen ground.

gem A general term for a precious or semiprecious stone, particularly after it has been cut and polished. It also applies to a particular variety of a mineral that is worn or used decoratively.

gemology The study of gemstones, including their description, origin, source, means of identification, and methods of evaluation. Br. spelling: *gemmology*.

genetic drift A gradual change with time in the genetic composition of a continuing population. It is a change that results from the disappearance of some genetic features and emergence of others; neither appears to be related to environmental effects.

genetics The study of heredity, specifically those aspects of the resemblances and differences of related organisms that are the product of interaction between genes and environment.

genotype That species on which the original description of a genus is primarily based. The term may also mean the genetic composition of an organism, in contrast to its physical characteristics.

genus A category in plant and animal classification between *family* and *species*.

geobarometry Determination by any method of the pressure at which geological processes occur. For example, the particular site occupied by aluminum in silicate structures is a function of the pressure of their formation. With an increase in pressure, aluminum shows preferred octahedral rather than tetrahedral coordination, thus making it possible to correlate the amount of aluminum in octahedral coordination with the pressure history of a rock.

geobotanical prospecting The visual study of the occurrence of plants as indicators of soil depth and composition, groundwater conditions, bedrock characteristics, and possible mineral or ore presence. Cf: *biogeochemical prospecting*.

geocentric Related to the earth as a center, or measured from earth's center.

geochemical anomaly See *geochemical exploration*.

geochemical cycle A path followed by an individual element or group of elements in the crustal and subcrustal region of earth. During the cycle, there occur processes of both separation of elements and elemental recombination. For the lithosphere, the cycle begins with the crystallization of magma at the surface or at depth. Decomposition of the igneous rock into sediments by surface alteration and weathering is followed by transport and deposition of the sediment, which will lithify and eventually metamorphose, until it melts and generates new magma. This idealized cycle could be interrupted at any point. Each element will be affected differently as the cycle progresses; e.g., during the weathering of an igneous rock, minerals containing iron, magnesium, and calcium will break down and be carried in solution, but quartz and feldspar will be moved and deposited as sediment. Sedimentary rocks formed of these sediments will be dominated by quartz and feldspar, whereas others will be dominated by calcium and magnesium, due to precipitation of the carbonates of both these elements. As partial melting of the sedimentary rock begins, elements will be separated according to melting characteristics. Volatiles escape into the atmosphere, and physical movement of the chemically separated bodies takes place.

geochemical exploration The application of methods and techniques of geochemistry to searching for mineral or petroleum deposits. The presence of a *geochemical anomaly* often indicates a deposit. A geochemical anomaly is a concentration of one or more elements in a rock, sediment, soil, water, or vegetation that differs significantly from the normal concentration. When the anomaly is in terms of abnormal concentrations of elements or hydrocarbons in soils, it is particularly applicable to mineral or petroleum deposits.

geochemical facies An area or region characterized by particular physicochemical conditions that affect the production and accumulation of sediments. The area is commonly distinguished by a characteristic element. In sedimentary environs, this is best illustrated in terms of electron concentration (eH) and proton concentration (pH). Certain related deposits exhibit very different mineralogical features because of different depositional environments. The sedimentary iron formations of the Lake Superior region, formed during the Precambrian, are classified according to the dominant iron mineral into four principal facies: sulfide, carbonate, oxide, and silicate. The respective eH-pH diagrams of each type can indicate the depositional environment, chemical region of stability, and conditions under which particular assemblages could have been formed.

geochemical prospecting *Geochemical exploration*.

geochemistry That branch of geology concerned with the quantities, distribution, and circulation of chemical elements in earth's soil, water, and atmosphere. Geochemistry has as its major concern the various classes of rocks and minerals, and their associated environments, as well as those conditions and processes that contribute to such variety. In summary, geochemistry applies the principles of chemistry to geological problems. As such, it overlaps many fields in geology such as mineralogy, petrology, and hydrology.

geochron An interval of geologic time that corresponds to a *lithostratigraphic unit*.

geochronologic unit *Geologic time unit*.

geochronology Measurement of time intervals of the geological past, accomplished in terms of absolute age dating by a suitable radiometric procedure, e.g., carbon-14 dating, relative age dating by means of sequences of rock strata or varved deposit analysis. Dating past events in terms of tree rings is called *dendrochronology*. Geochronology is concerned with areas such as orogenic events, sea-floor spreading rates, and intervals of glaciation. Methods based on sedimentation rates and the rate of increase in the salinity of sea water have been unreliable. See also: *dating methods*.

geochronometry The measurement of geologic time by the use of geochronologic methods, such as radiometric dating. Cf: *geochronology*.

geocosmology The discipline that deals with the origin and geological evolution of the earth, its planetary features, and the influence of the solar system, our galaxy, and the universe on the geological development of the earth.

geode, geodized A hollow, globular mineral body that can develop in limestone and lavas. Its usual shape is nearly spherical, from 2.5 to 30 cm in diameter. Prominent features are: 1. hollow interior; 2. a clay film between the geode wall and a surrounding limestone matrix; 3. an outer chalcedony layer; 4. a drusy lining (usually quartz) in the interior; 5. evidence of expansion. Geodes form by expansion from an initial fluid-filled (often with a salt solution) cavity. Chalcedony, the initial deposit, is

from a silica gel that isolates the salt solution. If the salinity of the surrounding water decreases, the internal salinity, through osmosis, creates outward-directed internal pressure. As a result of this, the geode will expand until equilibrium is attained. Finally, crystallization and dehydration of the silica gel occur, succeeded by cracking, shrinking, and the penetration of water-bearing dissolved minerals, which are deposited on the chalcedony wall. When the initial cavity is inside a fossil, such as the opening of a bivalve, the geode forms within this cavity, and the fossil is burst by the expanding geode. Geodized brachiopods are the result of such a process. Unfilled cavities within geodes of either type are called *vugs*. See also: *druse; amygdule*.

geodesic line The shortest distance between any two points on a surface.

geodesy The study of the external shape of the earth as a whole, which provides a geometrical framework into which other knowledge of the earth may be fitted and which, by itself, yields information about the internal construction of the earth. Fundamental to geodesy is the interpretation of the shape of the earth. Because no more than one quarter of the earth's surface is land, ordinary geometrical surveys cannot be applied to the planet in its entirety. However, the surfaces of the seas and shape of the earth are both controlled by the force of gravity; from this connection, a model referential figure of the earth, the *geoid*, is derived. See *geoid*. Because of the curvature of the earth and optical scattering by the atmosphere, direct measurements between points on the surface of the earth are limited to rays of some 50 km in length. The method followed in *geodetic surveying* is to make local geometrical measurements with respect to a reference surface that can be determined on a worldwide basis. The purely geometrical measurements need only have local validity; the worldwide reference surface, the geoid, is related to the gravitational field of the earth.

geodetic coordinates Values that specify the horizontal position of a point on a reference ellipsoid with respect to a particular geodetic datum; such values are generally expressed as longitude and latitude.

geodetic surveying See *geodesy*.

geodimeter Instrument designed to measure distance by the time required for light pulses to travel to and from a mirror located at a known distance from the geodimeter.

geofracture *Geosuture.*

geographic cycle *Cycle of erosion.*

geographic province An extensive region, all sections of which exhibit similar geographic features. Cf: *physiographic province.*

geography The study concerned with all aspects of the earth's surface. As such, it includes areal differentiation as shown by climate, elevation, soil, vegetation, population, land use, industries, and political divisions.

geohydrology A subdivision of hydrology, dealing with the waters beneath the earth's surface. Its main concern is with water associated with earth materials and with water-flow mechanics through the rocks. Although the terms *geohydrology, hydrogeology,* and *groundwater hydrology* are not clearly differentiated, all are in use.

geoid A model of the figure of the earth that coincides with the mean sea level over the oceans, and continues in continental regions as an imaginary sea level surface defined by spirit level. A geoid is perpendicular at all points to the direction of gravitational attraction, and approaches the shape of a regular oblate spheroid. However, it is irregular because of buried mass concentrations and elevation differences between sea floors and continents. The geoid is characterized by a constant potential function over its entire surface. This potential function considers the combined effects of earth's gravitational attraction and the centrifugal repulsion that results from earth's axial rotation.

geologic age The age of a fossil organism or geological event or formation, as referred to the geologic time scale and expressed as *absolute age* or in terms of comparison with the immediate environment (relative age).

geological oceanography *Marine geology.*

geologic column A composite diagram showing in columnar form the total succession of stratigraphic units of a given region; arranged so as to indicate the association of the units to the subdivisions of geologic time. See also: columnar section.

geologic hazard A geological phenomenon or condition, natural or manmade, that is dangerous or potentially dangerous to the environs and its inhabitants. Natural hazards include earthquakes, volcanic eruptions, and inundations. Ground subsidence due to overmining is an example of a manmade geologic hazard.

geologic map A map on which is shown the occurrence of structural features, the distribution of rock units, and their type and age relationship. Geologic maps are used in predicting where petroleum, water, iron ore, coal, and other buried, valuable materials may be found. Such maps are basic to deciphering the developmental stages of a mountain range, the steps in the evolution of fossil organisms, and climatic fluctuations.

geologic province An extensive area characterized by similar geological history and development.

geologic range *Stratigraphic range.*

geologic record The testimony of earth's history as evidenced by its bedrock, regolith, and morphology.

geologic thermometer See *geothermometer.*

geologic thermometry Determination or estimation by any technique of the temperatures at which geological processes take place. For current processes, such as volcanic eruptions, direct measurements can be made. For past geological occurrences, or inaccessible ones, indirect methods are used. Various rocks, minerals, or mineral groups can be used, since their presence or properties are known to be related to the temperature under which they are formed. Silicate minerals act as probes that record the evolutionary history of rocks. For example, the amount of substitution between compositional end-members and the mode of cation distribution, are functions of temperature. *Fluid inclusions*, can be used to indicate the formation temperature of the host grain. For magmatic systems (binary, ternary, quaternary), the temperature range of solubilities and precipitation can be determined from *phase diagrams* representing such systems. Certain isotope ratios can be used as thermal indicators; the ratio of the oxygen isotopes O^{16}/O^{18} in present-day shells is a function of the temperature of the ocean water in which they lived. The concentration of trace elements may also be used, since the amount of a particular trace element in a mineral is frequently a function of temperature. See also: *geobarometry.*

geologic time A term encompassing the time extending from the end of the formative period of the earth (i.e., after it was "established" as a planet) to the onset of human history; it is the part of earth's history that is recorded and described in the succession of rocks. Since no precise limits can be imposed, the term can only imply extremely long durations.

geologic time scale An arbitrary chronologic arrangement of geological events, such as mountain-building activity, usually presented in a chartlike form: the latest event and time unit are at the bottom, and the youngest at the top.

geologic time unit An uninterrupted interval of time in geological history, during which a corresponding *chronostratigraphic unit* was formed; i.e., a division of time delineated by the rock record. In order of decreasing magnitude, geologic time units are the eon, era, period, epoch, and age. Syn: *geochronologic unit; time unit.*

geology The study of the earth in terms of its development as a planet since its origin. This includes the history of its life forms, the materials of which it is made, processes that affect these materials, and products that are formed of them. See also: *geoscience; historical geology; physical geology.*

geomagnetic field The magnetic field associated with the earth. Although the field is generally dipolar on the surface of the earth, the dipole is distorted at locations removed from the surface because of interactions with the solar wind. As much as 6% of the geomagnetic field is non-dipolar in shape. This component of the field has its origins in the outer zone of earth's core, possibly at the core-mantle boundary. The theory that the core rotates more slowly than the mantle lends support to the idea that core-mantle interaction may be responsible for both non-dipolar and dipolar fields.

geomagnetic polarity reversal Alternation of earth's magnetic polarity, as indicated by alternating magnetic features of a sequence of rocks that are progressively younger. Such reversals are shown through measurements of the *remanent magnetization* of sedimentary or igneous rocks, and by the basaltic strips that parallel the mid-oceanic ridges. Within a period of perhaps less that 10,000 years the field may become very weak and then again increase to equal its former intensity, but with an opposite direction. The Atlantic Ocean floor consists of rocks advancing in age with increasing distance from the Mid-Atlantic Ridge, which is the site of new crustal generation. Rock of alternating magnetic polarity occurs as roughly symmetrical bands on each side of the ridge, and the absolute ages obtained are good indicators of the occurrence of geomagnetic reversals during the formation of ocean floor. Geomagnetic field reversals are now thought to be worldwide. A sequence of reversals has been established back to the Cretaceous. See also: *polar wandering; natural remanent magnetism; mid-ocean ridges; paleomagnetism; sea-floor spreading.*

geomagnetic pole   The location on the earth's surface where the continuation of a line along the resultant magnetic dipolar axis intersects with earth's surface. The north geomagnetic pole is in northwestern Greenland. The magnetic declination is the angle between a line through any point to the geographic North Pole and a line from that point to the geomagnetic north pole. See also: *geomagnetic fields; geomagnetic polarity reversal.*

geomagnetics   Branch of geophysics concerned with all aspects of the earth's magnetic field; its origin, changes through time, remanent magnetization of rocks, and magnetic anomalies. See also: *geomagnetic polarity reversal; natural remanent magnetism; magnetic anomaly.*

geomechanics   See *structural geology.*

geomorphic   Pertaining or related to the form of earth or its surface features; e.g., a *geomorphic* agent.

geomorphic cycle   See *cycle of erosion.*

geomorphogeny   That part of geomorphology that is involved with the origin and development of earth's surface forms and features.

geomorphology   The scientific discipline concerned with surface features of the earth, including landforms and forms under the oceans, and the chemical, physical, and biological factors that act on them; e.g., weathering, streams, groundwater, glaciers, waves, gravity, and wind. As parts of geomorphology, *geomorphography* is the description of geomorphic features and *geomorphogeny* deals with the origin and development of land features. Geomorphic maps, usually constructed with the aid of aerial photographs, indicate landforms of particular kinds or origins within a specific region. Cf: *geodesy.* See *cycle of erosion.*

geopetal structure   A structure that permits the up direction of a bed to be determined in highly deformed stratigraphic sections. Such structures are derived from rock characterized by *geopetal fabric.* Fossils having curved outlines (pelecypods, brachiopods, ostracods, trilobites) assume a concave-downward position in strata. They therefore form a bridge over the underlying sediment that subsequently can be partially or completely filled with sediment or sparry calcite. The partly spar-filled sub-bridge areas commonly serve as *geopetal structures.*

Geophone   Trademark for an acoustic detector that senses ground vibrations generated by seismic waves. Geophones are variously arranged on ground surfaces to record the vibrations produced by explosives in seismic refraction and reflection work. They are also used by the military as detection devices.

geophysical exploration   The use of appropriate geophysical methods to search for natural resources within the earth's crust, or to obtain information about subsurface structure for various civil engineering works. Geophysical prospecting techniques are also used in archaeological reconnaissance, where the ground is tapped and sounded for substructures such as walls and ditches. The measurement of gravitational and magnetic variations and anomalies is useful in oil and mineral prospecting. See *gravimetry*. In seismic prospecting, seismic waves, generated by an explosive charge, are used to determine underground structure. See also: *seismic shooting*. Variations in the electrical properties of rocks, soils, and minerals are used for detecting ore bodies. This is usually done by measuring the distance between electrodes placed in the ground. This method is useful for locating the depths of surface soil and rocks, water-bearing strata, and subsurface anomalies. Various sensors (infrared, microwave, or radio-frequency) provide relative surface information. Their application is based on the premise that all materials have particular signatures at different wavelengths. Since geophysical methods are indirect and require interpretation, they are most effective when used in conjunction with more direct methods; e.g., boring.

geophysical prospecting *Geophysical exploration.*

geophysics A section of earth science that employs the principles and methods of physics. The scope of geophysics touches on all physical phenomena that have affected the structure, physical conditions, and evolution of the earth. Its branches include the study of variations in earth's gravity field, seismology, the influence of earth's magnetic field on outer-space radiation, geomagnetism, thermal properties of the earth's interior, oceanography, climatology, meteorology, vulcanology, and deformation of the earth's crust and mantle, including warping of the crust and shifting and displacement of continents (continental drift). Techniques of geophysics include measurement of earth's gravitational and magnetic fields, elastic-wave measurement (by seismography), and

obtaining data on the radiation flux in space by means of space satellites.

geophysics, gravity The scientific discipline concerned with global gravity distribution and its significance. The objectives may be restricted geographically, as in the search for buried ore deposits, or may include large sections of earth's crust, as in the study of mountain belts, island arcs, or rift valleys. The basic investigative method in all cases involves obtaining gravity measurements and plotting the result on a gravity map. Gravity anomalies that appear on a map can be interpreted in terms of subsurface structures. Such studies are often the first phase of mineral or petroleum exploration. See also: *geophysical exploration.*

geophysics, magnetic See *geomagnetics.*

geophysics, solid-earth A scientific discipline involved with the origin, composition, structure, and physical properties of the entire planet. Solid-earth geophysics obtains its data from seismic studies of earthquakes and nuclear explosions, from gravity, magnetic, and heat-flow measurements, and from experimental study of the density and elasticity of rocks.

geopotential Combined differences between the potential energy of a mass at a given height and the potential energy of an identical mass at sea level. It is equivalent to the energy required to move the mass from sea level to the given height.

The difference between potentials at P_1 and P_2 is given by: $\int_0^H g \cdot dh$ where g is the value of gravity at height h above P_1 and H is the vertical distance between P_1 and P_2.

geoscience 1. *Geology.* 2. *Earth science.*

geosphere The inorganic world, contrasted with the world of life, the *biosphere*. It includes the *lithosphere* (solid portion of the earth), the *hydrosphere* (bodies of water), and the *atmosphere* (air).

geostatic pressure The vertical pressure exerted at a point within the earth's crust by a column of overlying material. For a completely granite crust, the geostatic pressure at a depth of 1.6 km is 350 kg/cm^2.

geostrophic motion Fluid flow that is parallel to the isobars (lines of equal pressure) in a rotating system; e.g., the earth. Such a flow is the result of a balance between the pressure gradient and the *Coriolis effect*. The velocity of the flow is proportional to the pressure gradient. For large-scale oceanic and atmospheric movements the geostrophic current generally represents the actual current, with about 10% error. Cf: *Ferrel's law*. See *Coriolis effect*.

geosuture The boundary zone between contrasting rock masses, probably extending as deep as the mantle; the contact between continental plates that have either spilt or collided.

geosynclinal couplet An *orthogeosyncline*.

geosynclinal prism An often extremely thick sediment accumulation that occurs in the downwarped part of a geosyncline. Its form resembles that of a long prism. Cf: *clastic wedge*.

geosyncline A basin-like or elongate subsidence of the earth's crust. Its length may extend for several thousand kilometers and may contain sediments thousands of meters thick, representing millions of years of deposition. A geosyncline generally forms along continental edges, and is destroyed during periods of crustal deformation, which often produces folded mountain ranges; e.g., the Appalachians of North America.

Geosynclines vary in their morphological features, and have been divided into subclasses based on form, rock type and derivation, and relationship to *cratons*, the stable interiors of continents.

Orthogeosyncline—The classic example lies between craton and ocean basin. It is divided into adjacent and parallel structures: the *eugeosyncline* and the *miogeosyncline*. *Eugeosyncline*—This forms at a distance from the craton, and is characterized by thick sediments with a high proportion of volcanic rocks. *Miogeosyncline*—This forms adjacent to the craton. Thinner sediments are found, and fewer or no volcanic rocks. Sandstone and limestone deposits are from continental erosion and the skeletal remains of marine life. *Parageosyncline*—This lies within a craton or stable area. Now subdivided into *exogeosynclines, autogeosynclines,* and *zeugogeosynclines*. *Exogeosyncline* —This lies along the cratonal margin, and collects sediments from an area outside the craton. Syn: *deltasyncline; foredeep*. *Autogeosyncline*—A *parageosyncline* without marginal uplift. An intracratonic basin. Essentially filled with carbonate sediments and some sands. *Zeugogeosyncline*—A *parageosynclines* with marginal uplift; an intracratonic trough. Contains sandy sediments derived from adjoining uplifted areas. Syn: *yoked basin*.

geotaxis The movement of an entire plant in response to gravity. Cf: *geotropism*.

geotectonic *Tectonic*.

geothermal energy Energy that can be extracted from the internal heat of the earth. Surface manifestations of geothermal energy are found in hot springs, geysers, and fumaroles.

geothermal gradient The change in temperature with respect to change in depth in the earth. The geothermal gradient, measured by means of holes bored into the earth, ranges upward from

5° C/km, with an average of about 30° C/km. The gradient depends on the heat flow in the region which, in turn, is related to the thermal conductivity of the rock.

geothermal heat flow The quantity of heat energy leaving the earth, measured in calories/cm^2/sec. Heat-flow measurements in igneous rocks have shown a linear correlation between surface heat-flow and heat produced in the rocks. The heat is due to the decay of uranium, thorium, and potassium. Syn: *heat flow.*

geothermometer Some feature of rocks, such as a mineral that forms within determined limits of temperature under specific pressure conditions. The presence of such a feature thus defines a range for the temperature of formation of the ambient rock. An example would be the filling temperature of *fluid inclusions.* Syn: *geologic thermometer.* See also: *geologic thermometry.*

geothermometry 1. Determination of the temperature of chemical equilibrium of a mineral, rock, or fluid. 2. Study of the earth's heat and its effect on physical and chemical processes. Cf: *geologic thermometry.*

geotropism The growth of a part of a plant in response to gravity. Cf: *geotaxis.*

geyser A type of hot spring that intermittently ejects sprays and jets of hot water and steam. Eruption is caused by the contact between groundwater and rock that is hot enough to create steam; the intermittent activity is due to the increase in the boiling point of water with pressure. If a geyser tube is long, crooked, and constricted in places, the water may first boil in only one part of the tube, thus forming steam and forcing out water that lies above this point. The pressure on the water remaining in the tube is reduced, and the boiling point is lowered. The next eruption will occur when the tube has filled with water and has again been brought to the boiling point. Such spout-

ing springs were first observed in Iceland; the type example was given the Icelandic name *Geyser.*

geyserite Syn. of siliceous sinter; esp. the opaline material deposited by precipitation from the waters of a geyser.

giant's kettle A cylindrical hole worn beneath a glacier by churning water or by rocks rotating in the bed of a meltwater stream. Syn: *glacial pothole; giant's cauldron.*

gibbsite A mineral, Al(OH)$_3$, the chief constituent of many bauxites.

gilsonite A variety of natural asphalt. Syn: *uintahite.*

gingko *Ginkgo.*

ginkgo A gymnosperm of the order Ginkgoales, which contains only a single living species (*Ginkgo biloba*). The known geologic range is from Late Permian to Recent, maximum abundance being reached during the middle Mesozoic. Its primitive characteristics have resulted in its being called a "living fossil." Ginkgoes, native to Japan and China, are also called "maidenhair trees." See also: *fossil plants.*

glacial abrasion See *glacial scouring.*

glacial boulder A large rock fragment that has been carried by a glacier. Cf: *erratic.*

glacial canyon A deep valley eroded (not cut) by a glacier. Such valleys are characteristically U-shaped in cross-profile.

glacial crevasse A crack or fissure in a glacier resulting from stress due to movement. Glacial crevasses may be 20 m (65 f.) wide, 45 m (150 f.) deep, and several hundred meters long. Longitudinal crevasses develop in areas of compressive stress; transverse crevasses develop in areas of tensile stress; marginal crevasses occur when the central portion of the glacier flows faster than the outer edges. Jagged ice pinnacles (*seracs*) may form where crevasses intersect at the glacier terminus. Glacial crevasses may be hidden by snow bridges, and may close up where the gradient is lower. See also: *cirque; bergschrund.*

glacial drift Rock material deposited by glaciers or glacier-associated streams and lakes. The term originated prior to glacial theory, when exotic stones and earth were thought to have been "drifted" in by water or floating ice. Drift material ranges in size from clay to massive boulders. The greatest thicknesses are found in buried valleys. Drift is divided into two types of material: *till*, which is deposited directly by glacial ice and shows little or no sorting or stratification; and *stratified drift*, well sorted and layered material that has been moved and deposited by glacial meltwater.

glacial epoch Any portion of geological time, from the Precambrian onward, in which the total area covered by glaciers greatly exceeded that of the present. Syn: *glacial age; ice age.*

glacial erosion Reduction of the earth's surface by processes for which glacier ice is the agent, including the carriage of rock fragments by glaciers, scouring, and the erosive action of meltwater streams.

glacial erratic *Erratic.*

glacial geology The study of the features and effects resulting from glacial erosion, and deposition by glaciers and ice sheets. The term also applies to the features of a region that has been subjected to glaciation. Syn: *glaciogeology.* Cf: *glaciology.*

glacial lake 1. A lake fed by meltwater, or forming on glacier ice because of differential melting. 2. A lake confined by a morainal dam. 3. A lake held in a bedrock basin produced by glacial erosion. 4. *Kettle lake.*

glacial mill *Moulin.*

glacial plucking An erosion process whereby fragments of bedrock are dislodged and transported by glaciers. The fragments may be plucked when meltwater in the bedrock joints refreezes and the resulting expansion forces blocks out of their places. The fragments may also be pried loose by the plastic flow of ice around blocks, which become part of the moving glacier. Glacial plucking is most pronounced on hilltops or knobs, and where bedrock is well jointed. It provides much of the rock material in *glacial scouring*, though erosion due to plucking is more extensive. Plucking is the dominant process in the formation of *cirques*, cliffs, glacial stairways, and other landforms resulting from glaciation. See also: *glaciation.*

glacial recession A decrease in the length of a glacier. Syn: *glacial retreat.*

glacial scouring Also called *glacial abrasion*, grinding, or rasping. An erosion process whereby a glacier abrades the bedrock over which it moves. The abrasive agent is the rock material dragged by the glacier ice. The ice itself does not erode the rock, since it is softer. Fine particles polish the bedrock to a smooth finish, larger particles produce striations or scratches, and boulders gouge deep cuts into easily eroded bedrock. All these features may be present at one time. Grooves and striations parallel the ice-flow direction. Glacial scouring usually erodes a smaller amount of material than *glacial plucking*. See also: *chatter marks.*

glacial striation 1. Fine-cut parallel or nearly parallel lines on a bedrock surface, inscribed by overriding ice. 2. Small grooves or scratches cut into a bedrock surface by rock fragments carried in a glacier, or cut on the transported rocks themselves. Glacial striations cover great areas in northern North America and Europe. They are usually associated with Pleistocene deposits, but also occur in Permo-Carboniferous, Ordovician, and Precambrian rocks. Syn: *glacial scratch*.

glacial valley See *glaciated valley*.

glaciated Said of a surface that was once glacier covered; esp. of a surface that has been modified by glacial action.

glaciated valley Also called *glacial trough*. A stream valley that has been glaciated, most often to a U-shaped cross-section, and which may be several hundred meters deep. It should be noted that glaciers do not cut the original valley, but modify or change the shape of an existing one. The formerly V-shaped valley is converted to a U-shaped one and, because a glacier cannot turn as easily as a stream can (because of its greater viscosity), the valley also becomes straighter. Smaller, shallower troughs are cut by smaller tributary glaciers. When the glaciers melt, they remain as *hanging valleys* on walls of the main glacial valley. Postglacial streams may form waterfalls from hanging valleys; e.g., Yosemite Falls. See also: *glaciation*.

glaciation The covering and alteration of the earth's surface by glaciers. Mountain glaciers produce erosional features such as *hanging valleys, cirques*, and *arêtes*, and depositional features such as *moraines*. Areas subjected to continental glaciation (ice sheets) are characterized by depositional features such as *eskers* and *drumlins*, as well as moraines. Lakes, *erratics, glacial striations, roche moutonnées, kettle holes*, and *kames* are common in both areas. Material from seasonal meltwaters is deposited into temporary lakes that form with the recession of continental ice sheets. This may produce *rhythmites*, layered deposits, developed by cyclic sedimentation, some of which are annual and can be used in dating and paleoclimate determination. (See *rhythmites*.)

Mountain glaciation usually increases topographic steepness and sharpness, whereas continental glaciation produces a more subdued landscape.

glacier A large mass of perennial ice showing evidence of present or past flow; it occurs where winter snowfall exceeds summer snow melt. Most of the ice of glaciers originates as fresh, light snowflakes, but it can also derive from rime ice and supercooled water vapor. The first requirement of glacier formation is the accumulation of newly fallen snow in hollows above the snowline. These accumulation areas are called *névé*. The snow crystals are first compressed into granular snow, a process that is most effective at temperatures not appreciably higher or lower than 0° C. After one winter, the granular snow reaches a density of 0.4 to 0.55 and is now called *firn*. As it ages, the firn increases in density and crystal size. When a critical density is reached, firn becomes ice; however, the time expended in this conversion is extremely variable. Periods of 200 years are not unknown.

When the weight of the amassed ice layers becomes great enough to overcome the internal resistance of the ice and the friction between ice and ground, motion is initiated and the ice becomes a glacier. The movement of glaciers consists of two components: *internal deformation* (creep) and *basal sliding*. Under the great pressure exerted by the weight of the ice, the ice crystals are rearranged into layers of atoms parallel to the glacier surface. The gliding of these layers over one another causes ice to creep. *Basal sliding* refers to the movement of the glacier over bedrock under the pull of gravity. Although glacier movement usually combines these two basic processes, the contribution of each varies greatly with local conditions. Polar glaciers exhibit no basal sliding, whereas the movement of certain temperate glaciers is almost totally due to this process, because of the lubricating effect of meltwater beneath them. The rate of both creep and basal sliding depends on thickness, slope, and temperature of the ice. Velocities also vary within sections of an individual glacier; e.g., the ice at the head and terminus moves more slowly than the ice in between. Englacial measurements show a velocity decrease from surface to bedrock.

Glaciers may be classified morphologically as being of three main types: 1. *Mountain* or *alpine glaciers* are confined to a path directing their movement. *Valley* and *cirque* glaciers are types of mountain glacier. *Valley* glaciers are longer than they are wide, and flow down a valley, at least in part. *Cirque* glaciers are short and wide, and originate with snow accumulated in cirques at valley heads. 2. *Ice sheets* and *ice caps* extend in continuous sheets that move outward in all directions. These glaciers are called *ice sheets* or *continental glaciers* if they are the size of Antarctica or Greenland (the only existing ice sheets), and *ice caps* or *ice fields* if they are smaller. 3. *Piedmont glaciers* are formed when valley glaciers issue from their confining channels and move out over level gound. *Ice shelves* are the floating parts of glaciers that have reached sea level. Smaller forms are *ice tongues*. Very large ice shelves (e.g., the Ross Shelf) may be formed of firn rather than true glacier ice. Glaciers are also classified according to thermal and dynamic characteristics. *Temperate* and *polar* glaciers are distinguished largely by the respective presence and absence of meltwater. A glacier may be defined as *active* if it is fed by a continuous ice stream from an accumulation zone; *passive* where the supply of snow is small or the slopes gentle; and *dead* if it no longer receives a supply of ice from an accumulation area. A dead glacier is not necessarily immobile.

glacier band A series of layers on or within a glacier, differing from adjacent material in color and texture. Such layers may consist of ice, snow, firn, rock debris, organic matter, or any mixture of these. See also: *ogive*.

glacier burst A sudden release of a great quantity of meltwater from a glacier or subglacial lake. Water accumulates in depressions within the ice margins, and at a critical stage erupts through the ice barrier, sometimes producing a catastrophic flood. Such accumulation and release may occur at almost regular intervals. *Jökullhlaup* (etymol: Icelandic) term for a glacier burst primarily used in Icelandic terrain, where the phenomenon is often associated with volcanic or fumarolic activity. Syn: *glacier flood*.

glacière 1. French generic term for all underground ice formation. It has been accepted by English writers and long recognized in American scientific journals. 2. Natural or artificial cavity in a temperate climate in which an ice mass remains unthawed throughout the year; an ice cave.

glacier flood *Glacier burst.*

glacier flow See *glacier.*

glacierization In British usage, approximately equivalent to glaciation, in the context of the gradual covering of land surface by glaciers or ice sheets.

glacier lake Water confined by the damming of natural drainage by a glacier or ice sheet.

glacier surge See *surging glacier.*

glacier tectonics The deformation of glacial masses and glacial sediment by the movement of ice, indicated by fractures and crevasses that appear in the upper ice. Fractured portions may develop as ridges perpendicular or parallel to the direction of movement. Successive bedding layers, commonly found in glacial ice, represent the residue of seasonal snowfalls mixed with rock debris. The clear patterns into which these layers are often folded helps to determine the general pattern of the glacial movement.

glaciofluvial Pertaining to glacier

meltwater streams or to deposits made by such streams. Syn: *fluvioglacial*.

glaciolacustrine Derived from or pertaining to glacial lakes, or the material deposited into such lakes, particularly deposits such as kame deltas.

glaciology The study of glaciers and of ice in all of its natural occurrences, including the formation and distribution of ice, its physical and chemical properties, glacier flow, and the interrelation between ice and climate. Glaciology can also refer to the structural character of glacial deposits, the materials of which are derived by the movement of ice and its differential pressures.

glass 1. A state of matter intermediate between the densely packed, well-ordered atomic arrangement of a crystal and the disordered atomic arrangement of a gas. Most glasses are supercooled liquids, i.e., they are amorphous, metastable solids with the atomic structure of a liquid. Glasses and liquids are distinguished on the basis of viscosity. 2. An amorphous substance resulting from the rapid cooling of a magma. Within the cooling period the strong cohesion between atoms of the magma inhibits crystal formation. Obsidian is an example of igneous glass. See also: *devitrification*.

glassy Textural term applied to igneous rocks or parts of rocks that completely lack crystalline structure as a result of rapid cooling. A glassy texture is unique to magmatic rocks, and is most common in volcanic rock. Glassy specimens may be composed almost entirely of massive glass, e.g., obsidian, or the glass may be vesicular, e.g., pumice. Syn. *vitric*. See also: *glass*.

glauberite A light-colored monoclinic mineral, $Na_2Ca(SO_4)_2$.

Glauber's salt *Mirabilite*.

glauconite A green mineral, essential-

ly a hydrous potassium iron silicate, closely related to the micas. Common in sedimentary rocks from the Cambrian to the present. Sandstones rich in glauconite (*greensand*) are especially prevalent in rocks of Cambro-Ordovician and Cretaceous age.

glauconitic sandstone *Greensand*.

glauconitization A replacement process common in formations containing large quantities of *glauconite* (a complex hydrous potassium iron silicate). The tests of many foraminifers are replaced in this manner. See *fossil; replacement*.

glaucophane A light blue prismatic or fibrous mineral of the amphibole group, $Na_2(Mg,Fe^{+2})_3Al_2Si_8O_{22}(OH)_2$. Found only in low-temperature, high-pressure metamorphic rocks. It is of interest to petrologists as a means of defining the metamorphic conditions under which the surrounding rock formed.

G layer See *seismic regions*.

gliding A deformation-related movement that takes place along lattice planes in crystalline substances. Gliding is of two types: *translation gliding* and *twin gliding*. In translation gliding, displacement takes place along preferred lattice planes without reorientation or rupture of the deformed parts. In twin gliding, the lattice of the displaced part of the crystal is symmetrically altered with respect to the lower, undisplaced part. There are a limited number of lines parallel to which movement can take place; these are the *glide directions*.

global tectonics Tectonics on a global scale, such as tectonic processes associated with the very large-scale movement of material.

globigerina A protozoan (foraminifer) with a calcareous shell. The initial shell is invisible to the unaided eye, but when the final shell is outgrown, the subsequent larger shells remain attached to the first shell. Eventually the foraminifer resembles a minute nautilus, which is then visible to the unaided eye. Globigerina live both on the sea bottom and in open water. They are the chief constituent of globigerina ooze. See *pelagic deposits*.

globular　*Spherulitic.*

Glomar Challenger　A research vessel used in the *Deep Sea Drilling Project*. It was specially designed to obtain long sediment cores by means of drilling into the ocean floor.

glory hole　Mining term for a large open pit from which ore is extracted.

Glossopteris　A late Paleozoic genus of fossil plants (tongue-ferns) found throughout the glaciated regions of the Southern Hemisphere, i.e., South America, Australia, and Antarctica. *Glossopteris* flora bore clusters of simple spatulate leaves. They produced winged spores, and some specimens have been found with reproductive organs attached to individual leaves. It is not certain whether these are seed vesicles or spore cases.

glowing avalanche　Syn: *Ash flow.*

glowing cloud　*Nuée ardente.*

gneiss　A foliated metamorphic rock commonly derived from granite, shale, diorite, mica schist, rhyolite, and other rocks by regional metamorphism associated with moderate to high temperatures and pressures. Gneiss is commonly composed of quartz and feldspar, with lesser amounts of mica; however, the term *gneiss* refers more to the particular texture, characteristic minerals present, and origin of a given rock. For example, a gneiss might be an augen gneiss, or a banded gneiss; one that is rich in hornblende is a hornblende gneiss; if there is evidence that it was metamorphosed from granite, it is a granite gneiss.

gneissic　Describing a rock that shows the texture typical of gneisses.

gneissoid　Said of a gneiss-like texture or structure that is not related to metamorphic processes; for example, the formation of a gneissoid granite by viscous magmatic flow.

goethite　A yellow, red, or brown mineral, FeO(OH), often the commonest constituent of limonite. It occurs as a deposit in bogs, and is the major constituent in *gossans*.

gold　A soft yellow mineral, the native metallic element, Au. It occurs principally as a native metal, but may also be alloyed with silver, copper, and other metals. Although a rare element, gold is widely distributed in nature. It is found as nuggets, leaves, grains, and in veins with quartz.

Goldschmidt's phase rule　A modification of Gibbs phase rule. It assumes that two variables are fixed externally and that, consequently, the number of phases in a system will not generally exceed the number of components. When used as a mineralogical phase rule, the variables are taken as *temperature* and *pressure*; the number of phases will be the *minerals*, and the system is the *rock* that is being formed.

goldstone　A translucent reddish-brown glass containing particles of copper or chromium oxide that display sparkling reflections. Goldstone originated in Venetian glass of the Renaissance as an imitation of *aventurine* and, thus, is also called *aventurine glass*.

Gondwana (Gondwanaland)　The southern supercontinent (protocontinent) named for the Gondwana system of India, dating to the late Paleozoic and early Mesozoic, and containing glacial tillites below coal measures. Similar rock sequences of the same age, containing identical fossil flora (*Glossopteris*) suggest an earlier connection of the southern continents of Antarctica, Africa, South

America, Australia, and India. Gondwana corresponds to the protocontinent *Laurasia* in the Northern Hemisphere. Both land masses are thought to have been derived by the splitting of *Pangaea*. See also: *plate tectonics*.

goniatite See *ammonoid*.

GOR *Gas-oil ratio.*

gorge A deep, narrow, steep-walled valley, sometimes carved by stream abrasion. It may also be a narrow passage between hills or mountains.

gossan; gozzan Also called "iron hat," a gossan is the oxidized and leached near- or above-surface portion of a mineral deposit containing sulfides, particularly iron-bearing ones such as pyrite and chalcopyrite. The term "gossan" derives from the Cornish for "blood," and refers to the rusty color resulting from oxidized pyrites. It basically consists of hydrated iron oxides (limonite) from which sulfur and copper have been removed by downward-percolating waters. Some highly insoluble minerals, e.g., gold, may be concentrated as a surficial residuum in a gossan. See *secondary enrichment*.

gouge 1. *Fault gouge.* 2. *Crescentic gouge.*

graben Etymol: Ger., "ditch." A crustal unit or block, generally elongate, that has been depressed relative to the blocks on either side. The bordering faults (on the longer side of the graben) are usually of near-parallel strike and steeply dipping. A true graben, in its initial surface form, is typically a linear structural depression. The flanking highland areas are commonly *horsts*.

grade 1. Degree of inclination with the horizontal of a road, railroad, or embankment. 2. The profile of a stream channel so adjusted that no gain or loss of sediment occurs. 3. Relative quantity or percentage of mineral content in an ore body. Syn: *tenor*. 4. A particular size

range of soil particles; e.g., *Alling grade scale*. 5. A classification of coal based on its content of impurities such as ash; higher grades of coal have fewer impurities. Cf: *rank; coal type*. 6. *metamorphic grade*.

graded 1. A geologic term referring to a sediment or rock containing particles of generally uniform size. Syn: *sorted*. 2. An engineering term that is essentially the opposite of the geological, i.e., a soil composed of particles of many sizes, or having a uniform distribution of particles from coarse to fine. 3. A land surface on which deposition and erosion are in such balance that a general slope of equilibrium is maintained is described as *graded*. Syn: *at grade*.

graded bedding A type of bedding in which each layer exhibits a gradual change in particle size, usually coarse at the base to fine at the top. Graded beds can be deposited by turbidity currents.

graded profile *Profile of equilibrium.*

graded stream A stream or river whose transporting capacity is in equilibrium with the amount of material supplied to it. Thus, a balance exists between degradation and aggradation. The downstream gradient of a graded stream is called a *graded slope*. It permits the most effective transport of load.

grade scale An arbitrary division of a basically continuous range of particle sizes into a sequence of grades for the purpose of standardization of terms.

gradient 1. Degree of inclination of a surface; steepness of slope. In quantitative terms, it is the ratio of the vertical to the horizontal. 2. *Hydraulic gradient.* 3. *Stream gradient.* 4. The change in value of one variable with respect to the change in another. See *thermal gradient*.

gradiometer Any instrument used to measure the gradient of a physical quantity; e.g., the gradient of a magnetic field.

grahamite 1. A black asphaltite with a high specific gravity, and a high fixed carbon content. 2. *Mesosiderite*.

grain 1. A mineral or rock particle less than a few millimeters in diameter. 2. Used in such expressions as "fine-grained." 3. A separate particle of ice or snow. 4. The linear disposition of topographic features in a region. 5. A term for the plane of splitting or separation in a metamorphic rock.

grainstone A clastic, grain-supported, mud-free limestone. Cf: *packstone; mudstone*.

grain-supported Referring to a sedimentary rock, esp. limestone, in which the quantity of granular material is sufficient to form the supporting framework, either wholly or in part.

granite A coarse-grained plutonic igneous rock in which quartz constitutes 10 to 50% of the felsic components, and the ratio of alkali feldspar to total feldspar is from 65 to 90%. Small amounts of mica (usually biotite) and hornblende are common in granite. The term granite is often applied to any holocrystalline, quartz-bearing plutonic rock. See also: *granitization*.

granite gneiss A gneiss derived from a sedimentary or igneous rock and having the mineral composition of a granite. Also, a metamorphosed granite.

granite porphyry A hyperabyssal rock characterized by sparse phenocrysts of mica, amphibole, or pyroxene in a medium- to fine-grained groundmass.

granitic and basaltic layers Terms referring to layers of the earth's crust, and often used as synonyms of *sial* and *sima*. The terms were proposed after the *Conrad Discontinuity* was postulated to explain abrupt differences in velocity, observed in some places, between seismic waves traveling through the earth's upper crust and those traveling through lower crustal layers. The velocities observed, 6.0 km/sec for the upper and 6.5 to 7.0 km/sec for the lower crustal layer, were generally appropriate to granite (or gabbro) and basalt, respectively, albeit the actual rock composition at these depths must be quite different. Because the Conrad Discontinuity is a local phenomenon, unlike the *Mohorovičić Discontinuity*, which is worldwide, many geologists discourage the use of the terms *granitic* (or *gabbroic*) *layer* and *basaltic layer*.

granitic layer See *granitic and basaltic layers*.

granitization An essentially metamorphic process whereby a pre-existing rock is converted into granite. The process seems to result in the addition of material to form feldspars, and in the removal of iron and magnesium from the original rock. However, the precise mechanism and extent of the process are still in question.

granoblastic A term for an isotropic fabric consisting of a mosaic of equidimensional anhedral grains. If inequant grains are present, they are randomly arrayed.

granodiorite A coarse-grained plutonic rock composed mainly of quartz, potassium feldspar, and plagioclase (oligoclase or andesine); biotite and hornblende are the usual mafic components. It is the approximate intrusive equivalent of *rhyodacite*. Granodiorite is similar to granite, except that the latter contains more alkali feldspar.

granophyre 1. A porphyritic rock of granitic composition having a crystalline granular groundmass. 2. A fine-grained granitic rock with micrographic texture. 3. A microscopic intergrowth of quartz and alkali feldspar.

granular disintegration A type of weathering in which rock masses, esp. coarse-grained rocks like granite, are broken down grain by grain. The process occurs in regions of great temperature extremes.

granular texture   A rock texture characterized by the aggregation of mineral grains of approximately equal size. The term may be applied to any type of rock, but is esp. applied to an equigranular, holocrystalline igneous rock whose particles range from 0.05 to 10 mm in diameter.

granule   1. A rock fragment of diameter 2 to 4 mm; larger than a coarse sand grain and smaller than a pebble. 2. A small, precipitated grain.

granulite   A metamorphic rock composed of equal-sized, interlocking grains. Also, a coarse granular metamorphic rock of the granulite facies.

granulite facies   The metamorphic facies in which basic rocks are represented by diopside + hypersthene + plagioclase. It is characteristic of deep-seated regional dynamothermal metamorphism at temperatures exceeding 650° C.

graphic   A texture of igneous rock; an intergrowth in which a single crystal of alkali feldspar encloses many small, wedge-shaped grains of quartz, creating an overall resemblance to writing. In *micrographic* texture, the size of the grains is visible only under the microscope. *Graphic granite* is a pegmatite characterized by graphic intergrowths of alkali feldspar and quartz.

graphite   A naturally occurring crystalline form of carbon, dimorphous with diamond. It is steel gray to black, soft and greasy to the touch. It most often occurs as crystals or flakes disseminated through metamorphic rocks such as gneisses, marbles, and schists. In some of these rocks it is found in veins large enough to be mined. It is probable that such graphite was derived from carbonaceous material of organic origin.

Graphite is used in pencils, electroplating, batteries, and electrodes; it is also used as a lubricant. Graphite is also called *plumbago*, because for centuries it was confused with galena, a black sulfide of lead.

graptolite   An extinct marine colonial organism, possibly related to the hemichordates, a primitive group of invertebrates. Graptolites first appeared during the Cambrian Period and survived into the Mississippian (Lower Carboniferous); their fossils are most frequently found as carbonaceous impressions in black shales. The typical graptolite consisted of one or more branches (stipes) bearing cuplike structures (thecae), the entire organism being covered with chitinous materials. Graptolites show a gradual development through time, and are very useful in stratigraphic correlation.

gravel   1. An unconsolidated accumulation consisting of particles larger than sand (diameter › 2 mm), i.e., granules, pebbles, cobbles, boulders, or any combination of these. Gravel is the unconsolidated equivalent of a *conglomerate*. 2. Common term for detrital sediments, composed mainly of pebbles and sand, found along streams or beaches.

gravimeter   A sensitive device for measuring variations in earth's gravitational field. One type consists of a weight suspended by a spring. Variations in gravity produce changes in spring length, measured by a number of mechanical and optical techniques. Some gravimeters operate on the principle that variations in the period of oscillation of a swinging pendulum are a function of changes in the gravitational field. Gravimetric surveys on oceans often involve the use of gyrostabilized platforms. Such surveys are useful in exploring natural resources and determining local anomalies produced by mineral deposits and salt domes. See also: *gravity anomaly; geophysical exploration.*

gravimetry   See *gravimeter.*

gravitational attraction   See *law of universal gravitation.*

gravitational constant   See *law of universal gravitation.*

gravitational gliding   *Gravity sliding.*

gravitational separation   The layering of oil, gas, and water in a subsurface reservoir according to their respective specific gravities. Also, the postproduction separation of these materials in a gravity separator.

gravitational water   *Free water.*

gravity anomaly   A deviation of the gravity value from the theoretical value, after a correction has been applied to the point of the earth's surface where the measurement was made. Gravity anomalies are the result of the earth's nonuniform density. Of the three types of gravity anomaly defined here, the *free-air* and *Bouguer anomalies* are the most common.

As an observer moves farther from the earth's center, the resulting decrease in gravity is called the *free-air effect.* The free-air gravity correction compensates for changes in surface elevation (to include points outside the *geoid*). Thus, the free-air anomaly, Δ y, represents the difference between the free-air gravity value (g_o) and the standard gravity value (γ), or $\Delta g = g_o - \gamma$, where $g_o =$ observed gravity + *free-air correction.*

A *Bouguer anomaly* remains after application of latitude and free-air corrections, and a correction (the *Bouguer correction*) that removes the effect of the mass of earth's material between the observation point and the geoid. The Bouguer anomaly is thus expressed as $\Delta g_B - \gamma$, where g_B is the Bouguer gravity value (observed gravity plus Bouguer correction), and γ is the standard gravity value.

Isostatic anomalies remain after correction for gravity reductions due to terrain.

Such corrections are of two principal types. One assumes compensation to be affected by horizontal density variations; the other, that compensation is influenced by variations in thickness of a constant-density layer. Gravity anomalies occur over mineral deposits, salt domes, and over caves where earth materials are absent or considerably less than the adjacent earth materials. They have been accurately correlated with global tectonic features; e.g., large regions of positive gravity anomaly are regions of current geological change, whereas principal regions of negative anomaly are either basins or areas of postglacial uplift.

gravity compaction   Compaction of a sediment as a result of overburden pressure.

gravity fault   *Normal fault.*

gravity geophysics   See *geophysics; gravity.*

gravity map   See *geophysics; gravity.*

gravity meter   See *gravimeter.*

gravity sliding   Also called *gravity gliding* and *gravitational gliding,* this is a downslope movement of a portion of the earth's crust, usually on the flanks of an uplifted region. This mechanism, activated by gravitational forces, seems to be responsible for many shallow faults. It is also thought to be closely associated with many major mountain systems such as the Alpine belt.

gravity tectonics   Tectonics in which the primary activating mechanism is believed to be downslope sliding due to gravitational effect.

gray mud   See *terrigenous deposits.*

graywacke   A term that has been redefined many times since its origin, and which is now usually applied to a very hard, dark-gray or greenish-gray, coarse-grained sandstone characterized by angular particles of quartz and feldspar embedded in a clayey matrix. Graywacke commonly exhibits the graded bedding characteristic of certain marine sedimentary facies, and so is believed to have been deposited by submarine turbidity currents. Cf: *wacke*. See also: *subgraywacke*.

great circle That circle defined by the intersection of a plane with the center of a sphere. The shortest distance between any two points on the sphere is along the arc of the great circle joining them.

Great Ice Age See *ice age*.

greenhouse effect The heating of earth's surface due to the presence, in the lower atmosphere, of carbon dioxide and water, which absorb long-wavelength terrestrial radiation, thus halting or hindering its escape from earth's surface.

green marble *Verd antique*.

green mud See *terrigenous deposits*.

greensand A greenish sand or sandstone; specif. an unconsolidated coastal plain marine sediment composed largely of dark green grains of glauconite. Syn: *glauconitic sand; glauconitic sandstone*.

greenschist A green, schistose, metamorphic rock whose color is due to the presence of chlorite, epidote, or ectinolite. Cf: *greenstone*.

greenschist facies The metamorphic facies in which basic rocks are represented by albite + epidote + chlorite + actinolite. The mineral assemblage is characteristic of low-grade regional metamorphism corresponding to temperatures in the range of 300° to 500° C. The greenschist facies assemblage of minerals partially to totally replaces primary magmatic minerals; e.g., ophitic pyroxene, calcic plagioclase. Relict igneous fabrics are often preserved.

greenstone A more or less overall term for any green, aphanitic, weakly metamorphosed igneous rock whose color is due to chlorite, actinolite, or epidote. Cf: *greenschist*.

greisen An aggregate of quartz and muscovite (or lepidolite) with accessory amounts of topaz, fluorite, tourmaline, rutile, cassiterite, and wolframite, formed by the metasomatic hydrothermal alteration of granitic rock.

Grenville orogeny A name for a major plutonic metamorphic and deformational event that affected a broad region along the southeastern border of the Canadian Shield during the Precambrian.

groove cast A long, straight, sometimes ridgelike structure on the underside of a sandstone bed. It is produced by the filling of a groove on the surface of an underlying mudstone. At least some of the initial grooves are thought to be produced by tools of early humans. Cf: *drag mark*.

grossular The calcium-aluminum member of the garnet series, $Ca_3Al_2(SiO_4)_3$. It is found mainly in crystalline limestone as a product of contact or regional metamorphism. Its color range is from white through golden to cinnamon. See also: *garnet*.

grossularite *Grossular*.

groundmass The fine-grained material that holds the phenocrysts in a porphyritic igneous rock. Syn: *matrix*. Also used for the matrix of sedimentary rock.

ground moraine See *moraine*.

"ground roll" A disturbance that often interferes with the clarity of reflections on seismic records. *Rayleigh waves* are thought to be the principal component of "ground roll."

groundwater 1. Subsurface water in the zone of saturation, including water below the water table and water occupying cavities, pores, and openings in underlying rocks. Syn: *phreatic water.* 2. In an unrestricted sense, all subsurface water as distinct from surface water. Syn: *subterranean water; underground water.*

groundwater barrier A natural or artificial obstruction, such as a dike, to the lateral movement of groundwater. There is a distinct difference in groundwater level on opposite sides of the barrier.

groundwater basin 1. An aquifer, not necessarily basin-shaped, having rather well-defined boundaries and more or less definite recharge and discharge areas. 2. A subsurface formation similar to a basin with respect to the collection, retention, and outflow of water.

groundwater divide An elevation along the *water table* from which groundwater flows away in opposite directions.

groundwater flow The natural or artificially produced movement of water in the zone of saturation.

groundwater hydrology See *geohydrology.*

groundwater level *Water table.*

groundwater reservoir *Aquifer.*

groundwater surface *Water table.*

group 1. The formal *lithostratigraphic unit* next in order above *formation*; it includes two or more geographically associated formations with notable features in common. A differentiated assemblage of formations is a *subgroup*, and an assemblage of related groups or of formations and groups is called a *supergroup*. 2. A term for a sequence of sedimentary beds or igneous rocks.

group velocity See *phase velocity and group·velocity.*

growth fabric The orientation of fabric elements of a rock that is characteristic of the manner in which the rock was formed and not dependent on the effect of stress and deformation.

growth ring *Tree ring.*

Guadalupian Lower series of the Upper Permian of North America.

guest A mineral introduced into and commonly replacing a pre-existing mineral or rock. Ant: *host.*

guide fossil A fossil that has actual or potential value in identifying the age of the strata in which it is found, or in indicating the conditions of its life surroundings.

gulch A deep, narrow ravine, esp. one marking the course of a stream or torrent.

gulf 1. A portion of an ocean or sea partly enclosed by land. 2. A deep narrow chasm or gorge.

Gulfian Upper Cretaceous of North America.

gully A small channel formed by running water or unconsolidated material.

gully erosion The erosion of soil by running water that forms clearly defined, narrow channels that generally carry water only during or after heavy precipitation. Cf: *sheet erosion.*

gumbotil A gray to dark-colored, leached, deoxidized clay representing the B horizon of fully mature soils. Gumbotil profiles develop under conditions that exist on flat and poorly drained till plains.

Gutenberg discontinuity The seismic velocity discontinuity at 2900 km, marking the mantle-core boundary within the earth. At this boundary the velocity of P waves (primary waves) is reduced, and S waves (secondary waves) disappear. It is believed to reflect the change from a solid to a liquid phase (solid mantle to molten outer core), as well as a change in composition. A molten outer core would block S waves entirely, since they cannot travel through liquids. See also: *shadow zone.*

guyot   Also called *tablemount,* a guyot is a seamount of more or less circular form and having a flat top. These structures are named in honor of Arnold Guyot, a Swiss-American geologist of the mid-nineteenth century. Guyots are thought to be volcanic cones whose tops have been truncated by surface-wave action. They occur in deep ocean, usually at depths exceeding 200 m, and are generally composed of alkali basalt, which differs from oceanic tholeiite in having a higher content of K, Na, and Ti. Guyots are found chiefly in the Pacific basin, where they occur in groups, as do the present volcanic island chains. See *basalt; island arc; seamount.*

gymnosperm   A member of the *Gymnospermae,* plants whose seeds are commonly held in cones and never enclosed in an ovary. Ex: cycad, ginkgo, fir, pine, and spruce. Range, Devonian to Recent. Cf: *angiosperm.*

gypsite   An earthy variety of gypsum found only in arid regions as an efflorescent deposit over an outcrop of gypsum.

gypsum   A common, widely distributed mineral, $CaSO_4 \cdot 2H_2O$, frequently associated with halite and anhydrite in evaporites. Curved, twisted crystal growths of gypsum found on cave walls are called *gypsum flowers.* The massive, fine-grained variety of gypsum is alabaster. Gypsum is used in making plaster of Paris, and as a retarder in portland cement.

gyre   A circular motion of water found in each of the major ocean basins. It is centered on a subtropical high-pressure region, and its motion is generated by the convective flow of warm surface water, earth's rotation, and the effects of prevailing winds. See also: *geostrophic motion.*

H

habit   1. *Crystal habit.* 2. The typical appearance of an organism.

hachure   n. One of several evenly spaced, parallel lines drawn on a topographic map to indicate relief features; the width of spacing between lines and the breadth of the lines indicate slope. v. To indicate by hachures.

hackly   Rough or jagged; such as a *hackly* fracture.

hadal   Pertaining to the deepest oceanic environment; in particular, to ocean trenches, where depths exceed 6000 to 7000 m. Sediments in this region accumulate very slowly, and organisms are relatively scarce. See *continental shelf.* See also: *abyssal.*

hairstone   A clear, crystalline quartz containing numerous fibrous or needlelike inclusions of other minerals such as rutile or actinolite. Syn: *needle stone.* See also: *Venus hair; sagenite.*

half-life   Time required for the decomposition of one-half of any given quantity of a radioactive element. It differs for various elements and isotopes, but is always the same for a particular isotope. It cannot be accelerated or decelerated by heat, cold, magnetic, or electric fields. A nucleus is regarded as stable if its half-life is greater than the estimated age of the earth (about 5×10^9 years). See also: *alpha decay; beta decay; radioactivity.*

halide   A compound in which the negative element is a halogen; e.g., HCl or, as a mineral example, halite, NaCl.

halite   A mineral, NaCl, that occurs

most commonly in bedded deposits. It is the most abundant of the evaporites, following gypsum and anhydrite in the sequence of precipitation of salts from seawater. Syn: *common salt; rock salt.*

halloysite   A clay mineral related to kaolinite. Halloysite has two forms: one with kaolinite composition, $Al_4(Si_4O_{10})(OH)_8$, and a hydrated form, $Al_4(Si_4O_{10})(OH)_8 \cdot 4H_2O$. The second type dehydrates to the first with the loss of interlayered molecules of water.

halmyrolysis   See *diagenesis.*

halo   1. A circular or lunate distribution about the source location of a mineral, ore, or petrographic feature. 2. A ring-shaped discoloration of a mineral that can be observed when the mineral is viewed in thin section. Most halos of this type come from radiation damage caused by the radioactive decay of mineral inclusions; e.g., uranium and thorium.

hammada   Etymol: Arabic. A bare rock surface in a desert region, all loose rock particles having been removed by wind. Cf: *erg; reg.*

hand level   A small, lightweight leveling instrument in which the spirit level and mirror are so positioned that the bubble may be viewed while sighting through the telescope.

hanging side   *Hanging wall.*

hanging valley   See *glaciated valley.*

hanging wall   Overlying side of a fault, orebody, or mine working; in particular, the wall rock above a fault or inclined vein. Syn: *hanging side.* Cf: *footwall.*

hard coal   *Anthracite.*

hardness   1. The resistance of a mineral to scratching; one of the properties by which minerals may be described. See *Mohs scale.* 2. A property of hard water, chiefly due to the presence of calcium and magnesium ions.

hardness scale   *Mohs scale.*

hardpan   A hard subsurface layer resulting from the deposition of silica, clay, oxides, and hydroxides, or carbonate salts in the layer of deposition. In humid areas, hardpans consist mostly of clay. *Caliche* hardpans are extensive in the western U.S., and their structure shows regular variation with age. Some hardpans are formed by accumulations of organic matter. Cf: *claypan, duricrust, iron pan.*

hard water   Water that contains large quantities of calcium and magnesium ions that prevent the ready lathering of soap, and which are responsible for the formation of boiler scale.

Harker diagram   See *variation diagram.*

harmonic folding   Folding in which the strata remain parallel or concentric with each other, and in which there are no abrupt changes in the form of the folds at depth. Ant: *disharmonic folding.*

Hartmann's law   Given two sets of intersecting shear planes, the acute angle between them is bisected by the axis of greatest principal stress, and the obtuse angle by the axis of least principal stress.

harzburgite   A *peridotite* composed mainly of olivine and orthopyroxene.

Hawaiian-type eruption   See *volcanic eruption, types of.*

HDR   *Hot dry rock.*

head erosion   *Headward erosion.*

headland   1. A projection of land, often with a cliff face that juts out from a coast into a sea or lake. Syn: *head.* 2. Promontory.

headwall   A precipitous slope at the head of a valley. The term is especially applicable to the rock cliff at the back of a cirque.

headward erosion   The lengthening of a valley or gulley, or of a stream, by erosion at the valley head. Syn: *head erosion; headwater erosion.*

heat capacity   The amount of heat required to increase the temperature of a body or system at constant pressure and volume by one degree. Usually expressed in calories per degree Celsius. Heat capacity differs for different materials. Syn: *thermal conductivity.*

heat conductivity   *Thermal conductivity.*

heat flow   *Geothermal heat flow.*

heave   1. An upward expansion of a surface that may be caused by such factors as clay swelling or frost action. 2. *Creep* in mines. 3. The horizontal component of displacement on a fault. Cf: *throw.*

heavy liquid   A liquid having a relative, adjustable density in the range 1.4 to 4.0, used in the analysis of minerals and to separate ore constituents into heavy and light fractions. Ex: *bromoform, methylene iodide, Clerici solution.*

heavy minerals   1. Dense accessory detrital minerals (e.g., tourmaline and zircon) of a sedimentary rock. These are separated from minerals with lower specific gravity, such as quartz, by means of high-density heavy liquids; e.g., bromoform. 2. Samples of crushed igneous or metamorphic rock that sink in a heavy liquid. Light minerals are those that float.

heavy oil   Crude oil that has a low *Baumé gravity* or *API gravity.* Cf: *light oil.*

heavy spar   *Barite.*

hectorite   A clay mineral of the montmorillonite group, containing lithium and magnesium.

height   1. The vertical distance above a reference, usually earth's surface; the

elevation above a specified level or surface. 2. An area that is significantly higher than the surrounding region.

helictite   A contorted, twiglike or branching cave deposit, usually of calcite. Unlike *stalactites*, helictites are often not vertical; many grow horizontally. Some helictites are formed where water seeps slowly through the bedrock and emerges through nearly microscopic fissures by capillary action.

helluhraun   Name used by Icelanders to describe *pahoehoe* lava flows when they occur in Iceland.

hematite   A mineral oxide of iron, Fe_2O_3, the principal ore mineral of iron. It occurs as individual crystals arranged in flowerlike formations (iron roses), but more commonly as massive granular formations. Massive formations that are soft and earthy are known as *red ocher*. Hematite is a common accessory mineral of many lavas: it is rare in plutonic rocks, but common in pegmatites and hydrothermal veins. Much of hematite is formed under sedimentary conditions through the diagenesis of limonite. It remains stable in a low-temperature metamorphic environment, where it often replaces magnetite.

hematization   The replacement of hard parts of an organism by hematite (iron oxide, Fe_2O_3). Cf: pyritization. Hematization generally results from iron deposition around a fossil. The shells of bryozoans, trilobites, brachiopods and crinoid fragments may be preserved in this manner. See *fossil; replacement.*

hemera   Etymol: Greek, "day." A geological time unit corresponding to the acme-zone, within which the greatest abundance of a taxonomic entity is found in a defined section. Period of time during which a certain type of organism is at its evolutionary apex. Pl: *hemera.* adj. *Hemeral.*

hemimorphic   Said of a crystal having a polar symmetry, so that the two ends have different forms.

hemimorphite   A hydrated zinc silicate, $Zn_4Si_2O_7(OH)_2 \cdot H_2O$, an ore of zinc. Single crystals are rare; it is more often found as fibrous radiating crusts, or mammillary masses. It forms in the oxidation zone of zinc sulfide and lead sulfide deposits.

Hercynian   *Variscan.*

Herkimer diamond   A local name for doubly terminated quartz crystals, named for the locality in New York State in which these are found.

heterochthonous   1. Said of a rock or sediment formed in a place other than that in which it now resides. Also said of fossils transported from their original site of deposition and re-embedded elsewhere. Cf: *allochthonous.*

heterogeneous equilibrium   Equilibrium that obtains in a multiphase system. Cf: *homogeneous equilibrium.*

heteromorphism   The crystallization of two magmas of almost identical composition into different mineral aggregates as a result of different cooling profiles.

hexacoral   A solitary or colonial coral characterized by septa in cycles of six or multiples of six; the septa appear to radiate symmetrically in all directions from the center. Range, Middle Triassic to present. See also: *anthozoan.*

hexagonal system   See *crystal system.*

hiatus   A gap or interruption in the stratigraphic record, as represented by the absence of rocks that would normally be part of a sequence, but were never deposited or were eroded away prior to the deposition of beds immediately overlying the break. The strata missing at a physical break represent the length in time of such a period of non-deposition. Cf: *lacuna.*

high-angle fault   A fault with a dip exceeding 45°. Cf: *low-angle fault.*

high-calcium limestone   See *limestone.*

highland   A mountainous region or high tract of land. Often used in the plural, as the Highlands of Scotland.

high plain   A term that implies an extensive area of level land away from any seacoast; e.g., the High Plains of the U.S., along the eastern side of the Rocky Mountains.

high quartz   *Beta-quartz.*

high-rank metamorphism   Metamorphism that proceeds under conditions of high temperature and pressure. Cf: *low-rank metamorphism; metamorphic grade.*

high tide   The tide at the maximum level reached during a tidal cycle. See also: *spring tide.*

hill   1. A natural, more or less isolated elevation of the land surface, usually considered to be less than 30 m from base to summit; the distinction between hill and mountain often depends on local terminology. 2. A region characterized by hills or by a highland; e.g., the Black Hills of South Dakota.

hill shading   A method of showing relief on a topographic map by the use of simulated sunlight and shadow. Syn: *relief shading.*

Hilt's law   The generalization that, at any location in a coal field, coal rank increases with depth in a vertical succession.

hinge   The locus of points (usually a line) of maximum bending or curvature along a folded surface. Syn: *hinge line; flexure.*

hinge fault   A fault on which the movement of one side rotates or hinges about an axis perpendicular to the fault plane (a "scissors effect"). Thus, the displacement is zero at the pivotal line and increases with distance from that axis. One component of the forces acting to cause this is *normal* and one is *reverse*. See also: *fault*.

hinge line 1. A boundary between a stable or stationary region and one experiencing upward or downward movement. 2. *Hinge*.

hinterland See *foreland*.

histogram A diagram, usually a vertical bar graph, in which the frequency of occurrence of some event is plotted against some other measurement, such as the particle-size distribution in sediments.

historical geology A major branch of geology concerned with earth and its life forms, from its origins to the present. As such, it involves studies in paleontology, stratigraphy, geochronology, paleoclimates, and plate tectonic motion. Cf: *physical geology*.

Histosol In the soil taxonomy of the U.S. Deptartment of Agriculture, a soil order that is more than half organic in its upper 80 cm. Histosols are usually saturated or nearly saturated most of the year, unless artificially drained.

hodograph A graphical representation of distance traveled versus time. A tidal hodograph is a plot of the mean tidal current cycle. In seismology, travel time is plotted as a function of distance from an earthquake epicenter. In oceanography, a special hodograph, the Ekman Spiral, is used to indicate the progressive departure of an ocean current from its initial direction.

hogback A linear, sharp ridge on steeply dipping sedimentary rocks, the opposite slope of which may be symmetrical in shape. Hogbacks are the resistant remnants of eroded and folded layers of rocks. Good examples can be found in the Bighorn Mountains of Wyoming and Black Hills of South Dakota. Hogbacks are similar to *cuestas*, but have steeper rock dips and sides that are more nearly symmetrical. See also: *cuesta*.

holoblast A mineral crystal that originates during metamorphism and is completely formed during that process.

Holocene The most recent epoch of geological time. Upper division of the Quaternary Period. See *Cenozoic Era*.

holocrystalline The texture of an igneous rock composed completely of crystals (i.e., having no glass) is said to be holocrystalline. A rock with such texture is also said to be holocrystalline.

holohyaline A term applied to an igneous rock composed entirely of glass.

holomictic lake A lake whose waters undergo complete mixing during periods of circulation *overturn*.

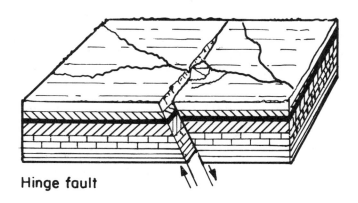

Hinge fault

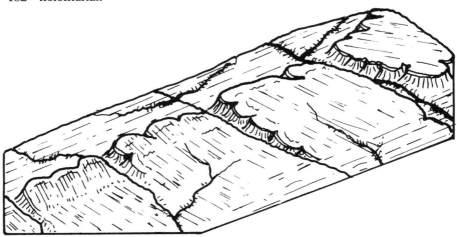

Transition from a hogback through a homoclinal ridge and cuesta to a mesa as controlled by varying rock dip

holothurian A member of the echinoderm class Holothurioidea; includes fossil and living forms; e.g., sea cucumbers. It has a cylindrical body with an anus at or near one end and mouth at or near the other. Feeding tentacles surround the mouth. Holothurians have little geologic importance. Range, Ordovician to Recent.

homeomorph 1. A crystal that resembles another in its crystal form and habit, but differs in chemical composition. Each crystal is said to be a homeomorph of the other. 2. An organism closely resembling another although deriving from different ancestors.

homeomorphism Crystalline substances that have a dissimilar composition but similar habit and crystal form exhibit *homeomorphism*. adj. *Homeomorphic; homeomorphous.*

homeomorphy A biological phenomenon in which species display general similarity but dissimilarity in detail. See also: *convergent evolution.*

homeostasis The tendency of a system, esp. the physiological system of higher animals, to maintain relatively stable internal conditions as a result of interacting physiological processes.

homoclinal ridge An erosional feature that develops in areas of moderately dipping strata; it is characterized by a significant difference in the length of its front and back slopes as well as in their steepness. Because a homoclinal ridge is less sharply defined than a *hogback*, and smaller in area than a cuesta, it is used as a term to describe those forms intermediate between these two. See *hogback.*

homoclinal shifting *Monoclinal shifting.*

homocline A structural state in which rock strata exhibit uniform dip in one direction; e.g., one limb of a fold. Cf: *monocline.*

homogeneous equilibrium Equilibrium that exists in a single-phase (typically liquid or gaseous) system. Cf: *heterogeneous equilibrium.*

homology The similarity but not complete identity between parts of different organisms, as a result of evolutionary differentiation of the same or corresponding parts derived from the same ancestor. adj. *Homologous*.

homoplasy The similarity or correspondence of organs or parts of different organisms that developed as a result of parallelism or convergence, rather than from a common ancestor. adj. *Homoplastic*. Cf: *homology*.

homopycnal inflow Flowing water that has the same density as that of the body it enters, resulting in easy mixing of the two systems. Cf: *hyperpycnal flow; hypopycnal flow*.

homotaxy Taxonomic similarity between stratigraphic or fossil sequences in separate locations. Two rock stratigraphic limits in different areas, which have a similar order of arrangement but are not necessarily contemporaneous, are called *homotaxial*. Also, two formations that contain fossils occupying corresponding positions in different vertical sequences are homotaxial.

honeycomb coral A compound coral whose corallites are so arranged as to resemble the structure of a honeycomb.

hoodoo A pillar, column, rock pedestal, or toadstool rock, often of bizarre shape, produced by rainwash and differential mechanical and chemical weathering. The particular agent involved seems to be related to the type of area in which the formation is found. Hoodoos are often massive, but some are delicate features 5 to 10 cm in height. The name appears to be a variant of *voodoo*.

hook A *spit* or long, narrow cape that curves sharply landward at its outer terminus; e.g., Sandy Hook, extending from the New Jersey coast into New York harbor.

Hook's law The statement that the strain in an elastic material is linearly proportional to the applied stress.

hopper crystal A type of skeletal crystallization often shown by rock salt; the crystal has grown faster along its edges than at the center, so that the faces appear to be depressed.

horizon 1. An interface indicating a particular position in a stratigraphic sequence. Theoretically, it is a surface with no thickness; in practice, however, it is usually a distinctive bed. It is not a synonym of *zone*. 2. *Soil horizon*. 3. In surveying, a line or plane used as a reference, generally relating it to a horizontal surface.

horizontal displacement *Strike slip*.

horizontal fault A fault having no dip. Cf: *vertical separation*.

horizontal separation In faulting, the distance, measured in any specified horizontal direction, between the two sections of a disrupted unit. Cf: *vertical separation*.

horizontal slip In a fault, the horizontal factor of the net slip.

horn See *arête*.

hornblende A silicate mineral, the most common member of the amphibole group, $(Ca,Na)_{2-3}(Mg,Fe^{+2},Fe^{+3},Al)_5$-$(Al,Si)_8O_{22}(OH)_2$. Dark green or black in color, sometimes acicular or fibrous in parallel or sheaflike aggregates. An important constituent of metamorphic rocks. Also found in mafic and ultramafic igneous rocks.

hornblendite A plutonic rock composed chiefly of hornblende.

horn coral A solitary species of rugose coral.

hornfels A fine-grained metamorphic rock composed of mineral grains of the same size. Porphyroblasts or relict phenocrysts may be present in the typically granoblastic fabric; relict bedding may also be found. Hornfels is formed by thermal metamorphism and solid-state crystal growth within a contact aureole surrounding an igneous intrusion.

hornfels facies An imprecisely defined term used to denote the physical conditions involved in, or the series of mineral assemblages produced by, thermal metamorphism at comparatively shallow depths in the earth's crust.

hornito Etymol: Span., "little oven." A small mound of spatter or driblet on the back of a lava flow (generally *pahoehoe*), formed by the gradual accumulation of clots of lava forced through an opening in the roof of an underlying lava tube. A *driblet spire* is a type of hornito resembling a thin column or spire. Hornito and *driblet cone* are synonymous terms. Cf: *spatter cone*. See *lava*.

horst Etymol: Ger., *horst*, one meaning of which is "top of a rock." An elongate block that has been raised relative to its surroundings. Its long sides are bounded by steeply inclined faults that generally dip away from each other. Lateral tension or stretching induced by vertical movements is believed to be the cause in the formation of most horsts. Cf: *graben*.

host A rock or mineral that is older than rocks or minerals introduced into it; a large crystal enclosing smaller crystals of a different species. Ant: *guest*.

host rock See *country rock*.

hot dry rock A potential source of heat energy within earth's crust, represented by rocks at depths less than 10 km, and temperatures above 150° C. Their heat energy is derived from magmas and conduction from the deeper interior. Abbrev: HDR.

hot spot A heat source deep within earth's mantle, 100 to 200 km across, and persistent for millions of years. Hot spots are thought to be the surface manifestation of a rising plume of superheated material from the mantle. They are associated with the formation of ocean ridges (*hot-spot volcanoes*) such as the Hawaiian ridge, but have no connection with *island arcs*. See also: *volcano*.

hot spring See *thermal spring*.

Hudsonian orogeny A period of metamorphism, plutonism, and deformation in the Canadian Shield during the Precambrian.

humic acid Black acidic organic matter occurring in soils, low-rank coals, and decayed plant substances.

hummock A term variously used to mean a knoll or mound above the general level of a surface, or a knob of turf or earth in alpine regions. Another meaning is that of a mound of broken, floating ice pushed upward by pressure.

hummocky moraine See *knob-and-kettle-topography*.

humus A dark organic material in soils, produced by the decomposition of animal or vegetable matter, and essential to the fertility of the earth.

Huygen's principle The statement that any particle excited by wave energy becomes a new point source of secondary waves.

hyaline Having glasslike transparency.

hyalite A colorless variety of opal whose appearance varies from glasslike transparency to translucent or whitish.

hyalocrystalline A textural term used for porphyritic rocks in which the volume occupied by phenocrysts is equal or nearly equal to the volume of groundmass. Cf: *intersertal*.

hyalopilitic A textural term for an igneous rock consisting of acicular microlites in a glassy groundmass. *Pilotaxitic* is the same texture without glass. When the amount of glass is decreased, the texture becomes *intersertal*; such texture also includes rocks without glassy interstitial material.

hybrid 1. A rock mass whose chemical composition is the result of the incorporation of matter from wall rocks (assimilation), or from inclusions physically ingested into the magma (contamination). 2. An individual whose parents are of different species.

hydrargillite *Gibbsite.*

hydrate n. A chemical compound produced by hydration (chemical combination of water with another substance), or one in which water is a constituent of the chemical composition. v. To effect the incorporation of water into the chemical composition of a compound.

hydration shattering A process of rock disintegration caused by the wedging or prying pressure of films of water on silicate mineral surfaces. Water drawn between the grains of rocks by a type of osmosis exerts a differential pressure of sufficient strength to dislodge and separate the grains. Such a process occurs without the aid or intercession of freezing or thawing.

hydraulic action The dislodging and removal of weakly resistant material by the action of flowing water alone.

hydraulic fracturing A process of injecting sand-charged liquids under such high pressure as to fracture a rock and inject the newly made crevices with coarse sand to keep them open.

hydraulic gradient The ratio between the difference of elevation and the horizontal distance between two points. The rate and direction of water movement in an aquifer are determined

by the permeability and the hydraulic gradient. Cf: *pressure gradient.*

hydraulic head The height of the exposed surface of a body of water above a specified subsurface point. Also, the difference in water level at a point upstream from a given point downstream.

hydraulic mining A method of mining in which high-pressure jets of water are used to remove earth materials containing valuable minerals.

hydraulics That branch of engineering science involved with the laws governing fluids in motion and their applications to engineering technology.

hydrocarbon Any of a class of organic compounds consisting of carbon and hydrogen only.

hydroclastic rock Any rock that has been broken by the action of water or deposited by it.

hydrofracturing *Hydraulic fracturing.*

hydrogen-ion concentration *pH.*

hydrogen sulfide A toxic gas, H_2S, having the odor of rotten eggs; it is emitted as a natural product of the decomposition of organic matter.

hydrogeochemistry The chemistry of surface and groundwaters, esp. as it refers to the relationship between the chemical features and quality of the water and the geology of an area. Cf: *biogeochemistry; lithogeochemistry.*

hydrogeology See *geohydrology.*

hydrograph A graph showing time-related variation in flow. The variation may be shown in terms of yearly, monthly, daily, or instantaneous change.

hydrolith A rock that is chemically precipitated from water in which it is dissolved; e.g., rock salt or gypsum.

hydrologic budget   An accounting of the inflow into, outflow from, and storage in a hydrologic unit such as an aquifer, soil zone, or lake.

hydrologic cycle   The constant interchange of water between the oceans, atmosphere, and land areas of the earth via transpiration and evaporation.

hydrology   The science concerned with the occurrence, distribution, movement, and properties of all waters of the earth and its atmosphere. It includes surface water and rainfall, underground water and the water cycle.

hydrolysis   In geology, a term generally implying a reaction between silicate minerals and water, either alone or in aqueous solution.

hydrolyzates   Sediments typified by elements that are easily hydrolyzed and which concentrate in fine-grained alteration forms of primary rocks. Thus, they are abundant in clays, shales, and bauxites. The elements found in hydrolyzates are aluminum and associated potassium, silicon, and sodium. Cf: *oxidates; resistates; evaporites; reduzates.*

hydrometer   A device for determining the specific gravity of a liquid.

hydrophone   A pressure-sensitive detector that responds to sound transmitted through water.

hydrosphere   The waters of the earth, as distinguished from the atmosphere, biosphere, and lithosphere.

hydrostatic level   The height to which water will rise in a well under a full-pressure head. It marks the potentiometric surface.

hydrostatic pressure   The pressure exerted by a homogeneous liquid at rest. This pressure exists at every point within the liquid and is proportional to the depth below the surface. At any given point in the liquid, the magnitude of the pressure force exerted on the surface is the same, regardless of the surface orientation, and is everywhere perpendicular to the surfaces of the container. Cf: *confining pressure; geostatic pressure.*

hydrothermal   Of or pertaining to heated water, its actions, or to products related to its actions.

hydrothermal alteration   Alteration of rocks by the interaction of hydrothermal water with preformed solid phases.

hydrothermal deposit   A mineral deposit formed from hydrothermal solutions at a range of temperatures and pressures. The minerals, which derive from fluids of diverse origin, are deposited in faults, fractures, or other openings. *Epithermal deposits* are formed within about 1 km of the earth's surface in the range of 50° to 200° C. These deposits are typically found in volcanic rocks; the chief metals are gold, silver, and mercury. Mesothermal deposits are formed at considerable depth in the temperature range of 200° to 300° C. Among the principal ore minerals in these deposits are pyrite, chalcopyrite, and galena. *Hypothermal deposits* are formed at great depth, in the temperature range of 300° to 500° C. Ore minerals of greatest importance in these deposits are cassiterite, wolframite, and molybdenite.

hydrothermal stage   That stage in the cooling of a magma during which the residual fluid contains abundant amounts of water and other volatiles. The limits of the stage are variously defined in terms of phase assemblage, temperature, and pressure.

hydroxide   An oxide composed of a metallic element and the OH ion.

hydrozincite   A minor ore of zinc, $Zn_5(CO_3)_2(OH)_6$, hydrozincite is found in the oxidized zones of zinc deposits as an alteration product of sphalerite. Syn: *zinc bloom; calamine.*

hydrozoan   Any coelenterate of the class Hydrozoa, including the hydra and the Portuguese man-of-war. There are marine and freshwater types, which may be solitary or colonial. Range, Lower Cambrian to Recent.

hygrometer   An instrument for measuring the water vapor content of the atmosphere.

hygroscopic   Having the property of absorbing moisture from the air.

hygroscopic water   Moisture confined in the soil and in equilibrium with that in the regional atmosphere to which the soil is exposed.

hyperfusible   Said of a substance capable of reducing the melting ranges in end-stage magmatic fluids.

hyperpycnal inflow   An inflow of water that is denser than the body it enters, resulting in the development of a turbidity current. Cf: *hypopycnal inflow; homopycnal inflow.*

hypersthene   A rock-forming mineral of the orthopyroxene group, $(Mg,Fe)SiO_3$. Enstatite is the pure magnesium end-member. Hypersthene is an essential constituent of many igneous rocks.

hypervelocity impact   The impact of an object on a surface at a velocity such that the magnitude of the shock waves created on contact greatly exceeds the static compressive strength of the target substance. In such an impact, the kinetic energy of the object is transferred to the target material in the form of profound shock waves that produce a crater much greater in diameter than the impacting object. The required minimum velocities needed for this effect vary for different materials.

hypidioblast   See *crystalloblast.*

hypidiomorphic   *Subautomorphic.*

hypocenter   *Seismic focus.*

hypocrystalline   A term describing the texture of an igneous rock having crystalline components in a glassy groundmass, the crystal-to-glass ratio being between 7:1 and 5:3. Syn: *merocrystalline.* Cf: *hypohyaline.*

hypogene   1. Formed by ascending fluids within the earth, as ore or mineral deposits. Cf: *supergene.* 2. Formed beneath the earth's surface; e.g., granite. Cf: *epigene.* 3. A synonym for primary deposits.

hypohyaline   A texture term for an igneous rock that has crystalline components in a glassy groundmass, in which the crystal-to-glass ratio is between 3:5 and 1:7.

hypopycnal flow   An inflow of water whose density is less than that of the body of water it enters. Cf: *hyperpycnal inflow; homopycnal inflow.*

hypothermal deposit   See *hydrothermal* deposit.

hypsographic   Pertaining to the study that deals with the measurement and mapping of earth's topography relative to sea level. A *hypsographic curve* is used to describe and calculate the distribution of prominences of land surface and sea floor. The elevations are usually referred to a sea-level datum.

hysteresis   1. In the elastic deformation of a body, hysteresis is a delay in the return of the body to its original shape. 2. The property that a substance is said to show when its magnetization is not reversible along the original magnetization curve.

I

ice   Water in the solid state, specif. the substance formed by the freezing of liquid water, thereby reducing the density.

ice age   A period of time during which glacial ice spreads over areas that are not generally ice covered. The Antarctic Ice Sheet and other extensive glaciers existed in high latitudes during the last 15 m.y. However, the Pleistocene was distinguished by the formation of large glacial masses in middle latitudes. The largest of these masses was the Laurentide Ice Sheet, whose area exceeded that of the existing Antarctic Ice Sheet. For this reason the Pleistocene is known as the *Great Ice Age* (cf: *Little Ice Age*); it spanned a period of 650,000 years. The most recent major glaciation began its retreat 18,000 y.b.p. In North America, it had left the U.S. about 11,500 y.b.p., but did not melt from Continental Canada until about 6000 y.b.p. See also: *glaciers and glaciation; interglacial period.*

iceberg   A large floating mass of ice detached from a glacier or ice shelf and carried seaward. When the terminus of a glacier moves into the sea, its lesser density subjects it to an upward force, which increases as the glacier reaches farther into the sea, and finally causes the forward end of the glacier to break off and form icebergs, a process called *calving*. Calving is less frequent in the extreme northern and southern hemispheres. The degree of submergence of an iceberg depends greatly on the iceberg's shape as well as on the density of the water and on the air content of the iceberg. The ratio of above-surface height to submerged height for a rectangular berg is 1:7, and for a rounded berg 1:4; however, the ratio of the total ice mass above water to the submerged mass is 1:9 for all icebergs.

ice cap   See *glacier.*

ice contact deposit   Stratified *drift*

that is deposited in contact with melting glacier ice; e.g., esker, kame, kame terrace.

ice crack   *Frost crack.*

ice field   See *glacier.*

ice island   A large tabular iceberg separated from an ice shelf and found in the Arctic Ocean. The surface commonly has a ribbed appearance when viewed from the air.

Icelandic-type eruption   See *fissure eruption; flood basalt.*

Iceland spar   A chemically pure and transparent variety of calcite perhaps the finest of which occurs in Iceland. It cleaves readily, and exhibits clear double refraction. This variety of calcite is found in vugs and cavities in volcanic rocks.

ice pack   *Pack ice.*

ice pole   The approximate center of the most consolidated portion of the Arctic pack ice and, therefore, not readily accessible by surface. Syn: *pole of inaccessibility.*

ice rafting   Transport of rock fragments and other materials by floating ice.

ice sheet   See *glacier.*

ice shelf   A sheet of very thick ice, most of which is afloat but is attached to land along one side. Ice shelves are fed by snow accumulations and the seaward projection of land glaciers. Cf: *shelf ice.*

ice wedge   Wedge-shaped, foliated ground ice formed in permafrost. It occurs as a vertical or inclined sheet, dike, or vein, and originates by the growth of hoar frost or by freezing of water in a narrow crevice.

ice-wedge polygon   See *nonsorted polygon.*

ichnofossil *Trace fossil.*

ichor A fluid thought by some to be responsible for granitization. The term initially bore the connotation of derivation from magma. Syn: *residual liquid.*

ichthyosaur An extinct, fishlike marine reptile. It bore a resemblance to modern dolphins, but had a longer beak. Range, Middle Triassic to Upper Cretaceous.

ideal cyclothem See *cyclothem.*

idioblast See *crystalloblast.*

idiogeosyncline A late-cycle geosyncline between mobile and stable areas of the earth's crust; its sediments are only slightly folded.

idiomorphic *Automorphic.*

idiotopic A term describing the fabric of a crystalline sedimentary rock in which most of the constituent crystals are euhedral.

idocrase *Vesuvianite.*

igneous breccia A breccia composed of fragments of igneous rock, or a breccia produced by igneous processes; e.g., volcanic breccia.

igneous rock One of the main groups of rocks that comprise the earth's crust. They constitute about 15% of the earth's surface, and are formed of molten material (*magma*) that flows up from the deeper part of the crust. When this material, which is composed largely of silicate minerals, solidifies beneath the surface, the resultant rock formations are referred to as *intrusive* or *plutonic*. Material that solidifies at or above the surface is called *extrusive* or *volcanic*. The classification of igneous rock is generally based on the following criteria:
  1. Texture. (a) *Pegmatitic*: an uneven, coarse texture. The mineral grains are usually of different sizes, but most are larger than 0.7 cm. (b) Phaneritic: the mineral grains are approximately the same size and large enough to be seen with the naked eye. (c) *Aphanitic*: a texture that is dense in appearance. The mineral grains are about the same size, but too small to be seen without a microscope. (d) *Vitric* or *hyaline*: no mineral grains are present; the rock is glassy in appearance.
  2. Color Index—or more precisely, lightness and darkness. Color is largely determined by the amounts of iron and magnesium present. See *mafic* and *felsic*.
  3. Mineral Content. The quantity of *feldspar* present and its type are among the determinants of the mineral character. See also: *sedimentary rock; metamorphic rock.*

igneous rock series A group of igneous rocks that are related by having been derived from the same general type of event (plutonic, volcanic), having certain textural or mineralogical features in common, and showing a continuous sequential variation from one end of the series to the other.

ignimbrite A term that includes *welded tuff* and nonwelded but recrystallized *ash flows*. It refers to rock that is formed by the consolidation of *pyroclastic material* deposited by *nuées ardentes* and ash flows. See also: *tufflava.*

Illinoisian Pertaining to the classical third glacial stage of the Pleistocene Epoch in North America. Also spelled: *Illinoian.*

illite A group of mica-like clay minerals of three-sheeted structure, widely distributed in argillaceous shales. They are intermediate in structure and composition between muscovite and montmorillonite. In illite, Mg, and Fe partly replace the octahedral Al of muscovite; a K layer is present in illite.

illuviation The accumulation in the subsoil of soluble or suspended material transported from an upper soil horizon by the process of *eluviation*. Iron and aluminum minerals, including clays, accumulate here. A soil horizon to which material has been so added is called an *illuvial horizon*.

ilmenite   A black or dark brown rhombohedral mineral, $FeTiO_3$, a major ore of titanium; it occurs in compact or granular aggregates. Ilmenite is a common accessory in plutonic rocks as a high-temperature segregation product. It also occurs in pegmatites and in metamorphic rocks such as gneiss and chlorite schist.

imagining   In the area of scientific intelligence, imagining is a technique that simulates a sequence of events and matches the end result of the simulation against the goal state. If the match succeeds, then the sequence is a valid explanation for the incidence of the goal. A combination of qualitative and quantitive simulations are used in imagining. Qualitative simulation is used in extending the sequence of events with numeric values for parameters, so that a quantitative simulation can be performed. The product of the quantitative simulation is compared with the quantitative goal state. In the case of geological interpretation, the quantitative simulation assembles diagrams that reflect the spatial effects of a sequence of geological events.

imbibition   The tendency of granular rock or any porous substance to absorb a fluid completely under the force of capillary attraction, without the exertion of any pressure on the fluid.

imbricate structure   1. A sedimentary structure in which gravel, pebbles, or grains are stacked with their flat surfaces at an angle to the major bedding plane. It is most obvious in gravels (gravel dips upstream) and conglomerates, but is also present in sands. Syn: *shingle structure; imbrication*. 2. A tectonic structure in which a series of lesser thrust faults that overlap, and are nearly parallel, are all oriented in the same direction, which is toward the source of stress.

immature   1. A landscape feature, such as a valley or drainage basin, that is still well above base level is said to be immature. 2. A term referring to either the texture or mineral composition of a coarse detrital rock, esp. sandstone; a sediment with at least 5% clay is said to have *textural immaturity*; an abundance of feldspar and lithic fragments denotes mineralogical maturity.

immature soil   *Azonal soil*.

immiscible   Said of two or more liquids that do not mix and which, when brought into contact, form more than one phase; e.g., oil and water.

impact crater   A depression formed by the impact of an unspecified body; it applies esp. to a crater on the earth or moon surface for which the identity of the impacting object is unknown.

impactite   A glassy to finely crystalline substance created by fusion of the target material as a result of heat generated by the impact shock of a large meteorite; impactite occurs in and around the crater. Syn: *impact slag*.

impermeable   A rock, soil, or sediment that cannot transmit fluid under pressure is said to be *impermeable*. Syn: impervious. Ant: *permeable*.

impoundment   The process of forming a lake or pond by some sort of barrier, such as a dam or dike. The body of water formed in this manner is also called an impoundment.

impregnated   A term applied to a mineral deposit in which minerals are epigenetic and diffused through the host rock. Cf: *interstitial*.

Inceptisol   In the U.S. Department of Agriculture soil taxonomy, a soil order typified by one or more horizons in which mineral substances other than amorphous silica or carbonates have been altered or removed, but have not accumulated to any great degree.

incised meander In a broad sense, an incised meander is a meander that has become deepened by renewed downcutting as a result of environmental change. It is usually bordered on one or both banks by vertical walls. Two types of incised meander are generally recognized: the *entrenched* meander, which shows little or no difference between the slopes of the two valley sides of the meander curve; and the *ingrown* meander, which exhibits obvious asymmetry in cross-profile, with undercut banks on the outside curve and prominent spurs on the inside.

inclination 1. Departure from the horizontal or vertical; also, the slope. 2. May be used as a syn. of *dip* in structural geology. 3. The angle of a well bore taken from the vertical at a given depth. 4. *Magnetic inclination.*

inclinometer 1. An instrument used for measuring the angle an object makes with the vertical; e.g., in engineering it would be used to measure inclination in a well bore. 2. An instrument that measures magnetic inclination.

included gas Gas that occurs in isolated, intervening spaces in either the *zone of saturation* or *zone of aeration.* The included gas may also be bubbles of air or some other gas that are enveloped in water in either zone. These impede the water flow unless the gas dissolves in the water.

inclusion An enclave of rock within an igneous rock that differs in fabric and/or composition from the igneous rock. Some inclusions are blocks of wall rock that have been incorporated into the magma body by *stoping*. Some are aggregates of early-formed crystals precipitated from the magma itself. Other inclusions might represent clots of heat-resistant residuum, left over from partial

melting processes deep within the crust where the magma originated. See also: *xenolith; autolith. 2. Fluid inclusion.*

incompetent See *competent.*

incompressibility modulus *Bulk modulus.*

incongruent melting Melting that involves dissociation or reaction with the liquid so that one crystalline phase is converted to another plus a liquid whose composition is different. different liquid. Most pure minerals with a fixed composition (albite, diopside, quartz) melt *congruently* to yield a liquid of precisely the same composition. However, some minerals melt incongruously; at its melting point, enstatite melts to a yield slightly more siliceous liquid plus forsterite crystals.

incongruent solution Dissolution that yields dissolved material in proportions different from those in the original fluid.

index fossil A fossil that identifies and dates those strata in which it is found. This especially applies to any fossil taxon that is morphologically distinct, fairly common in occurrence, broadly distributed (perhaps even worldwide), and confined to a narrow stratigraphic range. Graptolites and ammonites are two of the best index fossils. Cf: *characteristic fossil; guide fossil.*

index horizon A structural surface selected as a reference in studying the geological structure of an area. Syn: *index plane.*

index mineral A mineral that occurs throughout rocks of similar bulk composition and is exposed over a distinct area of the metamorphic terrane. Such a mineral develops under a specific set of temperature and pressure conditions, thus identifying a particular metamorphic zone.

index of refraction In crystal optics, a value expressing the ratio of the velocity of light in a vacuum or in air to the velocity of light within the crystal.

index plane *Index horizon.*

index zone A body of strata, discernible by its lithologic or paleontologic character, that can be followed laterally and which identifies a reference location in a stratigraphic section.

indicator 1. Some feature that suggests the presence of an ore or mineral deposit; e.g., a magnetic anomaly. 2. A plant or animal found in a specific environment, and which can therefore be used to identify such an environment. 3. An *indicator boulder.*

indicator boulder A glacial erratic of known origin used for determining the source area and transport distance for any given till complex. Its critical feature is a distinctive appearance, unique mineral assemblage, or characteristic fossil pattern. Indicator boulders are sometimes arranged in a *boulder train*, a line or series of rocks derived from the same bedrock source and extending in the direction of movement of the glacier. A *boulder fan* is a fan-shaped area containing distinctive erratics derived from an outcrop at the apex of the fan. The angle at which the margins diverge is a gauge of the maximum change in the direction of glacial motion.

indicator fan See *indicator boulder.*

induction log A continuous record of the conductivity of strata penetrated by a borehole, versus depth. See *well logging.*

induration 1. Hardening of rock or rock material by heat, pressure, or the introduction of some pore-filling material (cementation). See also: *lithification.* 2. Hardening of a soil horizon by chemical action to form a *hardpan.*

industrial diamond A general term for a diamond, synthetic or natural, that is used in work such as wire-drawing or drilling, and as a general abrasive for lapping and polishing. Any diamond that is too badly flawed to have value as a gem may have industrial use. Three natural varieties of such diamonds exist: ballas, bort, and carbonado. *Ballas* is composed of spherical masses of minute, concentrically arranged diamond crystals with poor cleavage. *Bort* is a gray to black massive diamond, the color of which is due to inclusions and impurities. The name is also used for highly flawed or badly colored diamonds unsuited for gem purposes. *Carbonado* is a black, opaque diamond having a slightly porous structure and no cleavage.

infiltration The movement of a fluid into a solid substance through pores or cracks; in particular, the movement of water into soil or porous rock. Cf: *percolation.* The *infiltration capacity* is the maximum rate at which soil under given conditions can absorb falling rain or melting snow.

influent adj. Flowing in. n. A stream that flows into a pond or lake (e.g., an inlet), or a stream that flows into a larger stream (e.g., tributary). Ant: *effluent.* Cf: *influent stream.*

influent stream A stream that contributes water to the zone of saturation; its channel lies above the water table. Syn: *losing stream.*

informal unit A body of rock or other feature that is referred to in casually descriptive but nondefinitive terms; e.g., "sandy beds," "mucky marsh." Cf: *formal unit.*

infraglacial *Subglacial.*

infrared The invisible part of the electromagnetic spectrum that comprises radiations of wavelengths from 0.8 to 1000 microns. Cf: *ultraviolet.*

infrastructure Structure produced in a plutonic environment at high pressure and temperature. It is characterized by plastic folding and the emplacement of granite and other magmatic rocks. Cf: *superstructure*.

ingrown meander See *incised meander*.

initial dip 1. *Original dip*. 2. The inclination of a bedded deposit due to compaction rather than tectonic deformation.

injection 1. *Intrusion*. 2. The forced insertion of sedimentary material into a crack, fissure, or crevice in an existing rock.

injection gneiss A composite rock formed when a deep-seated pluton intimately injects its wall and becomes interfingered with slabs of the host rock; such slabs or tongues are called *injection gneiss*. Although still in debate, there seems to be a somewhat general opinion that injection gneisses are formed by the forceful injection of magma along surfaces of schistocity, thus splitting the schist. Cf: *arterite*.

injection metamorphism Metamorphism characterized by the intimate injection of sheets and plumes of liquid magma in zones near plutonic rocks.

inland sea *Epicontinental sea*.

inlet A small, narrow "groove" in a shoreline, through which water penetrates into the land. 2. An inflowing stream, as into a lake. 3. Tidal inlet.

inlier A circular, elliptical, or irregular area of rock surrounded by younger strata. Inliers are found in a normal stratigraphic sequence wherever erosion has broken through the younger strata.

inorganic In chemistry, noting or pertaining to compounds that are not hydrocarbons. Cf: *organic*.

inosilicate Any silicate with a structure consisting of paired parallel chains of tetrahedral silicate groups, every other one of which shares an oxygen atom with a group of the other chain. The ratio of silicon to oxygen in such silicates is four to eleven (Si_4O_{11}). See *silicates*.

in place Said of a rock that occupies that position it had when formed, relative to surrounding masses; it was neither displaced nor separated from its source. Cf: *in situ*.

inselberg Etymol: Ger., "island mountain." Inselbergs are isolated residual uplands standing above the general level of the surrounding plains in tropical regions; they may be ridges, domes, or hills. Like *monadnocks*, they are remnants of the erosion cycle, but even though there may occasionally be great morphological similarity between the two, they are not the tropical equivalents of monadnocks, which are features of temperate zones. Typical inselbergs rise more abruptly from the plains than do typical monadnocks. The cause of this sharp transition appears to be anomalous weathering patterns related to the rock structure of a particular remnant and to the topography of the area. The sugarloafs of Brazil are renowned inselbergs. See also: *bornhardt*.

insequent stream A stream whose course cannot be ascribed to control or adjustment by any apparent surface slope, weakness in the earth's surface, or rock type. Because the stream course is random, its basin develops a dendritic *drainage pattern*. Syn: *insequent*.

in situ Said specif. of a rock, soil, or fossil when it is in the situation or position in which it was originally deposited or formed. Cf: *in place*.

in situ theory The theory that coal originates at the same location where the plants that gave rise to it grew, died and decayed. Ant: *drift theory*.

insolation In geology, insolation is the

geological effect of solar rays on earth's surficial materials; esp. the effect of temperature changes on the mechanical weathering of rocks.

intake Recharge.

interbedded Said of strata that are positioned between, or alternated with, other layers of dissimilar character; e.g., a lava flow interbedded with contemporaneous sediments. Syn: *interstratified*. Cf: *intercalated*.

intercalated A term applied to layered material that exists or is introduced between layers of a different type. It applies especially to layers of one kind of material that alternate with thicker strata of another material; e.g., beds of shell intercalated in sandstone.

intercept time In a seismic refraction survey, the intercept time expresses the time-distance relationship between two media separated by a horizontal discontinuity at a certain depth. In the equation

$$T_i = 2z \frac{\sqrt{V_1^2 - V_0^2}}{V_1 V_0}$$

T_i is the intercept time, V_0 and V_1 are the respective velocities of the two media, and z is the depth of the horizontal discontinuity.

interface 1. The boundary separating the top of the uppermost layer of a sediment from the water in which the sedimentation process is taking place. 2. The contact between fluids in a reservoir. 3. A seismic *discontinuity*.

interference ripple mark See *ripple mark*.

interfluve The comparatively undissected upland between adjacent streams flowing in the same direction.

interglacial adj. Pertaining to the time period between glaciations (ice ages). n. The warmer period of time that precedes and follows an ice age.

intergranular Said of the fabric or texture of an igneous rock characterized by an interlocking network of randomly oriented mineral grains. Such texture results from the formation of abundant nuclei of all crystalline phases during undercooling of the melt, combined with adequate crystal rate growths to consume all of the melt. This texture differs from *intersertal* texture by the absence of interstitial glass. Cf: *ophitic*.

intergranular movement A movement that involves displacements between individual mineral grains. If rocks that are composed of both mineral grains and crystals are subjected to stress, the individual crystals and grains may move independently, and individual grains may maintain their shape and size. Intergranular movement is a factor in glacier flow; it takes place within a glacier (only near the surface) when grains of ice rotate relative to each other and slide over each other. Cf: *intragranular movement*.

intergrowth The interlocking of grains of two different minerals because of their simultaneous crystallization.

interior basin 1. A depression from which no stream flows outward; an undrained basin. Cf: *closed basin*. 2. Intracratonic basin.

intermediate Referring to an igneous rock that is between *basic* and *silicic* in composition (or between *mafic* and *felsic*); its silica content is generally between 52 and 60%. See *silica concentration*.

intermediate-focus earthquake An earthquake whose focus is located between depths of about 60 and 300 km. Cf: *shallow-focus earthquake; deep-focus earthquake*.

intermittent stream 1. A watercourse that flows only at certain times of the year, such as when it receives water from springs. 2. A stream that may lose its water by evaporation, or absorption if its channel flows on highly porous rock. Cf: *ephemeral stream.*

intermontane Lying or located between mountains. Syn: *intermountain.*

internal waves Subsurface waves that occur between layers of different density or within layers where there are vertical density gradients. They can be present in any stratified fluid, and can be generated by atmospheric disturbances, shear flow, or tidal forces, and flow over an irregular bottom. The wave amplitude varies with depth, and is influenced by the density distribution of the fluid.

interpretive map As used in environmental geology, a map prepared for public use that classifies land according to its suitability for a particular function on the basis of its geological characteristics.

interrupted water table A water table that inclines sharply over a groundwater barrier; there is a pronounced difference in elevation above and below it.

intersertal A texture of igneous rock in which angular interstices within an interlocking network of grains are occupied by glass. This texture forms essentially in the manner of *intergranular texture,* but more rapid undercooling prevents complete crystallization of the melt.

interstade A warmer substage of a glacial stage, during which there is a temporary recession of the ice.

interstitial Said of a mineral deposit in which minerals occupy the pores of the host rock.

interstratified *Interbedded.*

interval velocity The distance across a specified stratigraphic thickness, divided by the time it takes for a seismic wave to traverse the distance. *Well shooting* is the most accurate procedure for velocity measurement. The interval velocity is the distance between successive detector locations in the well, divided by the difference in arrival times of the wave at two designated depths after a correction has been made for the fact that the actual wave is slanting rather than vertical. The *average velocity* is the total vertical distance divided by the total time.

intraclast A component of a limestone consisting of pieces of penecontemporaneous lithified carbonate rock from within the basin of deposition.

intracratonic Located within a stable continental region.

intracratonic basin A depression situated on top of a craton. Syn: *interior basin.*

intraformational Formed within the boundaries of a geologic formation, more or less contemporaneously with the sediments that contain it.

intrageosyncline *Parageosyncline.*

intragranular movement A movement in which displacement occurs within the individual crystals of a body by *gliding* and *dislocation.* Ice crystals that are oriented in a particular fashion are deformed by slippage along layers, without rupturing the crystal lattice. Cf: *intergranular movement.*

intratelluric 1. Of a phenocryst that grew slowly beneath the surface, prior to a second stage of rapid crystallization of the magma. 2. Said of something that is located, has formed, or has originated deep within the earth.

intrazonal soil A soil that more greatly reflects the effects of local conditions than of broad climatic patterns.

intrusion 1. The forceful injection of magma or a plastic solid into pre-existing rock. Intrusion can take place either by deformation of the involved rock (*batholith, dike, laccolith, sill*), or along some structural channels such as *bedding planes*, cleavages, or joints. Syn: *injection*. 2. Igneous rock formed in this manner. The term should not be applied to igneous-looking rock bodies that are the product of *metasomatic replacement*. 3. Injection of sedimentary material under abnormal pressure; e.g., emplacement of a *diapiric* salt plug. 4. Salt-water encroachment.

intrusive rock Igneous rock formed of magma that consolidates beneath the earth's surface. The texture of the intrusive rock depends partly upon the depth at which it has cooled. Rocks at greater depths cool more slowly, allowing the growth of large crystals, which results in a coarse texture characterized by clearly visible minerals; e.g., granite. Cooled at shallow depths, the rock tends to be more finely crystalline, and even aphanitic; e.g., syenite. See also: *intrusion*.

invariant equilibrium A phase assemblage for which there are zero degrees of freedom; none of the variables—temperature, pressure, or composition—can be changed without causing the loss of one or more phases.

invasion 1. An igneous *intrusion*. 2. *Transgression* of the sea over a land area.

inverse zoning In plagioclase, the change by which the outer parts of crystals become more calcic. Syn: *reversed zoning*. Cf: *normal zoning*.

inverted plunge A fairly common feature in excessively folded terranes, in which the inclination of the plunge of folds has been carried past the vertical so that the plunge is then less than 90° in a direction opposite its original attitude.

inverted relief A locality is said to have inverted relief if its geological structure and topography do not correlate; e.g., where the site of a valley later becomes a ridge.

ionosphere A region in the earth's atmosphere that extends from about 60 km to over 500 km altitude, and contains electrons and ions—produced by the ionizing action of solar radiation—in amounts sufficient to disturb the propagation of radio waves through it.

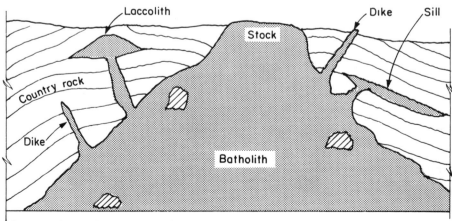

Cross section depicting types of igneous rock masses: sill, dike, stock, laccolith, batholith (not drawn to uniform scale)

iridescence The interference of light either at the surface of or within a mineral, thereby producing a series of colors as the angle of the incident light changes.

iron A heavy, chemically active mineral, the native metallic element, Fe. Native iron is rare in terrestrial rocks but fairly common in meteorites. Iron occurs in a wide range of ores in combination with other elements. The chief ores of iron consist mainly of the oxides of this element: hematite (Fe_2O_3); goethite (α -FeO(OH)); and magnetite (Fe_3O_4).

iron bacteria Bacteria that cause the deposition of iron. Iron-depositing bacteria are divided into three groups: 1. those that precipitate ferric hydroxide from solutions of ferrous carbonate, using the reaction energy and liberated carbon dioxide for their life processes; 2. those not requiring ferrous carbonate for vital processes, but able to cause the deposition of ferric hydroxide in the presence of iron salts; 3. those that break down iron salts of organic acids, using the organic acid radical as food and leaving ferric hydroxide as a deposit. Accumulations of iron developed through the agency of such bacteria are *bacteriogenic*. Cf: *sulfur bacteria*.

iron formation A chemical sedimentary rock consisting of alternating bands of quartz (including chert) and iron-bearing minerals, totaling at least 15%. Iron may be present as an oxide, silicate, carbonate, or sulfide. Most iron formation is of Precambrian age. Cf: *ironstone.* Many terms are essentially synonymous: *taconite; banded hematite; quartzite; banded iron formation; calico rock.*

iron hat *Gossan.*

iron meteorites (irons) One of the three major groups of meteorites. They consist mainly of iron, with up to 11% nickel. Because they are quite distinguishable from terrestrial rocks, they are found more frequently than other meteorite types. Irons are grouped according to their struc-ture, which can be revealed by the *Widmanstätten pattern*; there are three groups, grading into one another: *ataxites, hexahedrites,* and *octahedrites.*

iron pan A type of *hardpan* in which there is a considerable amount of iron oxide.

iron range A term used in the Great Lakes region of the U.S. and Canada for an area characterized by abundant iron formations. The term "range" implies a linear region rather than an elevation.

ironstone 1. A typically non-cherty, post-Precambrian sedimentary rock containing at least 15% iron. Ferriferous minerals found in ironstones include goethite, hematite, and pyrite. Ironstones are mostly unbanded, or coarsely banded and oölitic. Cf: *iron formation.* 2. *Clay ironstone.*

irrotational wave *P wave.*

isanomaly *Isoanomaly.*

island A body of land smaller than a continent and surrounded by water of an ocean, sea, lake, or stream. Islands may be formed in many ways, such as by submarine volcanic eruption, or where a mountainous coastal area submerges and peaks remain above water. Barrier beach islands are created off the shore of a mainland by waves and shore currents.

island arc A curved chain of volcanic islands, many of which are located along the circum-Pacific margins; e.g., the Aleutian Islands and the islands of Japan. Island arcs are commonly associated with deep-sea trenches (see *ocean trench*); they occur on the landward side of the trench and are generally curved, with their convex side facing oceanward.

iso- A prefix meaning "equal."

isoanomaly A line joining points of equal geophysical *anomaly*; e.g., magnetic anomaly, gravity anomaly.

isobar A line connecting points of equal pressure. An *isobaric surface* is one in which *every* point has the same pressure.

isobath A line on a chart or map that connects points of equal water depth. The term is also used for an imaginary line on a land surface, along which all points are at the same vertical distance above the upper or lower surface of the water table or an aquifer.

isochemical series An assemblage of rocks displaying the same bulk chemical composition throughout sequential, mineralogical, or textural changes.

isochore map See *isopach map*.

isochron In seismology, a line on a chart or map, passing through points at which the difference between the arrival times of seismic waves emanating from two reflecting surfaces is equal.

isochrone A line on a map or chart that joins all points at which an incident, event, or phenomenon occurs simultaneously; the points may also denote the same time value or time difference.

isoclinal fold Also called *isocline*. See *fold*.

isoclinic line A line on a chart connecting points of equal magnetic inclination.

isodynamic line *Isogam*.

isogal A contour line enclosing points of equal gravity value.

isogam A line joining points of equal field intensity. Syn: *isodynamic level*.

isogeotherm A line or surface within the earth, along which or over which temperatures are the same. Cf: *isotherm*.

isogonic line A line drawn through all points on the earth's surface having the same magnetic declination. Not to be confused with *magnetic meridian*. Cf: *agonic line*.

isograd A line on a map joining metamorphic zones of equal grade; i.e., the points at which metamorphism proceeded at similar values of pressure and temperature, as observed from rocks belonging to the same metamorphic facies. An isograd is delineated by the first appearance of an *index mineral*, or mineral assemblage, after which the next higher zone to it may be named. For example, the first appearance of sillimanite defines the lower-grade boundary of the sillimanite zone.

isogyre In crystal optics, a dark part of the interference figure resulting from extinction; it indicates the emergence of those light components having the same direction of vibration.

isohyetal line A line on a map connecting points that indicate locations receiving equal amounts of precipitation.

isolith An imaginary line connecting location points of similar lithology, and excluding rocks of dissimilar features, such as texture, color, or composition.

isometric system See *crystal system*.

isomorphism 1. (Crystallog.) The characteristic whereby two or more minerals closely similar in chemical composition crystallize in the same class. 2. Apparent likeness between individuals of different ancestry.

isopach map A map that shows the thickness of some unit, such as a bed or sill, throughout a geographic region by means of isopachs, which are contour lines representing equal thickness. An *isochore map* is based on the vertical distance between the top and bottom of particular strata. Equal drilled thicknesses of a specified subsurface unit are connected by *isochores*. The second type of map defines more accurately the present condition of the strata, rather than the thickness, which is mostly a function of the primary condition.

isopleth A general term for a line on a

chart or map that connects points of the same value; e.g., depth, elevation, population.

isopycnic adj. Of constant or equal density; this may refer to density in space (quantity or amount) or density in time (frequency).

isoseismal line A locus of points of equal earthquake intensity. If there were complete symmetry about the vertical through the earthquake focus, the isoseismal lines would be a family of circles with the earthquake epicenter as their center. But because of unsymmetrical factors affecting the intensity, the curves are often far from circular in shape. Indeed, an isoseismal curve of given intensity sometimes consists of more than one closed curve.

isostasy The condition of equilibrium whereby earth's crust is buoyantly supported by the plastic material in the mantle. See *Airy hypothesis; Pratt hypothesis.*

isostatic anomaly See *gravity anomaly.*

isostatic correction See *gravity anomaly.*

isotherm A line on a map or chart connecting points of equal temperature. Cf: *isogeotherm.*

isotope A particular atom of an element that has the same number of electrons and protons as the other atoms of the element, but a different number of neutrons; i.e., the atomic numbers are the same, but the atomic weights differ. Since the chemical properties of an element are determined by the atomic number, isotopes of an element have essentially the same chemical properties as the other atoms.

isotropic Said of a material whose properties are independent of direction. For example, an isotropic crystal will transmit light at the same velocity regard-less of its direction within the crystal. Ant: *anisotropic.*

isotropy The condition or state of having properties that are uniform in all directions.

isthmus A narrow strip of land that joins two larger bodies of land, and which is bordered on both sides by water.

itabirite A laminated iron oxide *iron formation* in which the original jasper or chert bands have been recrystallized into clearly discernable grains of quartz, and the iron is present in thin laminae of hematite or magnetite.

itacolumnite A flexible sandstone containing loosely interlocking grains of mica, chlorite, and talc.

IUGS classification An internationally adopted classification of plutonic rocks proposed in 1971 by the International Union of Geological Sciences. It is based on five modal proportions of minerals in five categories: quartz and other polymorphs of SiO_2; alkali feldspars, calcic plagioclase + scapolite; feldspathoids; all other phases.

J

jade 1. An extremely hard gemstone that may be either the pyroxene mineral *jadeite* or the amphibole mineral *nephrite*. Both forms are found in white and various shades of green, brown, yellow, and lavender. Jade of either mineral group is highly valued as a gemstone, especially among the Chinese, who for centuries have associated it with forces basic to life and fortune. 2. A term frequently applied to various hard green minerals.

jadeite A monoclinic mineral of the clinopyroxene group, $Na(Al,Fe^{+3})Si_2O_6$. It is found in compact waxy masses as metasomatic concentrations in serpentinized ultramafic rocks. Jadeite is one of the minerals referred to as *jade*. Burma is its principal source. *See also: jade; nephrite.*

jasper A compact, microcrystalline variety of quartz. Its colors are variable, including white, gray, red, brown, or black. Red varieties usually contain hematite.

jaspilite A rock consisting mainly of red jasper and iron oxides in alternating bands.

jet A dense black *lignite* formed from drift wood submerged in sea-floor mud. It takes a high polish and was once popular as a gemstone. Because it is a variety of coal, jet will burn.

jetty A pier or structure of stones, piles, or the like, projecting into the ocean or another body of water to protect a harbor or a pier, or to deflect currents.

Johannsen's classification A quantitative classification of igneous rocks formulated (1939) by the petrographer Albert Johannsen.

JOIDES Joint Oceanographic Investigation for Deep Earth Sampling; a program to obtain cores of sediments from the deep-ocean bottom.

joint A surface fracture (vertical or horizontal) in a rock without displacement. A *joint set* consists of a group of more or less parallel joints. A *joint system* consists of two or more joint sets with a characteristic pattern.

jökulhlaup *See glacier burst.*

jökull Etymol: Icelandic, "glacier." In glaciology, it is another term for ice cap, a continuous sheet of ice smaller than a continental glacier.

Joplin-type lead *See anomalous lead.*

J-type lead *See anomalous lead.*

jug A colloquial syn. of *geophone.*

Jurassic The period of geological time extending from 195 to 135 million y.b.p., occupying the middle portion of the Mesozoic Era between the Triassic and Cretaceous periods. Its name refers to the chalk sequence in the rock strata of the Swiss-French Jura Mountains. It is separated into lower (Lias), Middle (Dogger), and Upper (Malm) portions, which are further subdivided into a number of local divisions. Jurassic strata occur in all continents and reach great thicknesses in some regions. Marine Jurassic sediments are generally restricted to border regions of the present continents; only in North America and Asia did the oceans extend into the interior. During the Lower Jurassic, however, North America was largely emergent, while the seas advanced over large areas of the world. Sandstones, conglomerates, shales, and limestones were deposited during Jurassic time; clays and marly limestones were deposited in many geosynclinal areas. Among the economically important Jurassic deposits are *öolitic* and *sideritic* iron ores of marine deposits in central and western Europe, alluvial deposits of uranium ores and gold in United States, oil-producing rock strata along the North American Gulf Coast, coal in Siberia, and fuller's earth near Bath, England. The total absence of Jurassic sediments in places such as the east coast of North and South America suggests that the Atlantic Ocean had not yet fully formed. The Jurassic sequence of Madagascar in eastern Africa is thought to result from deposits in an ocean strait that was beginning to divide the super continent, *Gondwanaland.*

Among the invertebrates, dinoflagellates appear in the Middle Jurassic. Siliceous sponges reach their greatest abundance, and coelenterates, echinoids, and brachiopods are plentiful. Jurassic brachiopods are mainly the rhynchonellids and terebratulids. Bivalves show marked development, and many new gastropods appear at this time. Cephalopods are represented by *belemnites* and two genera of *ammonites* that had survived the Triassic ammonite extinction. The ammonite serves as the zone fossil for the Jurassic. Small, herring-like fish are the first teleosts with fully ossified vertebrae. Reptiles, especially the dinosaurs, are the dominant vertebrates; dinosaurs reach their maximum size during the Jurassic. Reptiles occupy sea and land, and develop flying forms that parallel the emergence of the first birds (*Archaeopteryx*). (Recent findings have possibly predated this by 75 m.y.) Mammals are sparsely represented by a small, ratlike form. Jurassic flora include the spore- and seed-bearing fern groups, club mosses (lycopsids), cycads, conifers, and *ginkgoes*. The first flora showing flower characteristics similar to *angiosperms* appear in the Jurassic.

The first effects of the Alpine orogeny were evidenced in southern Europe. In North America, the Nevadan orogeny produced the Sierra Nevada range during the Middle and Late Jurassic. Widespread plant growth and coal formation during this period indicate a moist, warm climate. Coral distribution shows that the equator lay about 20° farther north than at present. More uniform ocean temperatures are indicated by *oxygen-isotope fractionation* from belemnite shells.

juvenile gases Gases from the interior of the earth that are new and have not previously been at the surface.

juvenile water Water from magmatic sources within the earth, which has not previously existed as atmospheric or surface water.

K

kainite A monoclinic mineral, $MgSO_4 \cdot KCl \cdot 3H_2O$, a natural salt occurring in granular masses, used as a source of potassium and magnesium compounds.

Kainozoic Cenozoic.

kame Etymol: Scot., "crooked and winding" or "a steep-sided mound." A mound composed chiefly of stratified sand and gravel, formed at or near the snout of an ice mass, or deposited at the margin of a melting glacier. Kames, like *eskers*, occur in areas where large quantities of coarse material are available as a result of the slow melting of stagnant ice. Also, meltwater must be present in amounts great enough to redistribute the debris and deposit it at the margins of the decaying ice mass. Cf: *esker*. See also: *drumlin*.

kame terrace Terracelike body of stratified drift, left by a glacier against an adjacent valley wall after the ice has melted. See *kame*.

Kansan The classical second glacial stage of the Pleistocene Epoch in North America, after the Aftonian.

kaolin 1. A group of clay minerals that includes kaolinite. 2. A soft, white, high-grade clay (china clay), used in the manufacture of china and as a filler in paper.

kaolinite A phylosilicate clay mineral, $Al_2Si_2O_5(OH)_4$, formed by the alteration of alkali feldspars and other aluminum-bearing minerals; its largest deposits are beds of clay formed in lakes. It is polymorphous with dickite and nacrite.

K-Ar age method *Potassium-argon age method.*

karat A value indicating the proportion of pure gold in an alloy. Twenty-four-karat gold is pure or fine gold; 10-karat gold is 10/24 pure. Abbrev: K. Cf: *carat.*

karst A type of topography characterized by sinkholes, caves and caverns, *dry valleys,* and underground drainage. It is named after Karst, the type area on the coast of Yugoslavia. Development of karst topography requires the subsurface dissolution of some soluble rock, usually limestone (dolomite may suffice). The rock should be dense, highly jointed, and preferably thinly bedded, since permeability along joints and bedding planes is more favorable than mass permeability. With the exception of the Yucatan, karst regions occur in areas of moderate to abundant rainfall. *Sinkholes,* the most common topographic forms of karst terrane, are depressions varying in depth from slight indentations to 30 m or more. Fundamentally, they are of two major classes: those that developed downward by solution beneath the soil mantle, without physically affecting the surrounding rock, and those produced by the collapse of rock over an underground space. (See also: *doline.*) *Karst plains,* flat areas showing karst features, develop in regions of nearly horizontal limestone strata. A valley in karst that ends abruptly at the point where its stream disappears underground (the swallow hole) is called a *blind valley.* During periods of storm waters, a blind valley may become a temporary lake. A *karst valley,* although a type of blind valley, is not part of a typical karst plain, but rather is enclosed by clastic rocks. It is formed by the coalescence of several sinkholes, and represents a transitional stage between surface and underground drainage.

karst plain See *karst.*

karst valley See *karst.*

katamorphism Destructive surficial metamorphism in which complex minerals are broken down and changed to simpler, less dense minerals through such processes as oxidation, hydration, and solution. Cf: *anamorphism.*

katatectic layer A layer of solution residue, usually composed of gypsum and/or anhydrite in salt-dome cap rock.

katazone *Catazone.*

K-bentonite *Potassium bentonite.*

Keewatin A division of the Archeozonic rocks of the Canadian shield.

kernite A borate mineral, $Na_2B_4O_7 \cdot 4H_2O$, that occurs in great quantities beneath the Mojave Desert in a bedded series of Tertiary clays.

kerogen The solid bituminous substance in oil shales that yields oil when the shales are subjected to destructive distillation.

kerogen shale *Oil shale.*

kettle Also called *kettle-hole;* a depression in a glacial outwash drift, produced by the melting of a separated glacial ice mass that became buried either wholly or partly. Such stranded ice masses are thought to be the result of outwash accumulation atop the uneven glacier terminus. Kettles range from 5 to 13 km in diameter and up to 45 km in depth. Most kettles are of circular to elliptical shape, since melting ice blocks tend toward roundedness. When filled with water they are called *kettle lakes.* The kettles may occur in groups or singly; when large numbers are found together, the terrain, which appears as basins and mounds, is called *kettle and kame topography* or *knob and kettle topography.*

kettle hole See *kettle.*

kettle lake See *kettle.*

Keweenawan A provincial series of the Precambrian in Michigan and Wisconsin.

key A *cay*, esp. such small, low islands off the coast of Florida (Key West).

K-feldspar *Potassium feldspar.*

kick 1. *Arrival.* 2. A surge or thrust against the normal fluid circulation in an oil well, as a result of a well pressure that exceeds the pressure exerted by the drilling mud.

kidney ore A variety of hematite occurring in botryoidal to reniform shapes, with a radiating structure.

kieselguhr Syn: *diatomaceous earth.*

Kimberlite pipes See *diatreme.*

Kinderhookian Lowermost Mississippian of North America.

kinetic metamorphism A cataclastic metamorphism wherein rocks are shattered and deformed by external forces. The result is the formation of a new rock without alteration of the original chemical and mineralogical constituents.

kingdom The highest category in the sequence of classification of animals and plants that is subject to formal restriction in nomenclature.

kink band See *fold.*

klint An exhumed fossil *bioherm* or coral reef that appears as a knob or ridge because the enclosing material has been eroded away.

klippe See *nappe.*

knob-and-basin topography *Knob-and-kettle topography.*

knob-and-kettle topography An irregular assemblage of knolls, mounds, or ridges between depressions or kettles that may contain swamps or ponds. Such an undulating landscape is a type of end moraine that may result from slight oscillations of an ice front as it recedes. A section or area of knob-and-kettle topography that may have developed either along a live ice front or around masses of stagnant ice is called *hummocky moraine.*

knoll A small, rounded hill on a land surface, or a submerged elevation of rounded form rising from the ocean floor.

komatiite An igneous suite of Archean rocks that were among the first documented peridotitic magmas. *Spinifex fabric*, an array of crossed sheafs of platelike, feathery olivine, is a common feature of komatiites. Spinifex-textured komatiites range from peridotitic to basaltic composition.

kraton *Craton.*

KREEP An acronym for a basaltic lunar rock, first observed in Apollo 12 breccias and fines. It is distinguished by an inordinately high potassium content (K), rare earth elements (REE), phosphorous (P), and other trace elements.

kyanite A nesosilicate mineral of the triclinic system, Al_2SiO_5, often found in groups of light-blue crystals; less frequent colors are white, gray, and green with spots of color. Kyanite is trimorphous with sillimanite and andalusite. It occurs almost exclusively in rocks rich in aluminum and metamorphosed under high-pressure conditions: gneissic mica schists, amphibolites, and eclogites. It is important as a means of identifying the grade and type of metamorphism of the host rock.

L

labile 1. Unstable, apt to change. 2. Said of rocks and minerals that are easily decomposed.

labradorescence Brilliant flashes of color or iridescence displayed by some minerals when they are turned about in reflected light. It applies, in particular, to the light interference effect shown by labradorite as a result of the submicroscopic lamellae of feldspar.

labradorite A plagioclase feldspar common to gabbro and basalt; it has approximately equal proportions of calcium and sodium.

laccolith Etymol: Greek, *lakkos*, "pond" and *lithos*, "stone." (Also called laccolite.) A domelike concordant body of intrusive igneous rock. It arches the overlying rocks and sediment and has a floor that is more or less flat. Although they resemble inflated *sills*, laccoliths are massive rather then tabular, and may be *simple, composite* or *multiple*. Multiple laccoliths in the form of a series of stacked intrusions are sometimes called "cedar tree laccoliths." Cf: *phacolith*. See *intrusion; lopolith*.

lacuna 1. A chronostratigraphic unit representing a breach in the record, specif. the absent interval at a nonconformity. Cf: *hiatus*.

lacustrine 1. Related to, inhabiting, or produced by lakes; e.g., lacustrine environment. 2. A region populated by lakes is said to be *lacustrine*.

lag gravel 1. A residuum of coarse rock particles left on a surface, esp. desert, after finer material has been removed by deflation. See also: *desert pavement*. 2. Coarse material that is dragged along a stream bottom, thus "lagging" behind finer material. Syn: *lag; lag deposit*.

lagoon Essentially, a body of water that is more or less separated from its surroundings by some sort of barrier. It may be a coastal bay that does not exchange water with the open ocean, or a shallow body of fresh water cut off from a lake by a barrier. The shallow body of water enclosed within an atoll is also a lagoon.

lahar Etymol: Indonesian. A destructive landslide or mudflow of hot volcaniclastic material on the flanks of a volcano, formed when water from any source combines with the hot volcanic debris and slides downward, under its weight. The water source may be water ejected from a crater lake, rapidly melted ice or snow from subglacial eruptions, or very heavy rainfall. The descent of *nuées ardentes* into adjacent streams can cause lahars. Such mudflows are well known in Iceland, where subglacial volcanic eruptions occur. (See jökulhlaup.) *Tsunamis* may be initiated by sea-bottom lahars. In Indonesia, all mudflows are called *lahars*, but volcanologists restrict the use of the term to poorly stratified volcaniclastic beds.

lake Any inland body of nonflowing water. Its water source is rain or snow that falls directly into it, or is fed to it via springs and channels. A lake may also be a lake basin formerly or intermittently covered by water. Although a pond is generally considered to be smaller and shallower than a lake, some sections of the U.S. use the two terms interchangeably.

lake plain The nearly level surface marking the floor of a former lake. Because lakes are ephemeral land forms, their sites eventually are filled with inwash, or their outlets lowered by erosion. Also called *lacustrine plain*.

lamella A thin plate, scale, or layer. A structure composed of or arranged in lamellae is described as *lamellar*.

lamellar flow A movement of a liquid in which individual particles move in parallel sheets that slide over one another. Magma beneath the surface of the earth shows lamellar flow. Cf: *laminar flow*.

lamellibranch European term for *pelecypod*.

lamina The thinnest discernible layer in a sedimentary rock that differs from other layers in composition, particle size, or color; usually 0.005 to 1.00 mm thick. A structure is described as *laminar* if it is arranged in laminae; e.g., alternation of thin sedimentary layers of differing composition.

laminar flow Adjacent layers of fluid that move smoothly without mixing. Laminar flow occurs in conditions of low velocities, small channel sizes, and high viscosities. Cf: *turbulent flow*. See also *Reynolds number*.

lamination The formation of laminae or the state of being laminated; specif. the finest stratification, such as found in shale.

lamprophyre A group of mafic dike rocks. Although encompassing a range of compositions, lamprophyres exhibit in common a porphyritic fabric with mafic minerals as constituents of both phenocrysts and matrix. Feldspars and feldspathoids are confined to the aphanitic matrix. Lamprophyres tend to occur in dike swarms, and thus frequently intersect granite plutons in such a manner as to be in some way genetically related to granitic magmatism. However, their origin is as yet undetermined.

lamp shell See *brachiopod*.

land bridge A land area that forms a connecting route between continents or landmasses; it is often subject to temporary or permanent submergence.

Some of these bridges (e.g., the Bering Land Bridge) served as migratory routes for various plants and animals.

landform Any of the various features that make up the surface character of the earth. As such, it includes mountains, plains, valleys, rivers, and canyons.

landmass A term for a part of the continental crust above sea level.

Landsat An unmanned NASA satellite that orbits earth and transmits, to receiving stations on earth, spectral images in the 0.4 to 1.0 μm range.

landslide A general term for the downward movement, under gravity, of masses of soil and rock material as a result of a variety of processes. Types of landslides include rockfalls, mudflows, and slumps.

langbeinite An isometric evaporite mineral, $K_2Mg_2(SO_4)_3$. It is used as a source of potassium compounds in the fertilizer industry.

lapilli Pyroclastics with diameters in the general range of 2 to 64 mm. These ejecta, usually consisting of old lavas and scoriae, are thrown out in a completely solid state. Cf: *cinders*. See also: *pyroclastic material; pyroclastic rocks; accretionary lapilli*.

lapilli stone Pyroclastic rock composed of consolidated volcanic ejecta measuring 2 to 64 mm in diameter. See also: *pyroclastic material; pyroclastic rocks*.

lapis lazuli A crystalline rock composed basically of lazurite and calcite. It has a rich blue or blue-violet color and is used as a semiprecious stone and ornament. The same term is an old name for *lazurite*, and is used for the gem-variety mineral.

lapse rate *Thermal gradient*.

Laramide orogeny A series of Late Cretaceous and Early Tertiary mountain-building events that affected western North America. Its main effects appear centered along the eastern Cordilleran Geosyncline. Orogenic evidence for the Laramide orogeny consists of eastward-directed thrust faults and folds, initial vertical uplift accompanied by uncon-formities in the Central and Southern Rockies, and acidic plutonic intrusions (generally smaller than those of the Nevadan orogeny). It is also called the Laramide *Revolution*, since its effects were evidenced over a very long time, and is now considered a polyphase orogeny consisting of many deformation pulses varying in age and intensity through western North America. Events ascribed to the Laramide orogeny extend into the Oligocene.

larvikite An alkalic syenite with feldspar phenocrysts. Augite is the main mafic mineral, and apatite may be an accessory.

lateral accretion Materials deposited at the sides of channels, such as those where bed-load materials are being moved toward the inner sides of meanders. Cf: *vertical accretion*.

lateral erosion The action of a meandering stream as its course snakes from side to side, undercutting the banks; the process results in *lateral planation*.

lateral eruption See *volcanic eruption, sites of*.

lateral moraine See *moraine*.

lateral planation Reduction of the land adjacent to a river or floodplain to a plain or nearly even surface by the processes of lateral corrasion or lateral erosion by a stream.

laterite A soil residue composed of secondary oxides of iron, aluminum, or both. In regions of extreme weathering in-tensity, e.g., the Amazon River Basin, even kaolinite is unstable and silica is leached from the clay mineral, leaving an amorphous residue. Cf: *Latosol*.

latite A porphyritic extrusive rock with phenocrysts of potassium feldspar and plagioclase in approximately equal amounts, little or no quartz, and aphanatic to glassy groundmass; pyroxene and hornblende, if present, may be subordinate. It is the extrusive equivalent of *monzonite*. When the potassium feldspar content increases, latite grades into *trachyte*. If sodic or calcic plagioclase are present, it grades into andesite or basalt as the alkali feldspar content decreases. Generally considered synonymous with *trachyandesite* and *trachybasalt*.

Latosol A group of zonal soils typified by deep weathering and large amounts of hydrous-oxide material. They are found in forested, humid, tropical environments. Cf: *laterites*.

lattice-preferred orientation The preferred orientation of crystallographic axes or planes. In metamorphic rocks, it is produced by crystal gliding, and depends on the mineral structure and temperature, pressure, and stress during deformation. In igneous rocks, it is primarily a function of the original form of the crystals during flow or settling. Cf: *shape-preferred orientation*.

Laurasia The northern protocontinent of the Mesozoic Era, corresponding to *Gondwana* in the Southern Hemisphere. Together they comprised *Pangaea*, the original single-earth landmass. The present continents of the Northern Hemi-sphere have been derived from Laurasia by continental displacement. Laurasia in-cluded North America, Eurasia (exclusive of India), and Greenland. The main course of its breakup lay in the North Atlantic, extending somewhat into Hudson Bay. The name is derived from *Laurentia*, a paleogeographic term for the Canadian Shield and its surroundings, and Eurasia. See also: *plate tectonics*.

lava Molten rock (*magma*) that issues from openings at the earth's surface or on the ocean floor; the term also applies to the rock formed from the solidified material. Such openings may be located in craters or along flanks of volcanoes, or in fissures not associated with volcanic cones. (See *volcanic eruption, sites of.*) Molten lava reaching the surface has a temperature range of about 750° C to 1200° C. Its chemical composition and gas content vary with the particular volcano from which it is ejected; moreover, lava extruded at an early eruptive stage may differ from later extrusions. Lava generally contains varying proportions of oxides of Si, Al, Ca, Na, Mg, K, Fe, and water, together with smaller quantities of other elements. Based on chemical composition, lava is of three main types: *acidic* (rhyolite, obsidian), containing more than 65% silica; *basic* (basalt), containing less than 50% silica; and intermediate lava (andesite), containing between 50 and 65% silica. The viscosity of a lava depends mainly on its chemical composition and temperature. Silicic lavas are more viscous than basic flows. These factors partly determine the type of a volcanic eruption. See *lava flow; volcanic eruption.*

lava blister A small, hollow, relatively steep-sided swelling produced on the surface of a lava flow by gas bubbles trapped in the crust of the lava flow.

lava cascade A fall of fluid, incandescent lava, formed when a lava stream passes over an abrupt change in elevation.

lava cave *Lava tube.*

lava dome A term imprecisely used to mean: 1. a dome-shaped mountain formed by the extrusion of highly fluid lava, e.g., Mauna Loa; 2. a *Shield volcano*; 3. a *volcanic dome*.

lava flow A stream of lava issuing from a vent or fissure during a relatively quiet eruption or a nonviolent phase of an eruption; the term also applies to the solidified material so produced. The

velocity and dimensions of a flow depend essentially on viscosity. Thus, a basaltic lava tends to form fast-moving, thin, long flows, whereas the more viscous siliceous lava forms slow-moving, thick, short flows. Lava flows cool more rapidly than intrusive rocks, resulting in rock of a fine-grained to glassy texture. Flows may consist of such rocks as basalt, andesite, rhyolite, obsidian, pumice or scoria.

Morphologically, there are three principal kinds of lava flow: *pahoehoe, aa,* and block. *Pahoehoe* (Hawaiian: "ropy") flows, characterized by a smooth, glassy texture and ropy structure, develop from very fluid basaltic lava. As *pahoehoe* courses downslope it may change to *aa* (Hawaiian: "rough"), which has a chunky, broken, extremely rough surface, formed by a mantle of clinker-like or spiny fragments. The transformation is attributed to an increase in viscosity as a result of cooling, gas loss, and progressive crystallization. Although *block lava* is sometimes considered a type of *aa,* there are differences. Block lava is less vesicular than *aa* and lacks its spininess. It is more viscous and, although it may be basaltic, it is typically more siliceous. Flows associated with large volcanic cones along mountain-building belts (plate-collision zones) are generally block lavas. Flows that form shield volcanoes and basalt plains, and those along mid-ocean ridges, are typically pahoehoe and *aa.* See also: *pillow lava; lava; volcanic eruption.*

lava fountain A plume of incandescent lava sprayed into the air by gas rapidly escaping from magma as it reaches the surface. Fountains, which are characteristic of *Hawaiian-type eruptions,* usually range from about 10 to 100 m in height, but one in 1959 reached 400 m.

lava lake A lake of molten lava in a crater or other depression of a volcano; the term also refers to the solidified and partly solidified stages of lava.

lava plain See *flood basalt.*

lava plateau See *flood basalt*.

lava plug Another name for *volcanic plug*.

lava shield *Shield volcano*.

lava tube A tunnel-like space beneath the surface of a solidified lava flow, formed when still-molten lavas in the interior drain away after the formation of a surficial crust. Stalactites and stalagmites may form in these tunnels. Lava tubes are best developed in basaltic *pahoehoe* flows. They are rare in siliceous flows, rare or short in *aa* flows, and absent from *block lava*. Syn: *lava cave; lava tunnel*. See *lava; lava flow*.

lava tunnel *Lava tube*.

law of cross-cutting relationships A rock (esp. an igneous rock) is younger than any rock it transects.

law of faunal assemblages Similar assemblages of fossil fauna or flora indicate similar geologic ages for the rocks in which they are embedded.

law of original continuity A water-laid stratum, when formed, must continue laterally in all directions until it thins out as a result of non-deposition, or until it abuts the edge of the original basin of deposition.

law of original horizontality Water-laid strata are deposited nearly parallel to the surface, and thus nearly horizontally.

law of reflection The angle between an incident ray and the normal to a reflecting surface is the same as the angle between the reflected ray and that same normal; i.e., the angle of incidence equals the angle of reflection.

law of refraction When a wave traverses a boundary between two isotropic substances, the wave normal alters its direction in such a way that the sine of the angle of incidence between the wave normal and the boundary normal, divided by the wave velocity in the first medium, equals the angle of refraction divided by the wave velocity in the second medium. Syn: *Snell's law*.

law of superposition In any sequence of sedimentary strata that is not strongly folded or tilted, the youngest strata is at the top and the oldest at the bottom.

law of universal gravitation The statement that every body of the universe exerts a force of attraction upon every other body of the universe. Expressed as an equation, $F = G (m_1m_2/r^2)$ where m_1 and m_2 are the two bodies, and r is the distance between their centers. G, the gravitational constant, equals 6.670×10^{-8} dyne.

layering 1. In igneous or metamorphic rock, a tabular succession of different components, or the formation of layers in a particular rock, e.g., in plutonic rocks, as a result of the settling of crystals in a magma. 2. Stratification.

lazulite A bright blue monoclinic mineral, $MgAl_2(PO_4)_2(OH)_2$. See also: *lazurite*.

lazurite An ultramarine to violet blue mineral of the sodalite group, $(Na,Ca)_8(Al,Si)_{12}O_{24}(S,SO_4)$. It is the main constituent of *lapis lazuli*; not to be confused with *lazulite*.

leaching 1. Dissolution of soluble substances from a rock by the natural action of percolating waters. 2. Removal of mineral salts from an upper to a lower soil horizon by the action of percolating water.

lead A soft, heavy mineral, the native metallic element, Pb. Lead is found mostly combined with galena; it rarely occurs in native form. See also: *anomalous lead*.

lead-uranium ratio The ratio of lead-206 to uranium-238 and/or of lead-207 to uranium-235 that forms from the radioactive decay of uranium. See *dating methods*.

lee 1. The side of a prominence (hill, dune) that is sheltered from the wind. 2. *Lee-side* or *lee-seite*, the side of a hill in a glaciated region facing the direction toward which a glacier is moving. It is protected from abrasion, and is thus more jagged in appearance than the opposite side, the *stoss side*.

left-lateral fault See *strike-slip fault*.

left-lateral separation Displacement along a fault such that the side opposite the viewer appears displaced to the left. Syn: *left-handed separation*. Cf: *right-lateral separation*.

Lemberg solution An aqueous solution of logwood extract and $AlCl_6$ that causes calcite to stain violet but leaves dolomite unchanged.

lens An ore or rock body that is thick in the middle and thin at the edges. Something having this shape is said to be *lenticular*.

lentil 1. A lens-shaped rock body. 2. In North America, a subdivision of a formation that is impersistent; i.e., it thins out in all directions. Such impersistent subdivisions are also called *tongues*.

Leonardian Upper series of the Lower Permian in North America.

lepidoblastic A textural term for a schistose or foliated rock caused by parallel alignment during the recrystallization of platy minerals such as micas, chlorite, talc, and graphite.

Lepidodendron See *scale trees*.

lepidolite A mineral of the mica group, $K(Li,Al)_3(Si,Al)_4O_{10}(F,OH)_2$. Usually pink to lilac, but may also be pale yellow to grayish white; it occurs in pegmatites with other lithium-bearing minerals.

leucite A mineral of the feldspathoid group, $KAlSi_2O_6$, abundant in certain lavas. It characteristically occurs as white trapezohedral crystals in a fine-grained matrix.

leucocratic A term applied to light-colored rocks, especially igneous ones that contain between 0 and 30% of dark minerals; i.e., rocks of color index between 0 and 30. Cf: *mesocratic; melanocratic*. See also: *light mineral*.

levee 1. A manmade embankment along a watercourse to protect land from flooding. 2. A *natural levee*.

leveling Determining the relative altitude of different points on the earth's surface, generally by sighting through a leveling instrument. The term also means to find a horizontal line by means of a level.

lick *Salt lick*.

Liesegang rings Nested bands or rings produced by rhythmic precipitation within a fluid-saturated rock.

life assemblage *Biocoenosis*.

light mineral A rock-forming mineral with a specific gravity less than 2.80; this includes calcite, quartz, feldspars, and feldspathoids. The term is also applied to light-colored (leucocratic) rock-forming minerals, these generally being the same as those classified as light on the basis of weight.

light oil Crude oil with a high *API gravity* or *Baumé gravity*. Cf: *heavy oil*.

lignite Brown to black coal that has been formed from peat under moderate pressure. It is intermediate in coalification between peat and subbituminous coal, and dates chiefly from the Tertiary or the end of the Mesozoic. Although deposits of lignite are plentiful and quite readily available, its low calorific value, less than 10,000 BTU/lb., causes it to be inferior to bituminous coal. Cf: *brown coal*. See also: *coal*.

limb That area of a fold between

adjacent fold hinges; it may be planar or slightly curved. See: *fold*.

lime Calcium oxide, CaO; also hydrated lime, Ca(OH)$_2$. The term is often used incorrectly to mean *limestone*, or for calcium, as in "carbonate of lime."

limestone A sedimentary rock composed almost entirely of calcium carbonate (CaCO$_3$), mainly as *calcite*. *Organic limestones* consist of shell remnants or of calcite deposits precipitated by certain algae and bacteria; e.g., *coral limestone*. *Clastic limestones* are composed of broken fragments of shells or of calcite crystals. Although *chemically precipitated limestone* is today forming in warm, shallow seas, it is difficult to distinguish the role of inorganic precipitation from that of organic agents. Limestones differ greatly in color and texture, depending on the size of the shells or crystals they contain. The mineral *dolomite*, CaMg(CO$_3$)$_2$, is formed by the replacement of limestone. During the process of *dolomitization*, as the process is called, magnesium (Mg) is substituted wholly or in part for the original calcite (CaCO$_3$). The alteration process of limestone into dolomite proceeds, at least in some cases, in stages: High-calcium limestone—CaCO$_3\geq$95%. Dolomitic limestone—10 to 50% dolomite; 50 to 90% calcite. Magnesian limestone—$\leq$90% dolomite; $\geq$90% calcite. Dolomite—$\geq$90% dolomite; $\leq$1% calcite. See also: *dolomite; dolomitization*.

limnic 1. Pertaining to a body of fresh water. Cf: lacustrine. 2. Said of coal deposits formed inland in peat bogs, as distinguished from paralic deposits.

limnology The scientific study of fresh waters, in particular, ponds and lakes. It deals with the physical, chemical, meteorological, and biological conditions that pertain to such bodies of water.

limonite An amorphous, hydrated iron oxide, dark brown to black; a minor ore of iron, it occurs in earthy masses of various

forms. Limonite is of *supergene* origin, and is one of the chief constituents of the *gossan*. It often contains small amounts of hematite, clay minerals, and manganese oxides, and is used as the pigment yellow ocher. See also: *bog iron ore*.

limy Containing a significant amount of lime or limestone, e.g., a "limy soil," or containing calcite, e.g., "limy dolomite."

lineament A linear topographic feature that is thought to reflect crustal structure; e.g., fault lines, straight stream courses, aligned volcanoes. Cf: *lineation*. See *hot spot*.

linear momentum The product of a body's mass and velocity: momentum = mv. It is a vector quantity and its direction is that of the velocity. Since v represents the linear velocity of the body, the product mv is called its *linear momentum*.

lineation A general term for any rock feature showing a linear structure; e.g., parallel crystal arrangement, slickensides, boudinage, and fold axes. Cf: *lineament*.

liquefied natural gas Natural gas that has been cooled to about 160° C for shipment or storage.

liquefied petroleum gas A compressed hydrocarbon gas; usable as a motor fuel or in certain industrial processes. Abbrev: LPG.

liquid flow Movement of a liquid, usually one of low viscosity that involves laminar and/or turbulent flow. Cf: *viscous flow; solid flow*.

liquid immiscibility A process of *magmatic differentiation* in which the magma is separated into two or more immiscible phases which are, in turn, separated from each other.

liquid limit The boundary between semiliquid and plastic states of a sediment; e.g., a soil, determined by the moisture content (percent by weight of oven-dried soil) at which the soil will begin to flow if slightly disturbed. It is one of the *Atterberg limits.* Cf: *plastic limit.*

liquidus See *liquidus-solidus.*

liquidus-solidus In a temperature-composition diagram representing an equilibrium assemblage of crystals and melt, the temperature at which crystallization begins; i.e., the temperature above which the system is entirely liquid is the *liquidus;* the temperature below which the system is entirely crystalline is the *solidus.*

lithic 1. Pertaining to or made of stone. 2. Syn. of *lithologic.* 3. A sedimentary rock or pyroclastic deposit containing large amounts of fragments derived from previously formed rocks is described as *lithic.*

lithification 1. The conversion of unconsolidated sediments into a solid rock. It may involve a combination of such processes as compaction, cementation, crystallization, and desiccation. Laboratory experiments have shown that the presence of ductile fragments (e.g., schist) and absence of nonductile fragments (e.g., quartz) can produce lithification by *compaction* alone. The composition of the average sandstone (many fragments are nonductile) indicates that lithification must be achieved largely by the introduction of chemical precipitates as cements. Cf: *consolidation; induration.* 2. A type of coal-bed termination due to a lateral increase in impurities, wherein the bed changes gradually into bituminous slate or other rock.

lithofacies 1. A mappable subdivision of a specified stratigraphic unit differentiated from adjacent subdivisions by lithology. 2. The rock record of any sedimentary environment, including both physical and organic peculiarities. Cf: *lithotype.*

lithogeochemistry A branch of geochemistry dealing with the chemical properties of the lithosphere (rocks, soils, sediments). Cf: *biogeochemistry; hydrogeochemistry.*

lithographic limestone A dense, homogeneous, very fine-textured limestone; a *micrite.* Formerly used in lithography for engraving.

lithographic texture A texture of certain calcareous rocks characterized by uniform, clay-sized particles and a very smooth appearance resembling that of a lithographic stone.

lithohorizon A surface evidencing lithostratigraphic change, or having a particular lithostratigraphic character; highly useful for correlation. It is usually the boundary of a lithostratigraphic unit, but may be a thin marker layer within a lithostratigraphic unit. Cf: *biohorizon; chronohorizon.*

lithology The description and study of rocks, as seen in hand-specimens and outcrops, on the basis of color, grain size, and composition.

lithophile See *affinity of elements.*

lithophysae, lithophysal Lit: "stone bubbles." Rounded masses, generally a few centimeters in diameter, found in glassy and partly crystalline felsic volcanic rock. They commonly occur with spherulites, and consist of concentric shells separated by empty spaces, thus resembling a rose. (Cf: *orbicule.*) They are developed early in the consolidation of the surrounding volcanic rock. Lithophysal cavities are occasionally filled with secondary minerals to form geodes. Their origin has been ascribed to short periods of rapid crystallization alternating with gas expansion. Another theory considers them to be chemically altered forms of originally solid spherulites. Rocks containing these structures are said to have *lithophysal texture.* See *spherulite; orbicule.*

Lithosol An azonal group of soils that

generally develop on steep slopes. It is characterized by recent weathering.

lithosome A body of sediment deposited under uniform physiochemical conditions. It is the lithostratigraphic equivalent of a *biosome*.

lithosphere The solid outer shell of the earth, including the crust and uppermost rigid layer of the mantle. It is described as a strong or rigid zone, above the asthenosphere, or weak zone. In plate tectonics, a lithospheric plate is a segment of the lithosphere that moves over the plastic asthenosphere below. The earth's outer shell comprises eight large lithospheric plates and about two dozen smaller ones. They consist of continental and oceanic crust together with some of the underlying mantle. See also: *plate tectonics*.

lithostatic pressure *Geostatic pressure*.

lithostratigraphic unit See *stratigraphic unit*.

lithostratigraphy Stratigraphy concerned with the organization of strata into units based on lithologic character, and the correlation of these units.

lithotype A description of banded coal based on physical characteristics other than botanical origin. There are four lithotypes of coal as defined by its megascopic textural appearance. *Vitrain* occurs in thin horizontal bands, has a brilliant vitreous luster, and breaks with conchoidal fracture; *clarain* has a silky luster and shows platy fracture; *durain* is close-textured, showing a granular or matte surface when broken; *fusain* (Syn: *mineral charcoal*) occurs as patches or wedges and consists of powdery, somewhat fibrous strands. The differences in appearance of coal lithotypes result from differences in composition of the original plant material from which the coal was formed and in the degree of alteration, but the relationships are not well understood. Among coal geologists there is a widespread feeling that because of the subjectivity of the adjectives used, not much useful information is derived from such megascopic descriptions.

lit-par-lit Etymol: Fr., "bed-by-bed." A layered structure that occurs in rock, the laminae of which are penetrated by thin parallel sheets and lobes of igneous material. Cf: *injection gneiss; injection metamorphism*.

Little Ice Age A period from 1550 to 1850 characterized by extended cold seasons, heavy precipitation, and the expansion of glaciers. Of the many theories proposed for the occurrences of this interval, one of the more attractive ones seems to be that of its coincidence with a period of unexplained absence of sunspot activity. A collateral factor of the earth might have been a sharp increase in the albedo of vast areas that had been deforested just prior to the Little Ice Age; this could have worsened the conditions of the cold. Ice accretions during this period, however, were not great enough to alter world climate. See also: *ice age*.

littoral Designation of the shore area between tide marks. It may also describe the *benthic* zone extending from the high-water point on the beach to the edge of the continental shelf. See *continental shelf*. See also: *neritic zone*.

littoral current An ocean current that impinges on land and deviates in direction. A littoral current is also generated when waves break obliquely against a shore. Syn: *longshore current*.

llano See *savanna*.

LNG *Liquefied natural gas*.

loaded stream A stream that holds all the sediment it can carry.

local metamorphism See *metamorphism*.

lode A mineral deposit consisting of an entire zone of dissemination.

lodestone A variety of magnetite that is so strongly magnetic that it behaves as a natural magnet.

loess A fine-grained, loosely coherent blanket deposit composed of angular particles of quartz, feldspar, hornblende, mica, and bits of clay. Because it contains vertical tubules (remains of rotted-out grass roots), loess stands in nearly vertical walls, despite its weak cohesion. It is thought that loess is a windblown deposit of Pleistocene age. Most loess lies downwind from areas that were glaciated during the Pleistocene Period. The greatest loess deposits are found in China. In the U.S., loess covers parts of the Mississippi valley and of the Columbia Plateau.

loess doll A nodule or concretion of calcium carbonate found in loess; it bears a resemblance to a doll, a child's head, or a potato. Syn: *loess kindchen.*

Logan's Line A structural discontinuity located along the northeastern edge of the Northern Appalachians, between extensively deformed rocks on the southeast and undisturbed rocks on the northwest.

logging Any of various methods for obtaining a continuous record (a log) as a function of the depth of observations made on the rocks and fluids encountered in a well bore. See *well logging.*

longitudinal dune See *dune.*

longitudinal fault A fault whose structure parallels that of the general structural trend of the region.

longitudinal section See *cross-section.*

longitudinal stream A stream that flows in the direction of the strike of the underlying rocks.

longitudinal valley The valley of a subsequent stream that has evolved parallel to the general strike of the underlying strata.

longitudinal wave *P wave.*

longshore bar A low sand embankment formed primarily by wave action. It lies generally parallel with the shoreline at some distance from it, and is submerged during high tide.

longshore current *Littoral current.*

lopolith A concordant, typically layered igneous intrusion that is planoconvex or lenticular in shape. The central part is sunken because of the sagging of the underlying *country rock.* The term lopolith is used for both small tabular intrusions related to *sills,* and for very large intrusions whose dimensions are massive rather than tabular. The latter are sometimes referred to as *megalopoliths.* Lopoliths usually comprise more mafic rocks than those of other plutons, and most, particularly the megalopoliths, show alternating layers of light and dark minerals and have the appearance of sedimentary rock strata. It is thought that the origin of these banded structures must differ from that of the others. Cf: *phacolith.* See also: *laccolith; sill.*

lost river Referring to a stream that disappears underground in a *karst* region, or to a dried-up stream in an arid region.

Love wave See *seismic waves.*

low-angle fault A fault dipping less than 45°. Cf: *high-angle fault.*

low quartz *Alpha quartz.*

low-rank metamorphism Metamorphism that takes place under conditions of low to moderate temperature and pressure. Cf: *high-rank metamorphism; metamorphic grade.* See also: *metamorphism.*

low tide The tide at its lowest; the minimum level reached during the tidal cycle. See also: *neap tide.*

low-velocity layer correction *Weathering correction.*

low-velocity zone 1. *Weathered layer.* 2. The zone in the upper mantle of the earth in which seismic wave velocities are about 6% lower than in the outermost part of the mantle. Its depth is defined as being between 60 and 250 km. Syn: *B layer.* 3. A region within the core boundary, below a depth of 2900 km, that produces a *shadow zone* at the earth's surface.

LPG *Liquefied petroleum gas.*

L-tectonite See *tectonite.*

luminescence The emission of light by a substance or body after it is subjected to an external source of electromagnetic radiation of a different wavelength. See also: *phosphorescence; fluorescence.*

lunar basalt An igneous rock found as great flows in the *maria* of the lunar surface. Lunar basalts exhibit the same types of fabrics as terrestrial basalts but they differ compositionally. The major minerals in lunar basalts are titanium-rich clinopyroxene and very calcic plagioclase. All mare basalts contain iron-titanium oxides, usually ilmenite.

lunar crater A roughly circular depression in the lunar surface, relatively shallow and ranging in diameter up to hundreds of km; lunar craters may have resulted from meteorite impact, volcanism, or subsidence.

lunar playa A small area on the moon's surface in the ejecta blankets surrounding some of the lunar craters. It may be a small lava flow.

lunar regolith A thin grayish layer on the moon's surface, composed of loosely compacted fragmental material ranging in size from microscopic particles to blocks greater than a meter in diameter. It is thought to be the result of repeated meteoritic and secondary fragment impact extended over a long period of time. Syn: *lunar soil.*

lutaceous adj. Of a sedimentary rock formed from mud, or of the texture of such a rock; also, pertaining to a *lutite.* Cf: *argillaceous.* See *lutite.*

lutite Etymol: Latin *lutum,* "mud." General name for any fine-grained sedimentary rock composed of material that was once mud, such as mudstone or shale. More specifically, the proportion of silt contained should be between one-third and two-thirds of the total. The terms lutite, claystone, mudstone, siltstone, and shale often overlap in usage. Cf: *pelite.* (See also: *arenite; rudite.*) If the sedimentary rock consists of more than 50% detrital calcite particles of silt or clay size, it is called *calcilutite.* Cf: *calcisiltite.*

L wave *Surface wave.*

M

maar A shallow, flat-floored crater not associated with a volcanic cone. Maars appear to be the result of multiple shallow, explosive eruptions. Craters that intersect the water table may form lakes; many examples of these occur in the Eiffel area of Germany. It is possible that a significant number of maars are the surface manifestations of *diatremes.* See also: *volcanic crater.*

maceral A petrologic compositional unit of coal, identified microscopically by thin or polished sections. Macerals are related to coal as minerals are related to inorganic rock. Cf: *phyteral. See also: coal.*

macro- A prefix meaning "large" or "great." Syn: *mega.*

macrocrystalline Said of a rock texture in which the crystals are clearly visible to the unaided *eye* or with the use of an ordinary hand-lens. Cf: *microcrystalline.*

mafic A term in petrology, derived from *magnesium + ferric + ic,* and equivalent to subsilicic or basic. It pertains to compounds dominated by magnesium and iron rock-forming silicates, and also to some igneous rocks and their constituent minerals. It is the complement of *felsic.* In general, it is synonymous with *dark minerals.* Cf: *femic; salic.* See also: *mesocratic; melanocratic.*

magma Molten, mobile rock material that is a naturally occurring, high-temperature solution of silicates, water, and gases. Suspended solids such as crystals and rock fragments may or may not be included. Magma is generated deep within the earth's crust or upper mantle as a result of local melting of solid rock, and is the source of *igneous rocks.* Since the volatile components of a magma are usually lost during or after consolidation, the igneous rock formed cannot be said to entirely represent the original magma. During the process of *magmatic differentiation,* different kinds of igneous rocks are derived from an original magma. Magma that consolidates below the earth's surface is called *intrusive* or *plutonic rock;* that which emerges above the surface and solidifies is called *extrusive* or *volcanic* rock. *Lava* is magma that has reached the surface. Neither the source of the ingredients of magma nor the source of heat necessary to melt them is completely understood. The most plausible theory for the main source of heat seems to be that of decay of radioactive elements contained in these subterranean rocks. Slow dissipation of heat throughout the earth's crust maintains temperatures sufficient to melt

rocks even at moderate depths. See *magma chamber.*

magma chamber An underground reservoir of magma in the earth's crust, perhaps only 7 or 8 km underground, filled from magma sources deep within the mantle. Various surface volcanic materials are derived from the magma of these chambers through *magmatic differentiation.* These products, in general, have the same composition as the magma chambers. However, products that have solidified below the surface are coarsely crystallized in texture because of prolonged cooling. See also: *lava.*

Types of igneous rock

	Extrusive (erupted on surface)	Intrusive (solidified below surface)
About 50% SiO_2	Basalt	Gabbro
About 60% SiO_2	Andesite	Diorite
About 70% SiO_2	Rhyolite	Granite

magmatic differentiation Internal diversification processes that separate an initially homogeneous closed magma body into two or more daughter magmas of different chemical composition, thus developing more than a single type of igneous rock from a common magma. In partly crystallized magmas, the separation of crystals from melt (crystal-liquid fractionation) is a very effective means of magmatic differentiation. For example, as a basaltic magma cools, olivine and bytownite (plagioclase feldspar) are the first minerals to crystallize. If they react with the surrounding melt, olivine forms

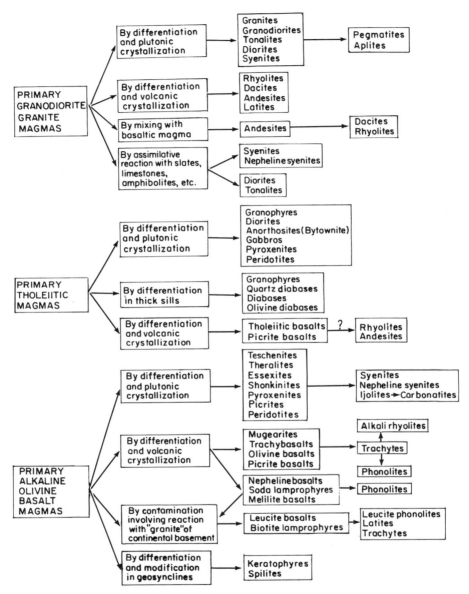

Magmatic differentiation

pyroxene, bytownite forms labradorite, and the resluting rock will be gabbro or basalt. If the early-formed crystals are segregated and isolated, the melt crystallizes to form a different rock. If other minerals are segregated before total crystallization, still other rock types will develop.

magmatic ore deposit A deposit of

minerals crystallized from a magma that has solidified into a particular rock mass. Through some process, such as magmatic segregation or magmatic differentiation, the minerals become concentrated and form an orebody.

magmatic segregation see *magmatic ore deposit.*

magmatic stoping See *stoping.*

magmatic system A conceptual body of magma composed of one or several components, which is used in model systems for the determination of crystal-melt equilibria.

magmatic water *Juvenile water.*

magmatism 1. The development and movement of magma, and its solidification into igneous rock. 2. The theory that much of earth's granite has been derived through crystallization from magma rather than through granitization, which results from a metamorphic process.

magnafacies A major, continuous belt of deposits that show similar lithology and paleontological profiles. It extends through several chronostratigraphic units and represents a depositional environment that prevailed with a shifting of geographic placement over time.

magnesia Magnesium oxide, MgO.

magnesian limestone See *limestone.*

magnesite A carbonate mineral of the hexagonal system, $MgCO_3$, usually found as compact, white, yellow or gray porcelain-like masses. It is formed by the alteration of ultramafic rocks through the action of waters containing carbonic acid. As a diagenetic mineral, it replaces calcite and dolomite.

magnetic anomaly Departure from the expected value of earth's magnetic field over a point of the earth's surface after adjustments have been made for local inhomogeneities. A magnetic anomaly generally indicates the presence of some object with either *remanent magnetism* or high *magnetic susceptibility* (the measure of the response of a material to a magnetic field) beneath the anomalous area. A change in rock type or the presence of a magnetic orebody may be indicated by such anomalies. Iron ore deposits are especially well-suited to magnetic exploration methods.

The continental pattern of anomalous magnetic values is, in general, irregular. The ocean is chiefly dominated by linear magnetic anomalies related to the mid-ocean ridge system. Basaltic submarine lava flows near the ocean floor, which typically show a high ratio of remanance to induced magnetization, appear to be responsible for most oceanic anomalies whose features are symmetrically arranged parallel with the axis, on both sides. According to the now widely accepted theory of *sea-floor spreading*, these parallel anomalies are underlain by zones of normally and reversely magnetized basaltic rocks which are created at the mid-oceanic ridge axis, and which acquire a *thermoremanent magnetism* as they cool. The magnetic polarity depends upon the polarity of the geomagnetic field at the time of cooling. Observed magnetic profiles are calibrated by matching with computer-simulated anomaly profiles formed by spreading during the Gauss geomagnetic normal polarity epoch, about 3370 m.y.a. In the assignment of ages to magnetic anomalies, it is assumed that the distinctive magnetic anomaly associated with this epoch can be noted in the simulated profiles. The spreading rate since then can thus be determined by measuring the distance of this anomaly feature from ridge crests. See also: *geomagnetic polarity reversal; mid-oceanic ridge; sea-floor spreading; geophysical exploration; paleomagnetism.*

magnetic declination The acute angle in degrees between the directions of magnetic and geographic north.

magnetic dip *Magnetic inclination.*

magnetic equator That imaginary line on the earth's surface where the magnetic needle remains horizontal; i.e., where the magnetic inclination is zero. Syn: *aclinic line.* Cf: *agonic line.*

magnetic field The region of magnetic influence that surrounds the poles of a magnet, or a moving charge.

magnetic field intensity The force exerted by a magnetic field on a magnetic substance. Syn: *magnetic field strength.*

magnetic field reversal See *geomagnetic polarity reversal.*

magnetic geophysics See *geomagnetics.*

magnetic inclination The angle at which magnetic field lines dip. Syn: *inclination; magnetic dip.*

magnetic meridian *Magnetic north.*

magnetic north The uncorrected direction indicated by the northerly directed end of a magnetic compass needle; it is the northerly direction of the magnetic meridian at any specified point. Syn: *magnetic meridian.* Cf: *true north.*

magnetic polarity reversal *Geomagnetic polarity reversal.*

magnetic prospecting See *magnetic anomaly.*

magnetic pyrites *Pyrrhotite.*

magnetic reversal *Geomagnetic polarity reversal.*

magnetic storm A disturbance of earth's magnetic field, thought to be caused by ionized particles ejected by solar flares. Magnetic prospecting is usually suspended during such periods, particularly if the disturbance is intense.

magnetic susceptibility The ratio of the strength to which a substance has been magnetized (induced magnetization) to the strength of the magnetic field causing the magnetization.

magnetite A strongly magnetic, shiny black mineral, $(Fe,Mg)Fe_2O_4$, the richest and most important ore of iron. It occurs as both compact and granular masses in a variety of rocks, and is plentiful in contact metasomatic conditions.

magnetometer Any of many types of instruments that measure the magnitude and changes of earth's magnetic field.

magnetosphere The outer region of the earth's magnetic field, in which magnetic forces predominate over gravity in the motion of charged particles. It is an asymmetric configuration compressed on the sunward side of the earth, and "pluming" far outward on the opposite side.

magnetostriction Elastic strain that accompanies magnetization.

magnitude Earthquake magnitude.

maidenhair tree See ginkgo.

malachite A banded, emerald-green monoclinic mineral, $Cu_2(CO_3)(OH)_2$, a minor ore of copper. It commonly occurs as a green film on other copper minerals and as botryoidal or reniform masses. Malachite is typically found in the oxidation zones of copper deposits, where it is produced by a sulfide-carbonate gangue reaction; it is valued as a gemstone and as a decorative mineral.

malpais Etymol: Span., "bad land." In the southwestern U.S., many early workers used this term for black rock areas and basalt; at present the term is exclusively applied to mesas or to any more or less elevated plateaus formed by rough lava flows. The term varies in connotation, depending upon the locality.

mammillary Having the form of breasts or portions of spheres. Said of some minerals such as limonite or malachite.

manganese nodule An irregularly shaped mass rich in manganese and iron that occurs on the ocean floor in areas of slow sedimentation, and also where red clay deposits are located. Nodules commonly found at the surface of pelagic sediments range from 0.5 to 25 cm, but generally average about 3 cm in diameter. Those from different parts of the ocean tend to have unique physical characteristics; however, all types consist of many minerals and crystallites, and lack an overall crystalline structure.

Nodule minerals are derived from the interaction of hot, upwelling, metal-bearing solutions at a plate boundary with organic sediments of the ocean itself. Precipitation from the mineral solutions about some center of accretion (e.g., a sand grain) is the apparent mechanism of nodule formation. The composition of manganese nodules shows definite regional variation. Mn/Fe ratios ‹ 1 are characteristic near the continents. In areas farthest from land, nodules show high nickel and copper content. In other regions, they are relatively rich in cobalt. In areas of concentration, the economic potential of manganese nodules can be significant.

manganite A hydrated manganese oxide mineral, $MnO(OH)$, of the monoclinic system. It is semiopaque with a slight metallic luster. Manganite occurs in low-temperature hydrothermal veins associated with calcite and barite.

mantle The zone within the earth that extends below the crust to the core. The upper boundary is marked by the *Mohorovičić discontinuity*, and the boundary between mantle and core by the *Gutenberg discontinuity*. The upper mantle extends to a depth of 400 km and is characterized by regional dissimilarities in seismic velocity profile. A *transition zone*, extending downward from about 400 to 1000 km, is characterized in general by a rapid increase of velocity with depth. The *lower mantle*, between 1000 and 2900 km, shows rather gradual seismic velocity increases with depth. Massive convection currents in the mantle are energized by the heat of decay of radioactive isotopes. Although it appears that this convective circulation is the agent for plate movement, much of it is not related to the movement or boundaries of the plates. (See *convection; convection cells.*)

A seismic P-wave velocity of 8.2 (± 0.2) km/sec in the uppermost mantle, together with certain broad mineralogical and petrological constraints, indicates some combination of olivine, pyroxene, garnet and, in restricted regions, amphibole. Peridotite (olivine-pyroxene) and eclogite (pyroxene + garnet) are the two principal rock types bearing these minerals. Geophysical evidence on the density of the upper mantle favors a density close to that of peridotitic composition. A study of diamond pipes in Africa and Siberia has revealed that the pipes carry inclusions of crustal rocks that are known to have intruded during rapid upward replacement. These pipes also contain inclusions of rocks not occurring in the vicinity, esp. peridotites and eclogites.

In many oceanic regions, the upper 70 km of the upper mantle shows an S-wave velocity of about 4.6 km/sec, which decreases suddenly to 4.2 km/sec at a depth of about 100 km. The zone of decreased velocity extends to about 150 to 200 km; a corresponding low-velocity zone for P waves is also believed to occur. It is thought that the observed magnitude of the velocity change suggests a small degree of partial melting in the region. It was demonstrated that the density in the transition zone (400 to 1000 km) increased much more rapidly with depth than would be expected in a uniform layer. After study, it was concluded that abnormal density changes in this zone were caused by a sequence of major phase changes. Studies of the density distribution in the lower mantle, and of shock-wave data on the densities of silicates at comparable pressures, indicate the density of the lower mantle to be about 5% higher than that of the composition that was assumed. The formation of a mineral assemblage that is intrinsically denser than the overlying assemblage is a possible explanation for this. See also: *crust; core.*

map projection A systematic method of drawing the earth's meridians and parallels on a flat surface. Some projections have equal-area properties, whereas others show conformal delineations in which, for small areas, the shape is essentially the same as the global representation. Insofar as flat maps of large areas must inevitably distort the global representation, the map projection best suited to the purpose of the map to be produced is also the one that minimizes the distortions. Most types of projection can be classified according to their geometrical derivation as *cylindrical, conic* or *azimuthal.* A few cannot be so classified or are combinations of these.

The problem of transposing meridians and parallels from globe to surface was solved in 1569 by the Flemish mathematician *Mercator,* by his well-known world map with the meridians vertical and parallels having increased spacing in proportion to the secant of the latitude.

The Mercator projection, which is of great value in navigation, belongs to the group of *cylindrical projections.* In these, the earth is treated as a cylinder in which parallels are horizontal lines and meridians are shown as vertical lines. For these projections, places of similar latitude appear at the same height. *Conic projections* are derived from a projection of the earth on a cone drawn tangent to the earth, and with the apex above either the North or South Pole. A *polyconic projection,* used in large-scale map series, regards each band of map as part of a cone tangent to the globe at the designated latitude. *Azimuthal* or *zenithal* projections represent a section of the earth as a flattened disk, tangent to the earth at a given point, as observed from a point at the center of the earth (*gnomonic projection*), or the opposite side of the earth's surface (*stereographic projection*), or from a point far distant in space (*orthographic projection*).

marble 1. A fine- to coarse-grained metamorphic rock consisting mainly of recrystallized calcite and/or dolomite. Colorful streaks in marble are due to inclusions of minerals such as mica, pyrite, epidote, quartz, feldspar, or iron oxide. Purer varieties, without inclusions, are valued by sculptors; veined varieties are used as ornamental stone. 2. In commerce, a general term for any crystallized carbonate rock, including true marble and particular types of limestone (orthomarble), that can be polished and used as ornamental or architectural stone. 3. *Verd antique.*

marcasite A pale-yellow sulfide mineral, FeS_2, dimorphous with pyrite. It occurs as massive aggregates in low-temperature hydrothermal veins and in sedimentary environments. Jewelry and trinkets made of marcasite were once popular.

mare (pl. *maria*) One of several large, level, low-lying areas on the lunar surface. The *maria* show fewer large craters than the uplands; they are composed of mafic or ultramafic volcanic rock.

marine abrasion 1. The erosion of a bedrock surface by the movement of sand that is agitated by waves. 2. The erosion of submarine canyons by the gravity-related downslope movement of sediments.

marine cave *Sea cave.*

marine geology That part of geology in which the ocean floor and its many features are studied. This includes submarine relief topography, petrology, the geochemistry of ocean floor rocks and sediments, and the effect of waves and seawater on the ocean bottom. Syn: *geological oceanography.*

marine terrace A relatively narrow coastal strip that is formed of deposited material sloping seaward. The slope just suffices to keep the material in slow seaward transit.

marl A friable mixture of subequal amounts of micrite and clay minerals. Cf: *marlstone.*

marlstone An indurated rock of about the same composition as *marl*; i.e., an earthy limestone; it is less fissile than shale.

marsh A type of wetland characterized by poorly drained mineral soils and by plant life dominated by grasses, sedges, reefs, and rushes. The latter feature distinguishes a marsh from a swamp, where trees are the dominant plant life. Marshes are common at the mouths of rivers, especially where large deltas have formed. The delta soil favors the growth of fibrous-rooted grasses, which bind the muds, thus hindering water flow and promoting the spread of both delta and marsh. Marshes may also be found in areas where depressions have been left by retreating glacial ice. *Salt marshes,* common along the east coast of the United States, are flat intertidal areas that are periodically flooded by sea water. Salt marsh grasses will not grow in permanently flooded flats. A *tidal marsh* is a marshy *tidal flat* bordering a sheltered coast that is regularly inundated during high tides. It is formed of mud and the roots of salt-tolerant plants. Cf: *bog; swamp.*

Marshall line *Andesite line.*

marsh gas *Methane* produced as a decomposition product of the decay of vegetable substances in stagnant water.

mascon A large, high-density mass concentration beneath the lunar surface.

massif A large elevated feature, usually in an orogenic belt, differing topographically and structurally from the lower adjacent terrain. It usually consists of more than a single isolated summit and exhibits no extensive level areas. The rocks of which a massif is formed are more rigid than those of its surroundings, and are usually older; they are generally thought to be rather complex structurally. Rocks forming a massif may be the mark of previous deformations and not necessarily related to its development as a massif.

massive In geology, a somewhat arbitrary term describing homogeneous rock bodies that have great bulk. *Batholiths* are examples of massive *plutons*. Massive bodies are also referred to as *non-tabular*. Cf: *tabular.*

mass transport The carrying of material in a moving medium, such as air, water, or ice.

mass wasting A general term for the bulk, downslope transfer of masses of rock debris under the direct influence of gravity. No other medium is an agent.

matrix The *groundmass* of an igneous rock, or the finer-grained material enclosing larger grains in a sedimentary rock. The term also applies to the background material in which a fossil is embedded.

mature 1. Pertaining to a particular point in the cycle of erosion, or to a stage in the development of a stream or shoreline. 2. A clastic sediment that has evolved from its parent rock and is now characterized by stable minerals.

M-discontinuity Syn: *Mohorovičić discontinuity.*

meander A curve or loop in a stream channel. A stream characterized by many such curves is a *meandering stream*. Many streams meander over part of their course and are relatively straight for other distances. Although some streams follow meandering courses cut in bedrock (see *entrenched meander*), most are on floodplains. Meandering streams have fairly light bed loads and banks that are much steeper than those of braided streams. Most are close to base level, and only by becoming more sinuous can they lower their gradients. Others not near base level become sinuous because of difficulty in carrying fine sediments when there is no bed load. As water rounds any bend, its inertia carries it toward the concave bank, causing a slight uptilting of the water surface; this tilt tends to make the stream rush toward the opposite bank as it flows downstream, and the curvature of the bank tends to reverse. Thus the outside bank of each curve is subject to erosion at the same time as the deficiency of water on the inner bank results in deposition. The combination of these two actions compels meander bends to migrate. See also: *oxbow lake*.

meander cutoff See *neck cutoff*.

mechanical twinning *Deformation twinning*.

mechanical weathering Weathering processes that are physical in nature and result in the fragmentation of rock without a chemical change; e.g., frost action, crystal growth, thermal expansion. Syn: *physical weathering*. Cf: *chemical weathering*.

medial moraine See *moraine*.

Medinan Obsolete syn. of *Alexandrian*.

mediterranean A term describing a sea that lies between continents, or within a continent. Cf: *epicontinental*.

Mediterranean suite See *Atlantic, Pacific, and Mediterranean suites*.

medium-grained Said of an igneous rock in which crystals have an average diameter in the range of 1 to 5 mm, or of a sedimentary rock in which individual particles average one-sixteenth to 2 mm in diameter.

megabreccia A very coarse breccia in which the larger fragments are hundreds or thousands of kilometers long. The area covered by a megabreccia is so extensive that it is measured in square kilometers. Although some are the result of landslides caused by rising fault blocks, the term megabreccia is a descriptive term without genetic implication. *Chaos* is a type of breccia consisting of small and large blocks of irregular shape. The type example is the Amargosa chaos of Death Valley, Calif., where most blocks are more than 60 m long and a few are 800 m long. It is believed to have been formed beneath a large thrust sheet by the imbrication of small thrust plates. *Mélange* describes a widespread coarse breccia, such as in Wales and Turkey. In Turkey, the Ankara Mélange consists of huge limestone blocks and some small blocks of graywacke. The origin of mélange is variously attributed to fragmentation due to compression, gravity sliding during sedimentation, or brecciation due to thrust faulting. *Olistostromes*, as found in the Appenines, are composed of greatly deformed black shales containing blocks of various rocks, ranging in age from Triassic to Miocene. They are believed to be the product of submarine landslides composed of blocks derived from uplifted areas.

megalopolith See *lopolith*.

mélange See *megabreccia*.

melanocratic A term applied to dark-colored rocks, particularly igneous rocks containing between 60 and 100% of dark minerals; i.e., rocks with a color index between 60 and 100. Cf: *leucocratic; mesocratic*.

mellorite *Fireclay mineral*.

melt In *petrology*, a fused liquid rock.

meltwater Water resulting from melting ice and snow. Its geologic effects include the depositing of drift, such as *eskers* and *kames*, and the formation of potholes. Glacial meltwater streams are often grayish-white because of "rock-flour" resulting from glacial abrasion.

member A minor rock stratigraphic unit comprising some portion of a *formation*. It may be a *formal unit, informal unit,* or *unnamed.* Cf: *lentil; tongue.*

Meramecian Lower part of the Upper Mississippian of North America.

Mercalli scale See *earthquake measurement.*

Mercator projection See map projection.

mercury A heavy, silvery-white, liquid mineral, the native metallic element Hg. Natural mercury is found as tiny globules in cinnabar or deposited in certain hot springs.

merocrystalline *Hypocrystalline.*

mesa Etymol: Span., "table." An isolated flat-topped hill bounded on at least one side by a steep cliff and having an extensive summit area. Mesas are erosional remnants that persist because of a protective cover of more resistant sedimentary rocks, esp. sandstones, or of lava flows or gravels. They range from 30 to 600 m in height, and from a few hundred meters to several kilometers in length. In the U. S., mesas are found in Arizona, Colorado, New Mexico, and Utah. Cf: *butte.* See *hogback.*

meso- A prefix meaning "middle."

mesocratic A term used for rocks, especially igneous ones that contain between 30 and 60% of dark mineral; i.e., whose color index lies between *leucocratic* and *melanocratic.*

mesonorm See *depth zone of emplacement.*

mesosiderites A miscellaneous group of *stony iron meteorites;* the metal content may exceed 40%, but does not form a continuous network.

mesothermal deposit See *hydrothermal deposit.*

Mesozoic Era The Mesozoic ("middle life") Era was a span of geologic time between 225 and 65 million y.b.p. It included the Triassic, Jurassic, and Cretaceous periods. Marine fossils form the world standard for the correlation of Mesozoic strata. Mesozoic rocks are widespread on every continent, and marine deposits from this era increase in area with successive periods. Triassic marine deposits consist of limestones and clastic and volcanic sediments. Continental deposits are mostly red shales, sandstones, and marls. Jurassic marine deposits proceed from reefoid and *stromatolitic* limestones in the early part to pelagic limestones toward the end. Jurassic sedimentation in North America is similar to that of Europe. Early Cretaceous sedimentation patterns are generally similar to those of the Jurassic. Increased marine deposits of the Middle Cretaceous reflect sea incursions, but a return to continental conditions is indicated toward the close of the Cretaceous.

The Mesozoic Era is known as the "Age of Reptiles," many groups of which populated the land and oceans. Certain amphibians such as the *labyrinthodonta* become extinct by the end of the Triassic, and are replaced by more advanced groups. Dinosaurs first appear in the Late Triassic and become quite suddenly extinct in the Late Cretaceous. Marine invertebrates include several important groups of fish. Mesozoic mammals, all of small size, represent several extinct orders. Among the invertebrate fauna of the Mesozoic, ammonites are most significant because of their rapid evolution, and are hence important as zone fossils; they become extinct by the end of the Cretaceous. Partly because of the evolution of siphons, bivalves achieve some dominance during the Triassic. Modern insects appear before the Cretaceous. The Mesozoic is often called the "Age of *Pteridophytes.*" Ferns, the *gymnosperm* order of cycads, *ginkgoes* and conifers, and the angiosperms are common at the end of the Triassic. Among primitive plants, cyanophytes (blue-green algae) are found as Mesozoic stromatolites.

The beginning of the Mesozoic saw the end of the Variscan orogeny in Europe and the Appalachian orogeny in North America. Volcanism was more intense and widespread in the late Mesozoic. Fossil evidence indicates that climatic changes were more equable than at present. At the onset of the Mesozoic, according to the concepts of continental drift, there were two major continental shield areas: *Laurasia* in the Northern Hemisphere and *Gondwanaland* in the Southern Hemisphere. These two supercontinents had joined by the Late Permian to form a universal land mass, *Pangaea*. This would imply that in Early Mesozoic time the Atlantic and Indian Oceans did not exist. Oceanographic data, as well as evidence from the distribution of terrestrial animals and the location of marine sediments, indicate the breakup of Pangaea no later than the Middle Jurassic.

mesozone See *depth zone of emplacement.*

meta-anthracite The highest rank of anthracite coal; it has a fixed carbon content of 98% or greater.

metabentonite See *bentonite.*

metacryst A large crystal in metamorphic rock, formed during metamorphism. Garnet and staurolite always occur as metacrysts in mica schist. Syn: *porphyroblast.*

metallogenetic See *metallogenic.*

metallogenic (Lit: "metal-forming.") A term used with the connotation of mineralization, especially of ore deposition. *Metallogenic epochs* are units of geologic time during which particular mineral deposits were formed. A *metallogenic province* is a region in which there occurs a particular series of minerals

or certain types of mineralization; e.g., the gold-quartz veins of California.

metamict Term applied to a mineral whose original crystal structure has been disrupted by radiation from radioactive elements (e.g., uranium or thorium) contained in that mineral, while the original external crystal form is retained. Consequently, a metamict mineral has no cleavage, and does not diffract X-rays. In many of these minerals, the crystallinity is restored by prolonged heating at a temperature below the decomposition point. Examples occur in thorite, zircon, and many other minerals.

metamorphic differentiation A process in which noticeably heterogeneous rock is created with layers or pods of contrasting mineralogical composition from an initially homogeneous body during metamorphism.

metamorphic grade The intensity of metamorphism, gauged by the difference between the parent rock and the metamorphic derivative. On a broad level it functions as an indicator of the P-T conditions that obtained when metamorphism took place.

metamorphic rock Any of a class of rocks that are the result of the alteration of pre-existing rocks in response to changes in heat, pressure, shearing, and the chemical effects of contained fluids. The pre-existing rocks may be sedimentary, igneous, or other metamorphic rocks. The relative effect of the physical conditions defines the type of metamorphism involved: 1. cataclastic—stress with low pressure; 2. contact—high temperature with low pressure; 3. low to high temperature; 4. metasomatic—a wide range of pressures and temperatures in the presence of solutions. The mineralogy and structure of a particular metamorphic rock reflect the type of metamorphism that produced the rock. Texture is a function of change in the interrelationship between grains grown in the solid state as opposed to liquid crystallization. The shape of new crystals may be regular or irregular, with crustal faces that depend on the mineral type and the presence or absence of stress. A *relic texture* is one that is inherited from the pre-existing rock.

Metamorphic rocks formed by *contact metamorphism* are associated with an igneous intrusion, and are limited in extent; pressure gradients are absent. A hornfels texture with equant grains is typical of such rocks. In sedimentary rocks, a sequence of minerals is formed mainly in response to temperature variations; e.g., andalusite, anorthite, and diopside.

The most abundant metamorphic rocks are those produced by *regional metamorphism*. Typical rocks of this type are schists and gneisses. Their coarse grain size implies a long-term metamorphism. Examination of a sequence of these metamorphic rocks shows that in addition to mineralogical changes, there is also a general pattern in the textures or structures of the rocks. For example, in the case of shale, low-grade metamorphism produces slate, which, in turn, becomes phyllite or a coarser laminated slate, and then schist. Continued metamorphism results in changes that produce gneiss, then granulite, and finally magnetite. In other sequences, quartz sandstone become quartzites, limestones become marbles, and basic igneous rocks become greenschist, amphibolites, and, ultimately, eclogites. The appearance of *index minerals* (biotite, garnet, staurolite, kyanite, or sillimanite) indicates constant metamorphic conditions. See also: *metamorphism*.

metamorphism Structural and mineralogical changes of any type of solid rock in response to physical and chemical conditions differing from those under which the rocks initially formed. Changes brought about by surface conditions such as compaction are usually excluded. Two metamorphic processes result from the effects of temperature, pressure, stress, and chemical environment: 1. Mechanical dislocation, where a rock is deformed, esp. as a result of differential stress; 2. Formation of a veir or mineral assemblage due to a nonequilibrium condition (caused by pressure and temperature changes) of the original assemblage.

Three types of metamorphism may take place, depending on the relative effect of mechanical and chemical changes. *Dynamic metamorphism* (or cataclasis) results chiefly from mechanical deformation, with little long-lasting temperature change. Such adjustments produce textures ranging from breccias to granulated rocks with foliation. In *contact metamorphism*, chemical, mineral, and physical changes in an earlier-formed country rock are produced by contact with magma that has migrated upward from depth. Metamorphism occurs primarily as a result of temperature increases under conditions of minor differential stress; *aureoles* are a common phenomenon. Because the extent of contact metamorphism is limited, the pressure is near constant. The resulting rocks are usually fine-grained, owing to the short duration of metamorphisis, and homogranular, because of lack of stress. *Regional metamorphism* results from a general increase of temperature and pressure over a large area. It is subdivided into different pressure-temperature conditions according to observed sequences of mineral assemblages. Regional metamorphism may thus include rock alteration occurring within a temperature range of $400°$ to $800°$ C and a range of pressures encountered at depths of about 10 to 35 km. See also: *retrograde metamorphism; hydrothermal alteration; metasomatism; metamorphic rocks*.

metaquartzite A quartzite formed by metamorphic recrystallization, as differentiated from an orthoquartzite, whose crystallinity is of diagenetic origin.

metasomatism A metamorphic process whereby existing minerals are transformed totally or partially into new minerals by the replacement of their chemical constituents. This occurs by the introduction of capillary solutions, originating either outside or within the rock, that are reactive with the existing minerals. Syn: *replacement.* In *contact metasomatism*, there is a mass change in the composition of rocks that are in contact with an invading magma. Certain fluid constituents from the magma combine with the country rock constituents to form a new series of minerals. *Pyrometasomatism* is the formation of contact-metamorphic mineral deposits at high temperatures; the process involves replacement of the enclosing rock by the addition or subtraction of materials.

meteoric water Water that occurs in or is derived from the atmosphere.

meteorite Any extraterrestrial solid mass that reaches the earth's surface. In the strictest sense, this would also include the dust particles that make up most of the material that survives passage through the atmosphere. However, the term is usually restricted to the exceptionally large fragments that fall to earth. Three classes of meteorites may be distinguished: 1. irons (*siderites*), composed entirely of metal (iron-nickel), are the ones most frequently found; 2. stony irons (*siderolites*) consist of both metal (iron and nickel) and silicate; 3. stones or stony meteorites (*aerolites*), composed mainly of silicate minerals, and the meteorites most frequently observed to fall. They are further divided into *chondrites* (containing inclusions) and *achondrites* (without inclusions).

Tektites are sometimes included as one of the meteorite classes. However, most recent studies indicate that they are re-formed earth materials resulting from meteoritic impact, and are not of extraterrestrial origin. See also: *tektite.*

methane An odorless, colorless, flammable gas, CH_4, the main constituent of *marsh gas* and the *firedamp* of coal mines; it is obtained commercially from natural gas.

methylene iodide A liquid compound, CH_2I_2; sp. gr. 3.32. Used as a *heavy liquid* in flotation processes of ore separation.

miarolitic 1. A descriptive term for plutonic rock bodies (e.g., granite) containing *vugs.* During crystallization of the intrusive body of magma, small gas pockets that may separate because of boiling are shaped by the crystalline surroundings and are thus crystallized. Cf: *druse.* In some of these miarolitic vugs, euhedral crystals subsequently grew into the gas pocket. 2. Petrological term for small, angular cavities in plutonic rocks into which crystals of the rock-forming minerals project. See *vug; vuggy.*

mica A group of monoclinic phylosilicate minerals characterized by their platy habit and perfect basal cleavage. Micas occur in a wide range of colors, as well as colorless and black. They are prominent rock-forming constituents of igneous and metamorphic rocks. Principal members of the mica group are muscovite, phlogopite, biotite, and lepidolite.

micaceous 1. Consisting of or pertaining to mica. 2. Resembling a mica, either in luster or in having the property of being easily split into thin sheets.

micrite 1. A term used for the dull, semiopaque to opaque, microcrystalline matrix of limestones, composed of chemically precipitated carbonate sediment with crystals less than 5 microns in diameter. Micrite is much finer-textured than *sparite.* 2. A limestone composed mostly of micrite matrix and with less than 1% allochems; e.g., *lithographic limestone.*

micro- A prefix meaning "small." When joined with a rock name, it usually signifies fine-grained.

microcline A mineral of the alkali feldspar group, $KAlSi_3O_8$. It occurs in red, yellowish, blue, green, or white, and usually shows tartan twinning. Microcline is found as compact aggregates in granitic pegmatite and metamorphic rocks formed at medium to low temperatures.

microcoquina See *coquina.*

microcrystalline A term applied to a rock texture comprising crystals that are visible only with the aid of a microscope.

microlite *Nickeline.*

microseisms In the broadest sense, ground motion that is not due to earthquakes or explosions; storms at sea are the main source. There is not yet a generally accepted theory of microseisms.

Mid-Atlantic Ridge That portion of the *mid-oceanic ridge* that lies within the limits of the Atlantic Ocean.

mid-oceanic ridge A continuous feature that extends through the Atlantic, Indian, Antarctic, and South Pacific Oceans, the Norwegian Sea, and the Arctic Basin. It is the greatest mountain range on the earth, with a total distance of over 40,000 km (35,000 miles). The *Mid-Atlantic Ridge* is that portion which lies within the Atlantic Ocean. Occasionally, the ridge reaches above the ocean surface as islands or reefs, as in Iceland, where the Mid-Atlantic Ridge has emerged.

According to *plate tectonics*, when two crustal plates separate, basaltic material wells up through the spreading center and produces a ridge. Rapid separation of the plates builds ridges with gentle slopes and broad elevations, such as those of the East Pacific Rise. The steep flanks of the Mid-Atlantic Ridge were formed by slow spreading. Mid-oceanic ridges are characterized by linear *magnetic anomalies* that are symmetrically distributed parallel to the ridge axis. Earthquakes at these extensional margins are shallow and follow the center of the rift. Volcanic activity is in the form of volcanic chains along the ridge axis. Many sections of the ridge show above-average heat flow (2 to 3 μ cal/cm^2/sec, where μ = 10^{-6}). Studies indicate that the submerged part of the mid-ocean ridge is composed predominantly of a low potassium tholeiite, and that only the highest parts are composed of alkaline basalts. See *plate tectonics, sea floor spreading.*

mid-ocean rift *Rift valley.*

mid-ocean rise *Mid-oceanic ridge.*

migmatite A highly complex rock that is generally an intimate mixture of igneous and metamorphic rocks; it is characterized by a banded or veined appearance. Migmatites are widespread, particularly near large granite masses. Their mineral composition is variable, but most contain abundant quartz and feldspar.

migration 1. Movement of gas and oil from their beds of origin into reservoir rocks via permeable formations or materials. 2. The change in position of the crest of a divide, directed away from a strongly eroding stream on a steep slope toward a less active stream on a gentler slope. 3. The gradual shifting downstream of a system of meanders, accompanied by widening of the meander course and enlargement of its curves. 4. The apparent movement of a dune due to continuous transfer of sand from its windward to its leeward side. 5. A term for the passage of plants and animals from one place to another over long periods of time.

Miller indices A set of three or four symbols used to define the orientation of a crystal face or internal crystal plane.

milligal The normal unit used in geophysical gravity surveying. It is equal to 10^{-5}m/sec^2.

mineral A naturally occurring element or compound of non-biological origin, having an ordered atomic structure and characteristic chemical composition, physical properties, and crystal form. Thus, coral would not be classified as a mineral. Moreover, if opal is considered to be amorphous, then it could not be categorized as a mineral. It is sometimes referred to as a *mineraloid*, and many mineralogists refer to it as *cryptocrystalline*, a term that would place it within the mineral family.

mineralization 1. The process by which valuable minerals are introduced into a rock, resulting in an ore deposit, either actual or potential. The term includes various types of this process; e.g., *impregnation, fissure filling, replacement.* 2. Process of fossilization in which inorganic materials replace organic features.

mineralizer A gas that is dissolved in a magma, which aids in the concentration, transport, and precipitation of certain minerals, and in the formation of particular textures as it escapes from the magma.

mineralogy The study of minerals, including knowledge related to their formation, composition, properties, and classification.

mineraloid See *mineral.*

mining geology The geological aspects of mineral deposits, with special regard to problems related to mining.

minor elements *Trace elements.*

Miocene The epoch of the Tertiary Period between the Oligocene and Pliocene Epochs. See *Cenozoic Era.*

miogeocline *Miogeosyncline.*

miogeosyncline See *geosyncline.*

mirabilite A yellow or monoclinic mineral, $Na_2SO_4 \cdot 10H_2O$, that occurs as a residue from playas and saline lakes; it is a source of sodium sulfate. Syn: *Glauber's salt.*

miscible Said of two or more liquids that, when introduced, will mix and form a single liquid. Cf: *immiscible.*

mispickel *Arsenopyrite.*

Mississippian A term introduced by Alexander Winchell in 1870 into American stratigraphic terminology for the Lower Carboniferous of the Mississippi Valley. Together with the term *Pennsylvanian* for the Upper Carboniferous, it has been accepted by the U.S. Geological Survey since 1953, though not used outside North America. Although the lower boundaries of the Lower Carboniferous and Mississippian are approximately the same, the upper boundaries vary. See also: *Mississippian-Pennsylvanian boundary; stratigraphic correlation.*

Mississipian-Pennsylvanian boundary The boundary between the Mississippian and Pennsylvanian Periods is placed at the surface of unconformity (discontinuous sedimentation). Because local events that produce the discontinuities do not occur contemporaneously, the base of the Pennsylvanian is not everywhere the same in North America. See also: *stratigraphic correlation.*

Missourian Lower part of the Upper Pennsylvanian of North America.

mixed-base crude A crude oil in which both naphthenic and paraffinic hydrocarbons are present in about equal proportions. Cf: *paraffin-base crude; asphalt-base crude.*

mixed crystal *Solid solution.*

mobile belt A belt of rock with consistent tectonic, metamorphic, and chronologic features, typical of Proterozoic terranes.

mobilization 1. Any process by means of which a solid rock becomes sufficiently plastic to permit the geochemical migration of particular mobile components. Cf: *rheomorphism.* 2. Any process that relocates and concentrates valuable rock components into an ore deposit.

mode The acutal mineral composition of a rock, expressed in weight or volume percent. Cf: *norm.*

modified Mercalli scale See *earthquake measurement.*

modulus of compression *Compressibility.*

modulus of elasticity The ratio of stress to strain within the range in which a material is deformed according to Hooke's law. Syn: *modulus of volume elasticity.* See also: *Young's modulus; modulus of rigidity; bulk modulus.*

modulus of incompressibility *Bulk modulus.*

modulus of rigidity Expresses the resistance to change in shape. It is given as $G = \tau / \gamma$, where G is rigidity modulus, τ is shear stress, and γ is shear strain.

modulus of volume elasticity *Modulus of elasticity.*

mofette A *solfatara* that is rich in carbon dioxide. See *fumarole.*

mogote A large residual hill of limestone associated with a karst-erosion landscape. The name derives from Spanish of western Cuba, where there are great numbers of mogotes. They are found only in regions of tropical or subtropical rainfall. Gigantic examples are characteristic of South China, and are the hills of fantasy that appear so frequently in Chinese paintings.

Moho Abbrev. for *Mohorovičić discontinuity.*

Mohole project National scientific project with the research objective of studying the earth as a planet, to be accomplished through sampling of all layers of the earth's crust and mantle by core drilling near Hawaii, to a total depth of approximately 9755 m. Phase I of Project Mohole, conducted in 1961, consisted of drilling into several hundred meters of the earth's crust off the California coast. Phase II was concerned with the problems created by drilling in deep oceans, and how to solve them. This work progressed from 1962 to 1965 but, in 1966, the project was terminated for lack of funding.

Mohorovičić discontinuity The boundary surface that separates the earth's crust from the mantle. It marks the level at which velocities of seismic P waves abruptly change from 6.7 to 7.2 km/sec (in the lower crust) to 7.6 to 8.6 km/sec (top of upper mantle), as a result of greater density of the mantle rock. The depth of the discontinuity ranges from about 5 to 10 km beneath the ocean floor to about 35 km below the continents. Its thickness is variously judged to be between 0.2 and 3 km. Sonic sounding has not detected its presence beneath the Mid-Atlantic Ridge, an indication that rocks beneath the discontinuity may be too hot for the transmission of elastic waves at velocities suitable to most of the upper mantle. The Mohorovičić discontinuity is named after its discoverer, Andrija Mohorovičić (1987-1936), a Croatian seismologist. Syn: *Moho; M-discontinuity.* Cf: *Gutenberg discontinuity.*

Mohs scale A standard of ten minerals by which mineral hardness may be rated. On a scale of one to ten, from softest to hardest: talc; gypsum; calcite; fluorite; apatite; orthoclase; quartz; topaz; corundum; diamond.

molasse Etymol: Fr., "soft." An association of conglomerates and sandstones deposited as alluvial fans and lacustrine deposits. Rocks of the facies are associated with most mountain ranges. The Coast Mountains of British Columbia and the California Coast Ranges show very good examples.

mold 1. An impression of a fossil shell or other organic structure made in the encasing material. An *external mold* shows the surface form and markings of the outer hard parts of a structure. The encasing material whose surface receives these molds is also called an external mold. An *internal mold* shows the form and markings of the inner surface of a structure. It is made on the surface of the material that fills the hollow interior of the structure. When a shell becomes filled with sediment, and the shell material is later removed by solution or erosion, the remaining hard core is called a *steinkern* (stone-kernel). Cf: *cast*. 2. A flute or other mark made on a sedimentary surface, the filling of which produces a *cast*.

Mollisol An order of soils that typically form under grass in climates that have a seasonal moisture scarcity.

mollusk An invertebrate of the phylum Mollusca, usually wholly or partially enclosed in a calcium carbonate shell secreted by a mantle covering the body. The shell may be external univalve, external bivalve, internal, or absent. Mollusks have a muscular "foot" that is modified in the different groups. They are highly diverse in body arrangement, and there is no standard molluscan shape or life pattern. Their habitat may be terrestrial, marine, or freshwater. The earliest molluscans (limpets and gastropods) date to Cambrian time. Classes of mollusks include *gastropods, cephalopods,* and *pelecypods.*

molybdate A mineral compound in which the MoO_4 radical is a constituent. Wulfenite, $PbMoO_4$, is an example.

molybdenite A bluish-gray hexagonal mineral, MoS_2, the principal ore of molybdenum. Molybdenite usually occurs as bladed or foliated masses, and is found in pegmatites and high-temperature pneumatolytic veins.

monadnock An isolated mountain or hill that rises above a lowland in a temperate climate, and which has been leveled almost to the theoretical limit (base level) by fluvial erosion. Such a lowland is called a *peneplain*, and represents the almost complete advance of the leveling process. Monadnocks remain as residual features, either because they are composed of more resistant rock or are farther away from the base level. Mt. Monadnock in New Hampshire is the classic example, from which these hills derive their name. Cf: *catoctin; unaka; inselberg.*

monazite A rare-earth phosphate, yellow to brownish red, generally occurring in disseminated granules. It is a common accessory in granites and gneisses, and is the main ore for thorium.

monoclinal shifting The slow downdip shift of a stream channel, as a result of two factors: the tendency of streams in an area of inclined strata to follow the strike of less resistant strata, and the tendency for erosion to proceed more rapidly along the steeper slopes of a cuesta. Cf: *migration.*

monocline A local steepening of the dip or grade of layered rocks in areas where the bedding is relatively flat. adj. *Monoclinal.* Cf: *homocline.*

monoclinic system See *crystal system.*

monogenetic 1. Derived from a single source, or resulting from a single process of formation; e.g., a volcano produced by a single eruption is called monogenetic. 2. Consisting of one element or material; e.g., a gravel composed of a single rock type. Cf: *polygenetic.*

monomictic 1. A term applied to a clastic sedimentary rock composed of a single mineral type. 2. Said of a lake with only a single yearly *overturn.* Cf: *oligomictic; polymictic.*

Monongahelan Upper Pennsylvanian of eastern North America.

monotropy The relationship between two different forms of the same substance where there is no transition point, since only one of the forms is stable, and the change from the unstable to the stable form is irreversible. Cf: *enantiotropy.*

montmorillonite A monoclinic clay mineral of the smectite group. Two noteworthy characteristics are its capacity to expand by absorbing liquids, and its potential for ion exchange when, under certain conditions, the cations sodium, potassium, and calcium, present between the structural layers, are replaced by other cations present in solution.

monzonite Plutonic rock with a composition intermediate between syenite and diorite. It contains approximately equal amounts of plagioclase and alkali feldspar, and little or no quartz; augite is often the main mafic mineral. Monzonite is the intrusive equivalent of *latite*.

moonstone A gem-quality alkali feldspar (*adularia*) that shows a silvery or bluish iridescence; an opalescent variety of orthoclase. Cf: *sunstone*.

moraine An accumulation of rock material (till) that has been carried or deposited by a glacier. It ranges in size from boulders to sand, and shows no bedding or sorting. Different kinds of moraines are distinguished: 1. *Ground moraine*—an irregular covering of *till* deposited under a glacier, and composed chiefly of clay, silt, and sand. It is the most prevalent deposit of continental glaciers; 2. *Lateral moraine*—debris derived from erosion and avalanches from the valley wall onto the glacier edge, and finally deposited as a long ridge when the glacier recedes. 3. *Medial moraine*—an enlarged zone of debris formed when lateral moraines join at the intersection of two glaciers. It is deposited as a ridge running approximately parallel to the direction of ice movement. 4. *Terminal* or *end moraine*—a ridgelike mass of glacial debris formed by the foremost glacial snout and dumped at the outermost edge of a given ice advance. It shows convex curving downvalley, and may form lateral moraines up the side. It may take the appearance of hummocky ground, perhaps with *kettles* and *knobs*. 5. *Recessional moraine*—secondary end moraine deposited during a temporary halt in a glacial retreat. Series of such moraines thus show the history of glacial retreat.

morphogenetic region A climatic zone in which prevailing geomorphic processes produce landscape features distinct from those of other regions that formed under different climatic conditions.

morphologic unit 1. A rock stratigraphic unit specified by its topographic features. 2. A surface that is recognized by its topograph.

morphology 1. *Geomorphology*. 2. The study of soil properties and distribution arrangements of soil horizons.

Morrowan Lower Pennsylvanian of North America.

mortar structure An apparently *cataclastic* structure resembling stones in mortar. It is found in gneisses and granites in which spaces between particles of feldspar and quartz are filled in with pulverized grains of the same minerals.

morvan The intersection of two erosional surfaces. The name is derived from the Morvan Plateau of northeastern France, which exhibits such a relationship. In the U.S., the Fall Zone peneplain of the eastern Piedmont Plateau is described as a morvan; a buried peneplain of probable Jurassic age is becoming exhumed, and its surface, when projected inland, intersects the late Tertiary Harrisburg peneplain at an angle.

mosaic 1. *Desert pavement*. 2. A petrologic texture in which the component mineral grains are approximately equant.

moss agate Chalcedony quartz with a moss- or fernlike inclusion. The inclusion is actually a *dendrite* of pyrolusite (MnO_2). Moss agate is often mistaken for a fossil. See also: *pseudofossil*.

moss animal *Bryozoan*.

moulin Etymol: Fr., "mill." A circular or subcircular cylindrical shaft extending down into a glacier; it is produced by the abrasive scouring action of swirling meltwater. Syn: *glacial mill.* See also: *giant's kettle.*

mountain Any part of the earth's crust that rises at least 300 m above the surrounding land and has a summit area smaller than that of a plateau. It can occur as a solitary formation, or be part of a *mountain range*, which is a single mass consisting of a succession of mountains or mountain ridges; these formations are closely related in position, formation, and age. A *mountain system* is a group of mountain ranges that show certain features in common; e.g., similarity in form and alignment, and a presumable origin in the same events. A complex connected series of several essentially parallel mountain ranges and mountain systems grouped together only on the basis of general longitudinal arrangement, or well-defined trend, is called a *mountain chain.* See also: *cordillera.* Mountains can be formed by earth movements, volcanic activity, or erosion.

mountain chain See *mountain.*

mountain glacier. *Alpine glacier.*

mountain range See *mountain.*

mountain system See *mountain.*

moveout See *normal moveout.*

mud 1. A mixture of water with clay- to silt-sized particles, ranging in consistency from semi-fluid to soft and plastic. 2. A fine-grained red, blue, or green marine sediment found mainly on the continental shelf; its colors are due to a predominance of various mineral substances. (See *terrigenous deposits.*) 3. The material of a *mudflow.* 4. *Drilling mud.*

mud crack An irregular polygonal pattern formed by the shrinkage of mud, silt, or clay in the process of drying. Mud cracks develop generally under conditions of long exposure to a dry, warm climate. Also called *shrinkage crack* or

desiccation crack. Cf: *mud-crack polygon.*

mud-crack polygon *Desiccation polygon.*

mud flat A relatively flat expanse of fine silt lying along a sheltered slope or expanse of an island. It is a muddy *tidal flat* devoid of vegetation, and may be covered by shallow water or alternately covered and uncovered by the tide.

mudflow A flow of fine-grained, water-saturated sediment in a stream channel. It is typical of semi-arid to arid regions where heavy rains transport earth material collected on hill slopes into pre-existing stream channels. Cf: *earth flow.* See also: *debris flow; lahar.*

mud pot A variation of a hot spring in which the channel way has become choked with mud that is broken down from the surrounding rock by acid volcanic gases dissolved in the water. Mud pots are frequently associated with geysers. Syn: *sulfur-mud pool.*

mudrock 1. A sedimentary rock composed of indurated mud, having the composition and texture of a shale but lacking fissility; also, a blocky or massive fine-grained sedimentary rock in which the amounts of clay and silt are approximately equal. (See also: *claystone; siltstone.*)2. A general term including clay, silt claystone, siltstone, shale, and argillite; it should be used only when the amounts of clay and silt are not known or cannot be accurately identified, or when it is desired to distinguish the entire group of finer grained sedimentary rocks from sandstones and limestones.

mudstone A mud-supported carbonate sedimentary rock containing less than 10% particles of clay and fine silt size; the original components are not bound together during deposition. Cf: *wackestone; packstone; grainstone.*

mud-supported Referring to limestone

in which there are not sufficient sand- and gravel-sized grains to form the supporting framework.

mullion structure 1. The large grooves on a fault surface that run parallel to the direction of displacement. 2. A structure consisting of a series of parallel columns or rods. Each unit may be several centimeters in diameter and several meters long. The columns are composed of folded sedimentary or metamorphic rocks. Syn: *rodding structure.*

multi- A prefix meaning "many" or "much."

multiple intrusion A term describing igneous intrusions whose mass consists of several emplacements of more or less identical chemical and mineral composition. (Cf: *composite.*) *Dikes* and *sills* are usually multiple; when dikes occur in large numbers they are called *dike swarms.*

multiple reflection A seismic wave that has been reflected more than once.

muscovite A mineral of the mica group, $KAl_2(AlSi_3)O_{10}(OH)_2$, may be white or yellow, or sometimes dark brown. Muscovite is one of the most common minerals in rocks, esp. pegmatite, granite, and low- or medium- to high-grade metamorphic rocks (green schist and amphibolite facies). Syn: *white mica.*

musical sand See *sounding sand.*

muskeg See *bog.*

m.y. *Million years.*

m.y.a. *Million years ago.*

mylonite A hard, fine-grained, chertlike rock with banded or streaky structure, formed by the extreme granulation of rocks that have been pulverized during faulting or intense dynamic metamorphism. It is typically related to faults (esp. thrust faults) or intense shear zones. Lenses of undestroyed parent rock may persist in the re-formed granulated groundmass. There is debate about whether new crystals form in the groundmass. When extremely fine-grained, mylonite may be described as a microbreccia with flow texture.

mylonitization The deformation of a rock by microbrecciation resulting from mechanical forces that are applied in a definite direction; there is no significant chemical reconstitution of the granulated minerals.

myrmekite An intergrowth of a plagioclase feldspar and quartz, generally replacing alkali feldspar; it is formed during the latter interval of consolidation in an igneous rock. The quartz occurs as blobs or as vermiform shapes within the feldspar.

N

nacreous 1. Pearly; having the luster of mother-of-pearl. 2. Of mother-of-pearl (nacre).

nappe Also called *decke*, a large, sheetlike body of solid rock that has moved a long distance (generally a mile or more) at low angles over the underlying rocks either by over-thrusting or recumbent folding. After long erosion, portions of a nappe may become isolated remnants called *klippen* (sing. *klippe*). A klippe is a *nappe outlier* and is distinguished from outliers of cuestas or plateaus in having younger rocks surrounding it. If, in the early stages of dissection of a nappe, erosion breaks through the overthrust sheet, exposing the rocks beneath the fault, a *window* (or *fenster*) is formed; the name applies because it is possible to look through the upper sheet to the lower.

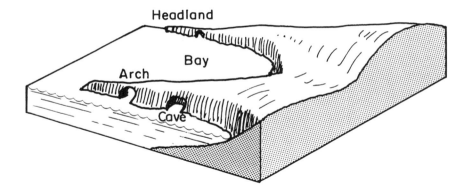

Waves attack a headland from all sides, forming <u>caves</u>. When two caves ultimately join, an <u>arch</u> forms. When the arch collapses, a <u>stack</u> results.

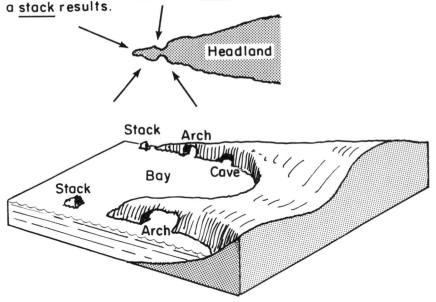

Formation of natural bridges and arches

native element Any element found in nature that is uncombined and in a non-gaseous state. Native metals include silver, gold, copper, iron, and mercury. Nonmetallic examples are carbon, sulfur, and selenium.

native paraffin *Ozocerite*.

natrolite A tectosilicate of the zeolite group, $Na_2(Al_2Si_3)O_{10} \cdot 2H_2O$.

natural arch See *natural bridges and arches*.

natural bridges and arches Archlike rock features formed by natural agents such as weathering and erosion. When lateral erosion by a stream enables it to cut through a meander neck, a bridge or arch will be formed if the upper level of the neck does not collapse. Rainbow Bridge in Utah, the world's largest natural bridge, was formed in this way. Some arches, such as those of Arches National Monument, also in Utah, are formed by weathering. In *karst* regions, which are characterized by tunnels created by the groundwater solution of limestone, the collapse of a tunnel roof eventually results in the formation of a bridge; Natural Bridge, Va., is the most famous example of a karst bridge in the U.S. Caves that develop on either side of a headland, such that they ultimately join together because of wave erosion, give rise to a sea arch. Upon collapse of the arch, the headland remains as a *stack*. See also: *stacks*.

natural coke See *coke*.

natural gas Hydrocarbons that exist as a gas or vapor at ordinary temperatures and pressures. It may occur alone or associated with oil. Like oil and coal, natural gas is an important fossil fuel. Although its origin is not fully understood, it is believed to have originated from the remains of microscopic life forms buried in sediments of prehistoric seas, where they eventually underwent decomposition.

natural levee An embankment of silt and sand built up by a stream along both its sides. The process is particularly effective during flooding, when the water overflows its banks and deposits its coarsest sediment.

natural remanent magnetism The permanent magnetism in rocks, resulting from the orientation of earth's magnetic field at the time the rock was formed. It provides the basic information for the paleomagnetic studies of *polar wandering* and *continental drift*. It can derive from several processes:

1. *Thermoremanent magnetism* (TRM), the most important process. When magnetic minerals forming in igneous rock cool through the Curie point (the critical point for magnetization to take place), and the crystals of the minerals align themselves in the direction of earth's magnetic field, a permanent record of the field orientation is made.
2. *Detrital remanent magnetism*, the result of the settling of small grains of magnetic minerals into a sedimentary matrix. It is thought that the grains arrange themselves in the direction of the geomagnetic field during deposition and before final consolidation of the rock.
3. Nonmagnetic minerals that are chemically altered to magnetic forms can acquire remanent magnetism in the presence of the geomagnetic field.
4. *Viscous magnetization* may cause already cooled igneous rocks to ultimately acquire remanent magnetism. The superimposition of later or earlier magnetization must be considered when determining remanent magnetization. See also: *paleomagnetism; polar wandering; geomagnetic polarity reversals*.

nautiloid Any cephalopod having straight, coiled, or curved shells whose septa (chamber walls) have a very wavy boundary (suture line) where they meet the outer shell. A siphuncle connects the chambers and adjusts the gases within them, permitting buoyancy. The last surviving genus of nautiloids is *Nautilus*; their peak was reached in the Ordovician and Silurian. Range, Cambrian to present.

neap tide A tidal pattern in which the difference in level between high and low tide is a minimum. Neap tides occur when the earth and sun are at right angles with respect to the moon; i.e. during the first and third quarter moon. Cf: *spring tide*.

Nebraskan Pertaining to the classical first glacial stage of the Pleistocene Epoch in North America; succeeded by the Aftonian interglacial.

Nebular Hypothesis A theory of the origin of the planetary system, according to which the system formed within a gas cloud that was part of the original sun. In the nebular hypothesis of Kant (1775), developed by La Place (1796), a rotating cloud of gas contracts, the outer region of the cloud becomes a cool gaseous disk in which the planets condense, and the innermost region contracts to form the sun. The problem of how such protoplanets condense in a Laplacian disk of gas is still unsolved. Moreover, according to this theory, the sun should rotate much more rapidly than is observed. Cf: *"tidal wave" theory.*

neck 1. A narrow strip of land that connects two larger areas, as an isthmus or cape. 2. A *volcanic neck.* 3. A mineral-bearing pipe. 4. A *meander neck.* 5. The narrow band of water that forms the part of a rip current where feeder currents converge.

neck cutoff A term for the *cutoff* that takes place in the final stage of meander-loop development when a meander neck is cut through. It results in shortening of the river course and local steepening of the stream gradient. Cf: *chute cutoff.*

negative shoreline *Shoreline of emergence.*

nekton A term including all animals able to swim effectively against the horizontal ocean current for prolonged periods. Seals, whales, tuna, herring, and squid are examples. See also: *plankton; pelagic.*

nektonic See *pelagic.*

nematath A "procession" of volcanoes which forms as a lithospheric plate drifts over a hot spot (thermal center). Periodic volcanic outpourings form a line of volcanoes, the ages of which are greater as the distance from the hot spot increases. By observing the location of succeeding cones as they converge into a ridge, it is possible to establish the absolute direction taken by the crust in that region.

nematoblastic A texture of metamorphic rocks in which an abundance of acicular or columnar grains are oriented in linear fashion, thus imparting a lineation to the rock.

Neogene Largely obsolete name for the Miocene and Pliocene Epochs of the Tertiary, when grouped together. Cf: *Paleogene.*

Neolithic In archaeology, the last division of the Stone Age, during which time agriculture developed. Syn: *New Stone Age.*

neomagma A magma formed by the fusion of pre-existing rock under conditions of plutonic metamorphism. Cf: *anatexis.*

neontology The science of existing organisms, as distinguished from *paleontology.* Approx syn: *biology.*

nepheline A hexagonal mineral of the feldspathoid group, $(Na,K) AlSiO_4$, generally occurring in compact granular aggregates of white, yellow, gray, or green. Its environment is in intrusive and volcanic rocks that are silica-undersaturated.

nepheline syenite A plutonic rock composed chiefly of alkali feldspar and syenite. It is the phaneritic intrusive equivalent of *phonolite.* Sodalite and cancrinite are common accessories.

nephelinite A porphyritic extrusive igneous rock, light gray in color, and often having a feltlike holocrystalline groundmass. It is composed primarily of nepheline and clinopyroxene, and is found in small flows associated with alkaline basalts.

nephrite An extremely tough, compact tremolite or actinolite. It is one of the two minerals referred to as "jade," jadeite being the other. Like jadeite, nephrite shows an almost full spectrum of colors, including white and black.

neritic deposits Sea-bottom deposits found chiefly at the edge of the continental shelf, but farther out than where *terrigenous* deposits would usually be found. They consist of material from the land and organic substances from shallow coastal waters, including the remains of crustacea, worm tubes, mollusks, etc. See *ocean bottom deposits.*

neritic zone Sedimentologists and stratigraphers use this term to include the shallow-sea environment from low water down to 200 m (100 fathoms). Oceanographers and ecologists prefer the term *littoral* for this zone, whereas geologists generally apply "littoral" to the intertidal and near-shore belt.

nesosilicate A silicate in which the SiO_4 tetrahedra occur as isolated units, i.e., they do not share oxygen ions as do all other silicate types. In nesosilicates, the tetrahedra are bound together by other ions. The ratio of silicon to oxygen is one to four. See *silicates.*

net balance The change in glacial mass from the time of minimum mass during one year to the time of minimum mass in the year following. Cf: *balance.*

net slip The distance between two previously adjacent points on either side of a fault, measured on the fault surface or parallel to it. Net slip defines both the direction and relative magnitude of displacement. Cf: *horizontal slip; vertical slip.*

Neuropteris A typical seed fern (Pteridospermae) that is a common fossil in certain Mississippian-Pennsylvanian strata. Oval leaflets alternate on either side of the stem. The length of the pinnules is 0.6 to 1.3 cm. See also: *fossil plants.*

neutral shoreline A shoreline whose basic character is independent of the sub-mergence of a former land surface, or the emergence of a surface that was formerly under water. It includes alluvial and outwash planes, volcanoes, and coral reefs.

neutron-gamma log See *well logging.*

neutron log See *well logging.*

Nevadan orogeny Originally applied to a mountain-building event in the Sierra Nevada region of eastern California in the very late Jurassic. The term is now used for a series of orogenic pulses in the western *Cordilleran Geosyncline* of western North America. These pulses extended from Late Jurassic to Middle Cretaceous. Early phases of the Nevadan orogeny are noted in the folding and thrust-faulting on the western slope of the Sierra Nevadas and in the Klamath Mountains. Mid- and late phases are represented by massive batholiths in southern California, the Sierra Nevadas, and British Columbia. The Nevadan orogeny may be the result of underthrusting of the western portion of the North American plate by oceanic crust along a former oceanic trench found in the western part of the *Cordilleran Geosyncline.*

névé See *firn.*

new global tectonics A general term introduced in 1968 for *global tectonics,* and which would incorporate the concepts of sea-floor spreading, transform faults, continental drift, and subduction of the crust and uppermost mantle, and would apply them in the same context as that with which they are used to analyze the relative motions of crustal segments as delineated by major seismic belts.

New Red Sandstone The red sandstone facies of the Permian and Triassic systems, particularly well developed in northwestern England. New Red Sandstone has been widely adopted, at least in Britain, as a single Permo-Triassic system. Cf: *Old Red Sandstone.*

New Stone Age *Neolithic.*

Niagaran Middle Silurian of North America.

niccolite *Nickeline.*

nickeliferous Containing nickel.

nickeline A copper-red nickel arsenide, NiAs, one of the chief ores of nickel. Nickeline occurs with other arsenides and sulfides in vein deposits. Syn: *niccolite.*

Nicol prism A device for obtaining plane-polarized light. It consists of a rhombohedron of optically clear calcite, cut and cemented together in such a way that the ordinary ray produced by double refraction in the calcite is totally reflected, while the extraordinary plane-polarized ray is transmitted. It has largely been replaced by sheet polarizers such as Polaroid.

Niggli's classification A classification of ore deposits comprising two main divisions: plutonic and volcanic. The former has the categories orthomagmatic, pneumatolytic to pegmatitic, and hydrothermal.

niter A white mineral, KNO_3, found in arid regions and as surface efflorescence in caves. Syn: *saltpeter.* Cf: *soda niter.*

nitrate A compound characterized by the $(NO_3)^{-1}$ radical. Soda niter $(NaNO_3)$ and niter (KNO_3) are examples. Cf: *carbonate; borate.*

nitride A compound containing two and only two elements, of which the more electronegative one is nitrogen.

nivation Processes acting on high mountain slopes that enable a snow mass to persist and alter the bedrock or *regolith.* Nivation includes *creep, frost action, solifluction,* and sheetwash, all of which take place on the snow-patch edge or beneath it. *Glaciation* is initiated when the snow that accumulates in such niva-tion hollows becomes thick enough to move downslope as glacial ice.

noble metal Metals that do not enter readily into chemical combinations with nonmetals; e.g., gold, silver, platinum. Such metals are resistant to oxidation or to attack by corrosive agents. Cf: *base metals.*

node 1. The point on a fault at which the apparent displacement changes. 2. That point on a standing wave at which there is no vibration.

nodular Composed of or having the form of nodules; e.g., a nodular ore.

nodule A term that conveys the idea of a rounded mass or lump; there is no connotation of size. It can be applied to a lump of mineral aggregate that is normally without internal structure and contrasts with its embedding matrix. Cf: *concretion.* The term is also used to describe the concretionary lumps of metals found on the ocean floors; e.g., *manganese nodules.*

noise In seismic studies, noise is all recorded energy not derived from the explosion of the shot.

non-artesian groundwater *Unconfined groundwater.*

nonconformity See *unconformity.*

nonflowing artesian well An artesian well in which the hydrostatic pressure is not great enough to elevate the water above the land surface; the water does, however, rise above the local water table. Cf: *flowing artesian well.*

nonplunging fold A fold whose hinge line is horizontal.

nonsorted circle See *patterned ground.*

nonsorted net See *patterned ground.*

nonsorted polygon A form of *patterned ground* showing a dominantly polygonal mesh, the units of which are not bordered by stones, as are *sorted polygons*. The structure typically results from the filling in of fissures that commonly mark its borders. The outlining fissures of a *desiccation polygon* (also called a *mud-crack polygon*) contain vegetation. A *fissure polygon* is marked by intersecting grooves that form a slightly convex pattern. It is typical of areas of the northwestern Canadian lowlands. When the outline is formed by intersecting frost cracks, the configuration is called a *frost-crack polygon*. An ice-wedge polygon is a large feature averaging 10 to 110 m in diameter, and typically bordered by intersecting ice wedges. It is found only in permafrost regions, and is formed by the contraction of frozen ground. See also: *patterned ground*.

nonsorted step See *patterned ground*.

nonsorted stripe See *patterned ground*.

norite A coarse-grained intrusive igneous rock. Its essential components are plagioclase and labradorite-bytownite in a content greater than 50%.

norm The theoretical mineral composition of a rock, given in terms of *normative mineral* molecules, that has been determined by specific chemical analyses for classification and comparison of rocks. adj. *Normative*. Cf: *mode*.

normal dip *Regional dip*.

normal fault A fault in which the *hanging wall* is displaced downward relative to the *footwall*; i.e., displacement or *net slip* is in a vertical plane down the inclined plane of the rupture. The angle of dip is usually 45 to 90°. Syn: *gravity fault; slump fault*. Cf: *thrust fault*. See also: *fault*.

normal horizontal separation *Offset*.

normal moveout The increase in arrival time of a seismic reflection incident as a result of an increase in the distance from the shot to the geophone, or due to the dip of the reflecting interface. *Moveout* is the difference in reflection times, as observed from adjacent traces of a seismic record; in particular, when the difference is due to dip of the reflecting interface.

normal polarity 1. A natural remanent magnetization that closely parallels the existing geomagnetic field direction. (See also: *geomagnetic polarity reversal*). 2. A configuration of earth's magnetic field with the magnetic pole located near the geographic north pole. Cf: *reversed polarity*.

normal stress That component of stress that is perpendicular to, i.e., normal to, a given plane. The stress may be either compressive or tensile.

normal zoning In plagioclase, that change by which crystals become more sodic in their outer parts. Cf: *reversed zoning*.

normative mineral A mineral whose presence in a rock is theoretically possible on the basis of particular geochemical analyses, but which may not actually be included. Syn: *standard mineral*.

norm system *CIPW normative composition*.

nucleation The initiation of crystal growth at one or more points in a system; e.g., a melt.

nuclide A particular species of atomic nuclei with a specific Z (atomic number) and A (atomic weight).

nuée ardente Etymol: Fr., "glowing cloud." A highly mobile, turbulent, gaseous cloud, sometimes incandescent, erupted from a volcano. The lower component of a nuée ardente contains ash and other *pyroclastics*, and is comparable

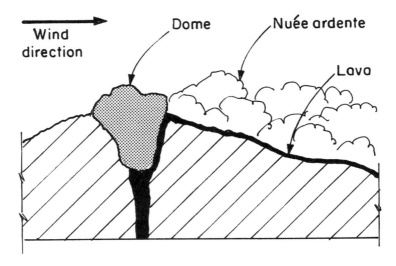

Nuée ardente

to an *ash flow*; the terms are occasionally used synonymously. The typical surface-hugging movement of a nuée ardente can be caused by a lateral blast of volcanic material. When a solidified *lava dome* obstructs the vent of a volcanic crater, pressure builds up beneath the obstruction until some escape route is found or made. A famous example of a nuée ardente was that which occurred a few minutes after Mt. Pelée, Martinique, erupted in 1902; it annihilated 30,000 inhabitants. Syn: *glowing cloud*. See *ignimbrite; ash flow*.

nummulite One of a family of very large Foraminifera having discus-shaped or coin-shaped multichambered cells. Nummulites were extremely abundant in the seas of the Mediterranean region during Eocene and Oligocene time, and their tests contributed largely to the *nummulite limestones* that are widely distributed in southern Europe, northern Africa, and the Himalayan region, where the Eocene is still spoken of as the *nummulite period*. Although less common in the American Eocene, such shells are abundant in the upper Eocene of Florida.

nunatak Etymol: Eskimo, "lonely peak." An isolated peak of bedrock that projects above an icecap or glacier and is surrounded by ice. Nunataks can be seen along the coast of Greenland.

O

oblique fault A fault that trends oblique to, rather than perpendicular or parallel to, the direction (strike) of the dominant rock structure. Syn: *diagonal fault*. Cf: *oblique-slip fault*.

oblique joint *Diagonal joint*.

oblique-slip fault A fault on which the displacement has had a *net slip* down and to the side. Thus, movement occurs diagonally downward and upward, with both strike and dip components. This type

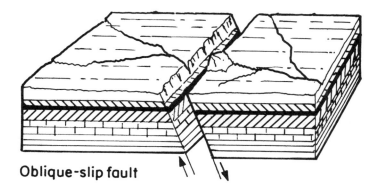

Oblique-slip fault

of fault is related only to the type of motion, and not to the original attitude of the rocks. Syn: *diagonal-slip fault.* Cf: *oblique fault.* See also: *fault.*

obsequent Said of a feature, geologic or topographic, that does not resemble or coincide with the *consequent* feature from which it evolved; e.g., an *obsequent stream.*

obsequent fault-line scarp See *fault-line scarp.*

obsequent stream A watercourse that flows in a direction opposite to that of the original inclination of the land; that is, contrariwise to the *consequent stream,* and usually, but not always, opposite to the dip of the sediment. Syn: *obsequent.*

obsidian Volcanic glass, usually black, but sometimes brown or red, characterized by conchoidal fracture. Most obsidian is formed from a rhyolitic magma. See *glass.*

obtuse bisectrix See *bisectrix.*

ocean basin The part of the earth's crust that lies beyond the continental margin, including abyssal plains and hills, oceanic ridges, volcanic islands, and trenches. Some oceanographers include the slope and rise of the continental margin.

ocean-bottom deposits Conditions of sedimentation beneath the shallower waters of the continental shelf are different from those below the deep waters of open ocean. Because of these conditions, ocean-bottom deposits are considered in three basic groups: 1. *Terrigenous deposits*—found on the continental shelf; these are mainly materials from the land; e.g., shallow water sands and muds. 2. *Neritic deposits*—farther out along the continental shelf; these consist of land (terrigenous) materials mixed with organic substances from shallow coastal waters. 3. *Pelagic deposits*—beyond the direct influence of land. These comprise biogenic and nonbiogenic materials whose origin is almost solely the sea.

oceanic crust The crustal rocks that underlie the ocean basins, referred to as the sima layer (silica and magnesia); about 5 to 10 km in thickness. Cf: *continental crust.* See also: *crust.*

oceanography The application of various disciplines to the study of the oceans and seas of the earth. Such disciplines include marine biology, marine geology, chemistry, physics, and meteorology.

oceanography, optical A branch of physical oceanography that embraces the study of seawater's optical properties and the underwater field of natural light from the sky and sun. Optical measurements are useful in oceanographic problems, such as diffusion mixing and the characterization of water masses, as well as in marine biology.

oceanography, physical That science involved with describing the physical state of the sea and its variations in space and time, and with formulating the physical processes occurring in it.

ocean trench A long, narrow, submarine depression with steeply inclined sides. The deepest ocean trenches all have about the same maximum depth—approximately 10,000 m—and are characterized by strongly negative *gravity anomalies*. Deep oceanic trenches occur along *subduction zones*, which are the sites of the earth's most intense earthquake activity. Although they are located in the three major oceans, trenches lie predominantly around the Pacific periphery. Some are arcuate in form, some rectilinear, and others show sharply angular changes in trend. See also: *plate tectonics*.

ocellar The texture of an igneous rock in which aggregates of small crystals are arranged about larger crystals in such fashion that eyelike structures, similar to the eye of a cat, are formed.

ocher A powdery red, brown, or yellow iron oxide used as pigment; e.g., brown or yellow ocher (limonite) and red ocher (hematite); also, any of various clays deeply colored by iron oxides. Cf: *umber; sienna*.

Ochoan Uppermost Permian of North America.

octahedral cleavage See *cleavage*.

octahedrite *Anatase*.

offset 1. The horizontal component of displacement along a fault, measured perpendicular to the interrupted horizon. Syn: *normal horizontal separation*. 2. In reflection shooting, the horizontal displacement of the reflection point from the vertical plane of the profile. The displacement is used as a correction for converting slant time to vertical time. 3. A spur or branch from a range of mountains or hills.

offshore bar 1. *Longshore bar*. 2. A term often imprecisely applied to both *barrier beach* and *barrier island*.

offshore beach *Barrier beach*.

offshore oil Oil that resides in the largely mostly marine sedimentary rocks that are the submerged extensions of the continents.

ogive Any dark, curved, convex downslope band visible on a glacier surface. Typically, an ogive is composed of debris-laden ice. These bands are also called *Alaskan bands, dirt bands*, and *Forbes bands*.

oikocryst A matrix of host crystal through which smaller crystals (*chadacrysts*) of different minerals are distributed as poikilitic inclusions.

oil field Two or more *oil pools* on a single geologic feature.

oil pool A subsurface accumulation of petroleum that can yield crude oil in quantities that are economically favorable. The oil occurs in the pores and crevices of rock, and is not a "pool" in the ordinary sense of that word.

oil sand A general term for any porous stratum that contains petroleum or is impregnated with hydrocarbons. See also: *gas sand; tar sand*.

oil shale See *oil sources*.

oil sources Oil is found in permeable rocks confined by virtually nonpermeable ones and in structural positions that are determined by hydraulic laws. Most oil pools occupy the crests of elongated anticlinal folds. Less common are the *salt-dome* fields. It appears to be the opinion of most geologists that petroleum originates exclusively in sediments. Almost all oil pools are located in sedimentary rocks. Some rare exceptions could have received oil by migration from adjacent sedimentary rocks. Most pools are hundreds or thousands of meters removed from igneous or metamorphic rocks. There are also many oil pools in or near thick accumulations of marine or deltaic sedimentary rocks that include a high proportion of shales. These shales contain as much as 80% organic matter, mainly carbon compounds derived from plant and animal remains; they were probably deposited as marine ooze, as judged from the content of residues in oozes. Continued bacterial alteration of such residues would partly eliminate oxygen. If, then, some oozes accumulated in oxygen-depleted stagnant basins, the hydrogen-carbon residue might not combine with oxygen to re-form water and carbon dioxide. Although the presence of carbon and hydrogen might "set the stage" for the next geologic scene, the pressure-temperature conditions and other factors are not defined, and so, the actual transformation of organic material to oil is still not well understood.

Little free oil has been found in sedimentary rocks of freshwater origin, despite great quantities of *"oil shale"* (a shale containing no free oil, but from which oil can be distilled) in Eocene lake beds of Utah and Wyoming. Oil and coal are not the respective liquid and solid alteration products of peat, even though both sometimes occur in the same sedimentary sequences in mid-continent. Insofar as oil is absent from Holocene strata, although similar compounds have been extracted from them, the oil-forming process seems to be extremely slow.

oil trap Any barrier that impedes the upward movement of oil or gas, and allows either or both to accumulate. A dis-

tinction is usually made between traps that result from some deformation, such as folding or faulting (e.g., an anticlinal trap), which are called *structural traps*, and those in which certain lithologic peculiarities are accountable. These are called *stratigraphic traps* or *lithologic traps*. An example would be the grading of a porous section into a nonporous section. See also: *shoestring sands*.

oil-water contact The boundary layer or surface between an oil accumulation and the underlying "bottom water."

old age A term applied to the final stage in the cycle of erosion, in which the surface has been brought to almost base level. It may also refer to that stage of stream development in which sediment builds up and water flow decreases.

Old Red Sandstone A thick sequence of Devonian rock, nonmarine in origin, that occurs in northwest Europe and represents the Devonian System in Great Britain and other parts of northwest Europe. Great masses of sand and mud accumulated in structural basins between the ranges of the Caledonian Mountains. The resulting sediments are poorly sorted and quite variable, consisting of red, green, and gray sandstones and gray shales; fossil fauna and flora also occur in the sequence. Cf: *New Red Sandstone.*

Old Stone Age *Paleolithic.*

oligo- A prefix meaning "few" or "a little."

Oligocene Epoch of the Tertiary Period between the Eocene and Miocene Epochs. See Cenezoic Era.

oligoclase A plagioclase feldspar ranging in composition (albite and anorthite) from $Ab_{90}An_{10}$ to $Ab_{70}An_{30}$. It is common in igneous rocks of intermediate to high silica content, and is present as a major constituent in granodiorite and monzonite rocks.

oligomictic 1. A lake that circulates only at times of inordinately cold weather is described as *oligomictic.* 2. Said of a clastic sedimentary rock comprised of only a few rock types. 3. Said of a conglomerate that is composed almost entirely of a single type of fragment. Cf: *monomictic; polymictic.*

oligotrophic lake A lake characterized by a deficiency in plant nutrients and generally by large amounts of dissolved oxygen in the bottom layers; the bottom deposits have only small amounts of organic matter.

olistostrome See *megabreccia.*

olivine An olive-green to yellowish, or brown through alteration, orthorhombic inosilicate, $(Mg,Fe)_2SiO_4$. The term refers to forsterite and fayalite, which form a continuous series. Olivine is a common rock-forming mineral, typical of ultramafic and mafic intrusive or volcanic rocks; it is abundant in gabbros and basalts. In metamorphic rocks, olivine occurs in high-temperature regional or contact-metamorphosed dolomitic limestones.

omphacite An inosilicate of the pyroxene group, usually occurring in light-green grains. It is found with garnet as a typical mineral of eclogites. Also found in massive seams.

oncolith A limestone which, in thin section, reveals a filamentous structure characteristic of algal encrustations. Oncoliths have an organic origin, and form in marine waters of normal salinity; they are found frequently in modern carbonate environments.

onion-skin weathering *Spheroidal weathering.*

ontogeny The development of an individual organism through various stages. Cf: *phylogeny.*

onyx A variety of chalcedony similar to *banded agate*, in that it consists of alternating bands of different colors; the bands of chaldedony, however, are straight and parallel, unlike those of banded agate. Onyx, esp. that with alternating black and white layers, is used in making cameos. Cf: *agate; sardonyx.*

onyx marble A compact, usually translucent, variety of calcite (rarely, aragonite) similar in appearance to true onyx, esp. parallel-banded travertine, which is used as decorative or architectural material. It is usually deposited from cold-water solutions, often in the form of stalactites and stalagmites in caves. Sometimes sold as "Oriental alabaster." See also: *cave onyx.*

oölite A sedimentary rock, usually a limestone composed mainly of small accretionary bodies cemented together. These bodies, called *ooliths*, resemble fish eggs.

ooze See *pelagic deposits.*

opal A substance variously referred to as mineral, cryptocrystalline, mineraloid (if classified as non-crystalline), or a mineral gel. Its chemical composition, $SiO_2{\cdot}nH_2O$, indicates the content of an indefinite amount of water, usually 3 to 9%. Opal occurs in most colors from transparent to almost opaque. *Precious opal* may have various colors, and a characteristic play of colors. *Fire opal* is a variety of precious opal, reddish in hue, but not necessarily iridescent. Opal is commonly precipitated at low temperatures from silica-rich solutions. It also replaces the skeletons of many marine organisms. *Wood opal* replaces the fibers in fossilized wood without destroying texture or detail.

opalescence A milky, lustrous appearance of a mineral, such as of a moonstone or opal. Cf: *play of colors.*

opalized wood *Silicified wood.*

open-cut mining Surficial mining in

which the valuable substance is exposed by removal of overlying material. Coal, iron, and copper are often worked in this way. Syn: *strip mining.*

open fold See *fold.*

open packing See *packing.*

ophicalcite A marble containing serpentine.

ophiolite Masses of mafic and ultramafic igneous rocks, such as basalt, gabbro, and olivine-rich rocks, whose structure and composition identify them as segments of ocean crust pushed into the continents by plate collisions. Ophiolites found embedded within certain deformed mountain belts represent the last vestiges of oceans that disappeared as continents approached each other. By dating these ophiolite zones, geologists are able to arrive at a chronology for continental motion and for the opening and closing of former oceans. See also: *plate tectonics.*

ophitic A texture of an igneous rock in which randomly oriented plagioclase crystals are surrounded by a single large crystal of pyroxene. Cf: *poikilitic.*

optical calcite Crystalline calcite whose clarity makes it suitable for optical use. It is usually *Iceland spar.*

optic angle *Axial angle.*

optic axis A direction in an anisotropic crystal along which there is a no double refraction.

orbicule, orbicular Large, spherical, onionlike masses with distinct concentric shells. The usual size range is between 2 and 15 cm, but some are 3 m or more in diameter. They are variously distributed through phaneritic rocks of silicic to basic composition. Shells within individual masses are sharply defined (cf: *lithophysal*), and each differs from its immediate neighbors in texture or composition. The minerals of most orbicules are

the same as those of the host rock. The concentric structure seems to be a result of rhythmic crystallization around centers that sometimes hold a xenolithic nucleus. The formation of orbicules usually occurs at early stages of rock consolidation. *Orbicular* describes rocks containing orbicules. See also: *lithophysal; spherulite.*

order A category between class and family in the classification sequence of plants and animals.

Ordovician A period of time extending for 65 million years, from 500 to 435 million y.b.p. The name was proposed for rocks in the Arenig Mountains of North Wales, formerly inhabited by an ancient tribe, the Ordovices. The geological time scale for the Ordovician is based solely on the distinctive *graptolite* fossils contained in the strata. The rocks themselves have no unique characteristics to distinguish their boundaries from the younger Cambrian or older Silurian. Ordovician strata are recognized in the Northern Hemisphere continents, in the Andes, and in Australia, but are absent in the Antarctic and in greater India and Africa. Terrestrial sediments cannot be detected in the Ordovician system, since no land organisms are known. Marine successions are of two types: geosynclinal—thick, deformed, and metamorphosed; and *shelf areas*—thinner, little deformed, and unmetamorphosed. Of the geosynclinal type, the Ordovician rocks of Wales and the Lake District include graywackes and volcanics. The central and southern Appalachians of North America are partly formed of a thick, limestone-shale sequence containing tongues of red sandstones and graywackes. The Great Basin of the western U.S. shows a limestone-sandstone-shale succession. Shelf areas are overlain with thin sediment series, mostly limestone.

Upper Ordovician limestones of such areas in North America typically contain corals and large cephalopods. In contrast with the Cambrian, Ordovician rocks contain more advanced trilobites and brachiopods. Ostracods, crinoids, and the bryozoan groups become abundant in the lower Ordovician. Graptolites, the most important index fossil for this system, are found in all types of sedimentary rock thoughout the Ordovician. Shelf areas of northern North America in the Late Ordovician are characterized by corals, large nautiloid cephalopods, and gypsum-bearing rocks. All these imply warm water and a moderate climate, at least in these areas. Fragmented fossils of the first vertebrates, jawless freshwater fish (ostracoderms), appear in North America but not yet in Europe. Broken fossils of the earliest land plants have been found in Poland and the southeastern United States. Both red and green algae types are known in the Ordovician. Although no fossils of freshwater plants have been identified, the presence of ostracoderms, whose food must have been freshwater vegetation, implies their existence at this time. The upper boundary of the Ordovician System is based on the changes in the brachiopod, trilobite, and graptolite fauna. The lower boundary is variously defined in different countries. In the U.S. the Canadian Series is taken as earliest Ordovician. Although the base of the Ordovician system was originally defined by the Arenig Series, a later erroneous modification that included all the Tremadoc Series in the Ordovician was adopted by many countries.

The beginnings of the Taconian orogeny in eastern North America, and the Caledonian orogeny in northwest Europe, are evident in the Middle Ordovician. Both events were accompanied by widespread volcanicity. There are clear signs that the Sahara was covered by an ice sheet during the Ordovician, implying that North Africa was near the South Pole. This has been accepted as evidence that the continental masses were in motion long before the breakup of Pangaea in the Jurassic. See *continental drift.*

ore A naturally occurring material from which a valuable mineral can be extracted. The word commonly connotes metalliferous minerals, but may be expanded to include any natural material used as a source of a nonmetallic substance. A mass of rock or other material housing enough ore to make extraction economically feasible is called an *orebody.*

ore control Any tectonic, geochemical, or lithologic feature thought to have influenced the formation and placement of ore.

organic reef A *reef* or *bioherm.*

organic rock A sedimentary rock consisting essentially of plant or animal remains. Cf: *biogenic rock.*

original dip *Primary dip.*

orogenic belt A crustal region formed by earth's activity. There are two major groups of orogenic belts that form the continental crust: widespread, relatively undeformed accumulations of sedimentary or volcanic rocks; and long, deformed belts of sedimentary, igneous and metamorphic rocks. The first group is not found everywhere on the continent but, wherever present, always overlies the second group. The second belt group comprises the bulk of the continental crust. Examination of the rocks in the belt shows that many older belts are similar to ones formed in the more recent geological past, and also resemble the belts now being formed by tectonic activity. Thus, a means of extrapolating to past geological time is provided.

orogeny A term literally meaning "the formation of mountains." In its original usage, orogeny was the process responsible for the development of mountainous terrain, and for deformation of rock within the mountains. The term as now used means the process by which features such as thrusting, folding, and faulting within fold-belt mountainous regions were formed. The formation of mountain systems is termed a part of *epeirogeny.* It is now widely accepted that orogeny is a result of compressive forces acting on crustal segments. See also: *plate tectonics.*

orpiment A bright, lemon-yellow to orange mineral, As_2S_3, that occurs in veins of lead, sulfur, and gold ores associated with realgar and, sometimes, cinnabar. Orpiment is deposited in hot springs, and also found in metamorphic dolomites.

ortho- In petrology, a prefix which, when used with the name of a metamorphic rock, indicates deviation from an igneous rock; e.g., *orthogneiss.* The prefix may also indicate the primary origin of a crystalline sedimentary rock; e.g., *orthoquartzite.*

orthoclase A white, gray, or pink mineral of the potassium feldspar group, $KAlSi_3$, found as compact granular masses. Orthoclase is a common rock-forming mineral and an essential component of many intrusive rocks that are formed at medium to high temperatures and cooled slowly: granites, granodiorites, syenites. Cf: *plagioclase; anorthoclase.*

orthogeosyncline See *geosyncline.*

orthographic projection See map *projection.*

orthomagmatic The main stage of crystallization of silicates from a typical magma, during which as much as 90% of the magma may crystallize. Syn: *orthotectic.*

orthoquartzite A clastic sedimentary rock composed almost totally of silica-cemented quartz sand; it is a quartzite of sedimentary origin. The rock is distinguished by scarcity of heavy metals, lack of fossils, and prominent cross-beds. Syn: *quartzarenite.* Cf: *metaquartzite.*

orthorhombic system See *crystal system.*

Osagean Upper part of the Lower Mississippian of North America.

oscillation ripple mark See *ripple mark.*

osmosis The tendency of a liquid to pass through a semipermeable membrane into a solution where its concentration is lower, thereby equilibrating the conditions on either side of the membrane. Cf: *dialysis.*

ostracod A small, bivalved crustacian of subclass *Arthropoda,* found mostly in oceans but also in fresh water; also called mussel shrimp. The carapace, which has pits, lobes, and spines, is hinged at the upper margin. Ostracods are divided according to features of the carapace, which is shed regularly. They range from Cambrian to present, and are usually of microscopic size, although larger forms are known. Ostracods are significant as index fossils in limestone, marine marls, and shales. Certain types are often used as indicators in subsurface petroleum exploration.

outcrop Cru l rock that is exposed at earth's surface, or bedrock that is surficially covered by deposits such as alluvium.

outgassing The removal of occluded gases, generally by heating. Outgassing, in which gases and water vapor were released from molten rocks was responsible for the formation of earth's atmosphere.

outlier An area of rock surrounded by rocks of older age. Cf: *inlier.*

outwash Gravel and sand deposited by meltwater streams on land. *Outwash plains* are produced by the merging of a series of outwash fans or aprons.

overfold See *fold.*

oversaturated 1. A synonym of silicic. 2. *Silica-oversaturated.* Cf: *undersaturated; unsaturated.* See *silica saturation.*

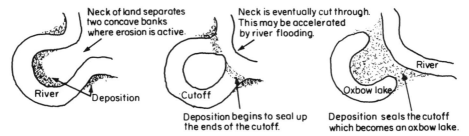

Formation of oxbow and oxbow lake

overthrust *Thrust fault.*

overturn Circulation, esp. in spring and fall, of waters of a lake or sea, whereby surface water sinks and mixes with bottom water; it is the result of density differences due to changes in temperature. See also: *turnover; circulation.*

overturned fold See *fold.*

oxbow See *meander.*

oxbow lake An arc-shaped body of standing water located in river bends (meanders) that are abandoned whenever a stream crosses the neck between bends, thus shortening its course. The abandoned meander itself is referred to as an *oxbow.* After the channel is bypassed, the ends of the channel are rapidly silted and eventually become a crescent-shaped lake. See also: *meander.*

oxidates Sediments composed of the oxides and hydroxides of iron and manganese that have crystallized from aqueous solution.

oxidation The process of combining with oxygen.

oxide A mineral compound of oxygen and one or more metallic elements; e.g., H_2O.

oxidized zone The upper portion of an orebody that has been modified by a solution of surface waters bearing oxygen, soil acids, and carbon dioxide. It extends from the surface to the water table. Primary ore minerals in this zone react with this solution to produce various secondary minerals containing oxygen. Chemical alteration of sulfide minerals within the zone of oxidation results in the formation of acids, which then enable meteoric water to act as a solvent. For example, the weathering of pyrite, FeS_2, produces ferric sulfate, $Fe_2(SO_4)_3$, and sulfuric acid, H_2SO_4. The ferric sulfate reacts with sulfides of copper, lead, and zinc to form soluble sulfates of these metals. The sulfuric acid dissolves carbonates such as the gangue minerals, calcite, and dolomite. As a result of these processes, buried deposits of copper and copper sulfide will change to malachite, $Cu_2CO_3(OH)_2$, or azurite, $Cu_3(CO_3)_2(OH)_2$, under increasingly oxidizing conditions, i.e., toward the surface of the earth. See *secondary enrichment; protore.*

Oxisol A soil order characterized by mixtures of quartz, kaolin, and organic matter, and lacking distinctive horizons. These are deeply weathered soils found in tropical to subtropical regions.

oxygen-isotope fractionation See *dating methods.*

ozocerite A dark-colored paraffin wax that occurs in irregular veins. There are many varieties. Syn: *fossil wax; native paraffin.*

P

Pacific suite See *Atlantic, Pacific, and Mediterranean suites.*

Pacific-type coastline A coastline that runs generally parallel to the dominant trend of the land structures, such as mountain chains. The margins of the Pacific Coast cordillera of North America and the island area of the western Pacific are ideal representations. See also: *Atlantic-type coastline.*

pack ice Any area of sea ice formed by the jamming together of fragments of floating ice.

packing The spatial arrangement of grains or crystals in a rock. It is a function of packing density and packing proximity. If uniform, solid spheres are packed as loosely as possible so that porosity is a maximum; this arrangement is known as *open packing.* If the spheres are packed as closely as possible so that porosity is a minimum, the arrangement is known as *close packing.* An aggregate with *cubic packing* has a porosity of about 47%. This is the "loosest" systematic arrangement of uniform spheres in a clastic sediment or crystal lattice. The unit cell is a cube whose eight corners are the centers of the spheres. *Rhombohedral packing* is the "tightest" systematic arrangement of uniform, solid spheres in a crystal lattice or clastic sediment. The unit cell is composed of six planes that pass through the centers of eight spheres positioned at the corners of a regular rhombohedron. An aggregate with such packing has the lowest porosity (25.95%) possible without grain distortion. The average porosity of a *chance-packed* aggregate of uniform spheres is a bit less than 40%. This is a random combination of grains packed in ordered fashion and grains packed haphazardly.

packstone A clastic limestone supported by its own grains, but also containing some calcareous mud. Cf: *mudstone; grainstone; wackestone.*

pahoehoe A Hawaiian term for lava flows having a smooth, undulating, or ropy surface. Varieties include corded, elephant-hide, entrail, and festooned. Syn: *ropy lava.* Cf: *aa.* See also: *lava flows.*

palagonite A yellow to orange-brown mineraloid that is optically isotropic and often concentrically banded. Unstable basaltic glass can be replaced by many minerals by means of hydration and devitrification. More massive bodies of basaltic glass, that form in subaqueous environments, such as the rinds of pillow lavas, are replaced by palagonite. Palagonite further alters to a clay that is often mistaken for mudstone. See also: *palagonite tuff.*

palagonite tuff A consolidated deposit of glassy basaltic ash in which most of the constituent particles have been altered to palagonite. The term is confusing, insofar as these are not pyroclastic tuffs. (Cf: *tuff.*) To compound the confusion, the term palagonite is used in Iceland to cover all kinds of clastic formations: fragmented subglacial basaltic lavas, glacial deposits, genuine tuffs, and alluvial deposits.

paleo- A prefix meaning old or ancient; e.g., *paleowinds.* Also *palaeo-.*

paleobotany See *fossil plants.*

Paleocene Lowermost division of the Tertiary Period. See *Cenozoic Era.*

paleoclimatology The study of climate through geologic time, extending back to the Precambrian. Basic data are derived from the analysis of sedimentary rocks, their contained fossils, and other sedimentary features. These indicate certain climatic factors, such as rainfall, land and ocean temperatures, and the effects of variations in these. More indirect evidence comes from meteorology, geophysics, and astronomy. Much research in paleoclimatology is focused on explaining the ice ages, the irregular advances and retreats of ice sheets during glacial periods, and the dry hot periods that have prevailed at certain times. Warm periods can be largely explained by continental drift, but ice ages have not yet been totally or satisfactorily explained. See also: *paleocurrents; paleoecology; paleogeography.*

paleocurrents Currents of water or wind, active in the geological past, whose direction can be inferred from sedimentary structures such as crossbedding, ripple marks, and *striations. Dune* type and heavy mineral distribution are also used to determine paleocurrent direction.

paleoecology The study of all aspects of relationships between fossil organisms and their ancient life environments, in particular, their relation to the sediment. These relationships are also treated by paleontology and biostratigraphy. Much paleoecological work depends on comparative studies of modern and ancient forms.

Paleogene Infrequently used term for the Paleocene, Eocene, and Oligocene Epochs of the Tertiary, when grouped together. Cf: *Neogene.*

paleogeographic map A map showing reconstructed physical geography at a specified time in the geologic past; it includes the distribution of land and seas, depth of the seas, geomorphology of the land, and climatic belts.

paleogeologic map A map that shows the areal geology of a land surface at some time in the geologic past; e.g., showing an unconformity as it was before it was covered with deposits.

Paleolithic In archaeology, the time characterized by the appearance of human life and manmade tools. Syn: *Old Stone Age.* Cf: *Neolithic.*

paleomagnetism The study of changes of the geomagnetic field during geological time. The basic assumption is that some rocks, at the time of their formation, acquire a permanent record of the earth's magnetic field because of the presence of certain iron-bearing minerals. (See *natural remanent magnetism.*) Paleomagnetism has provided substantial evidence to support the theory of *continental drift* and the occurrence of *polar wandering.* It has been used to locate the past positions of land masses in relation to each other. Studies of alternating magnetic patterns on either side of mid-ocean ridges (see *geomagnetic polarity reversals*) have supported the hypothesis of *sea-floor spreading.*

paleontology The study of life forms that existed in past geological periods, as represented by fossil plants and animals. Cf: *neontology.* See also: *historical geology.*

paleosols Soils of past geologic ages, usually buried by more recent materials.

paleotectonic map A map intended to represent geological and tectonic features as they were at some time in the geologic past, rather than the totality of all tectonics of a region.

paleotemperature Surface temperature during the geological past, as determined indirectly through the use of different minerals, rocks, fossils, and geochemical isotope ratios. For example, certain corals can live only within a particular temperature range. The presence of iceberg-transported glaciated boulders in water whose present temperature does not fall below 10° C indicates much lower paleotemperatures. Because the isotope ratio of ^{16}O to ^{18}O is partly temperature-dependent, its measurement in carbonate shell matter of marine organisms indicates the ambient water temperature at the time of shell secretion.

Paleozoic The era ranging from 600 to

230 million y.b.p. The Lower Paleozoic includes the Cambrian, Ordovician, and Silurian periods. The Upper Paleozoic includes the Devonian, Carboniferous (Mississippian and Pennsylvanian), and Permian periods; it was preceded by the Precambrian and followed by the Mesozoic. The first widespread fossils, esp. those with hard parts, occur in the Lower Cambrian: algae, arthropods (ostracods and trilobites), brachiopods, sponges, coelenterates, worms, mollusks, and echinoderms. Graptolites are important early Paleozoic *index fossils*. The earliest fish remains occur in the Ordovician. Trilobites and graptolites become extinct in the Upper Paleozoic. Corals, crinoids, and sponges increase in abundance; the first terrestrial flora appears in the Devonian. Three major orogenies occurred during the Paleozoic in North America; the *Taconic* in the latter Ordovician to Lower Silurian, the *Acadian* in the Devonian, and the *Allegheny*, which was probably in the latest Paleozoic and perhaps up to the earliest Tertiary. The greatest of all submergences took place during the Ordovician. Fully half the North American continent was covered with marine waters at the time of maximum submergence.

paleozoology That branch of paleontology in which both vertebrate and invertebrate animals are studied.

palimpsest The fabric or texture of a metamorphic rock in which remnants of a pre-existing fabric or texture are still preserved. Cf: *relict*.

palingenesis Formation of a new magma by in situ melting of pre-existing magmatic rocks. Cf: *anatexis; neomagma*.

palisade An extended rock cliff or line of cliffs rising sharply from the border of a river or lake; e.g., the Palisades along the Hudson River of New York and New Jersey.

Palisades disturbance A time of orogeny that is thought to have closed the Triassic Period in Eastern North America. The disturbance is named for the Palisades of New York and New Jersey, because part of a diabase sill was injected at this time.

pallasites *Stony iron meteorites* consisting of grains of olivine embedded in a matrix of nickel-iron (10 to 20% Ni). The nickel-iron is a coherent mass enclosing separated stony parts.

paludal Referring to a *marsh*.

palustrine Pertaining to substances growing or deposited in a marsh or *paludal* environment.

palygorskite A fibrous clay mineral, $(Mg,Al)_2Si_4O_{10}(OH)\cdot4H_2O$; that forms lathlike crystals. It occurs in sediments from *playa lakes* and desert soils, and can be a component of fuller's earth. Syn: *attapulgite*.

palynology The study of pollen and spores, and their dissemination. Ancient pollens have proved most helpful in certain stratigraphic analyses and in paleoecology.

panfan *Pediplain*.

Pangaea A supercontinent that existed about 300 to 200 m.y.a. This primitive land mass included most of the continental crust of the earth. About 320 million y.b.p. the protocontinents, *Gondwana* and *Laurussia* (Laurasia's ancestral landmass), converged to form Pangaea. It is believed that at some stage Pangaea split into Gondwana to the south and Laurasia to the north, with the *Tethys* between. The present continents were derived from the fragmentation of these two land masses, and moved north and west of their original position. Pangaea was surrounded by an ancestral universal sea called *Panthalassa*, part of which is the present Pacific Ocean. The Atlantic and Indian oceans formed after the breakup. Paleomagnetic readings,

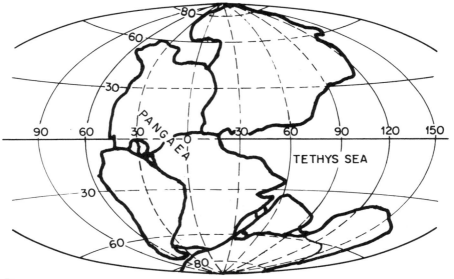

Pangaea

geological evidence, and fossil distribution strongly suggest that plate tectonic activity had already started about 2 billion y.b.p.; it is probable, then, that Pangaea was not the first such supercontinent, but may have had two or three predecessors. The concept of the supercontinent, Pangaea, was part of the theory of *continental drift* proposed in 1912 by Alfred Wegener, a German climatologist. The theory remained controversial until the 1960s, when supporting evidence led to its increased acceptance. See *plate tectonics*.

panning A technique of prospecting for heavy metal by rinsing placer of crushed vein material in a pan. Lighter material is washed away, and the heavy metals remain in the pan.

Panthalassa The ocean that surrounded Pangaea before its fragmentation.

parabolic dune See *dune*.

paraconformity See *unconformity*.

paraffin-base crude Crude oil that will yield large quantities of paraffin during distillation. Cf: *asphaltic-base crude; mixed-base crude*.

paraffin series The methane, CH_4, series; its general formula is C_nH_{2n+2}.

parageosyncline See *geosyncline*.

paralic 1. A term applied to environments of marine margins such as littoral basins and lagoons. 2. Said of coal deposits formed along sea margins, as contrasted with *limnic* deposits.

parallel drainage pattern See *drainage patterns*.

parallel folding See *fold*.

paramagnetic Having a magnetic permeability slightly greater than 1, and a small positive magnetic susceptibility.

Platinum has a permeability value of 1.00002. A paramagnetic mineral placed in a magnetic field shows a magnetization in direct proportion to the field strength; magnetic movements of the atoms have random directions. Cf: *diagmagnetic; ferromagnetic.*

paramorph A pseudomorph having the same composition as the original crystal.

parasitic cone See *volcanic eruption, sites of.*

parent element The radioactive element from which daughter products are given off by radioactive decay.

parent magma The magma from which a particular igneous rock was formed or from which a mineral crystallized; it may also be the source of another magma. Sometimes used as a syn. of *primary magma.* See also: *magmatic differentiation.*

particle size The general dimensions, such as average diameter or volume, of particles or grains that make up a sediment or rock. The gauge of particle size is based on the premise that particles are spheres.

particle velocity In wave motion, the velocity with which an individual particle moves in a medium.

paternoster lakes A chain of small lakes occupying a rock basin in a glacial valley. Their name derives from the similarity of their arrangement to beads on a rosary.

patterned ground A term for various forms, more or less symmetric, that are characteristic of but not restricted to surficial material subject to extensive frost action. *Striped ground*, a pattern of alternating stripes, is produced on a sloping surface by frost action. If the pattern is oriented down the steepest available slope, lack of the alternating bands of fine

and coarser material comprising the pattern is called a *sorted stripe.* The finer component is a *soil stripe*, and the coarser, a *stone stripe*; if the material is very coarse, it is a *block stripe.* A sorted stripe never forms singly; an individual stripe may range from a few centimeters to 2 m wide and exceed 100 m in length. *Sorted circles* and *sorted polygons* are dominantly circular or polygonal shapes having a border of stones surrounding finer material. The diameters of both range from a few centimeters to 10 m. *Sorted net* is a form whose mesh is intermediate between that of a sorted circle and a sorted polygon. A steplike form with a downslope border of stones containing an area of finer material is called a *sorted step.* Forms of patterned ground in which the border of stones is absent, and in which vegetation characteristically outlines the pattern, are referred to as *nonsorted.* Thus, there are nonsorted forms of stripe, step, circle, polygon, and net. See also: *nonsorted polygon.*

peacock coal An iridescent bituminous or anthracite coal. It occurs in the upper levels of mines where water seepage through rock fractures deposits a film of iron oxide.

peacock ore An informal name for a copper mineral having a lustrous, tarnished iridescent surface.

peak zone *Acme-zone.*

pearlite *Perlite.*

peat A deposit of semicarbonized plant remains in a water-saturated environment such as a bog or a meadow with spongy soil. As dead plants fall into the bog or swamp waters, anaerobic conditions permit partial but not total decomposition of the remains. Successive accumulations result in the compaction of material at the bog bottom, and eventually the entire bog may be covered. A vertical section of a peat bog shows the stages of development: living moss, dead moss, laminated peat, and solid peat. Peat that has been cut into blocks and dried is used as fuel in some European countries. Although its carbon content (50 to 60%) and calorific value (7200 to 10,800 BTU) are fairly high, it is an inferior fuel because of its smoke and heavy ash. Peat is considered to be an early stage in the development of coal. The humid climate of the Carboniferous fostered the growth of the great tree ferns and created vast swamp areas that became the first peat beds. Under pressure, peat dries and hardens to become lignite. See also: *coal.*

pebble A rock fragment, generally rounded, with a diameter 4 to 64 mm. Its size is between that of a granule and a cobble.

pedestal rock See *rock pedestal.*

pediment A very slightly inclined erosion surface that slopes away from a mountain front; typically formed by running water. The bedrock may be exposed or thinly covered with alluvium. Pediments are sometimes mistaken for bajadas; they are most conspicuous in *basin-and-range* type arid regions. Syn: *rock pediment.* Cf: *bajada.* See also: *pediplain.*

pediplain A broad, flat, thinly alluviated rock surface formed by the joining of adjacent pediments, usually in arid or semi-arid climates. It is postulated that the pediplain represents the final stage of the arid erosion cycle. Syn: *panfan.* Cf: *pediplane.*

pediplane Any degradational piedmont surface produced in arid climates, and either exposed or covered with a thin blanket of contemporary alluvium.

pedogenesis Soil formation. The study of the morphology, origin, and classification of soils is *pedology.*

pegmatite A very coarse-grained igneous rock, typically found around the margins of large, deep-seated plutons, usually extending from the pluton itself into the surrounding country rocks. Pegmatite bodies are lenticular or podlike in shape. Although diorite and gabbro pegmatites occur, syenite and granite ores are the most common pegmatites. Many pegmatites have an internal zonation of fabric and composition. Most are mineralogically simple, comprising mainly quartz and potassium feldspar, with lesser amounts of muscovite and tourmaline. Other pegmatites have rather high concentrations of gaseous constituents. Pegmatites represent the last portions of magma to crystallize.

pelagic 1. Pertaining to the water of the ocean environment. 2. A term describing the mode of life of those marine organisms living without direct dependance on the sea bottom or shore. This includes the free swimming (*nektonic*) organisms such as fish, and floating forms (*planktonic*), such as jellyfish and sargassum weed, that are carried by the ocean currents. 3. A term referring to the sediments that cover the floor of open sea, in distinction of the *terrigenous deposits* of coastal areas. See *continental shelf.*

pelagic deposits Deep ocean-bottom deposits beyond the direct influence of the land, composed of materials that originate almost exclusively in the sea and are fine enough to be carried in suspension. Pelagic deposits consist of four primary biogenic oozes and non-biogenic red clay. Great quantities of manganese oxide nodules are occasionally found in association with the oozes.
1. Calcareous deposits: *pteropod ooze*—occurs principally in tropical regions at depths of 1000 to 2500 m. The shells are delicate and readily dissolve at greater depths; *globigerina ooze* is found at depths of 2000 to 4000 m. About two-thirds of the Atlantic sea floor and more than one third of the total sea floor are covered with globigerina ooze. These

oozes are unusual at depths exceeding 4000 m, since the $CaCO_3$ of the shells dissolves in sea water under great pressure.

2. Siliceous Deposits: *diatom ooze*—found at depths of 1100 to 4000 m in Antarctic regions and in the southern and far northern Pacific. Although diatoms themselves are not restricted to these regions, their shells are so abundant here that other skeletal remains are obscured; *radiolarian ooze* is found at depths greater than 5000 m in isolated areas of the tropical Pacific and Indian Oceans. At depths greater than 4000 m, $CaCO_3$ is more soluble than silica, and hence, only siliceous skeletal remains survive in sediments deposited below these depths. Siliceous oozes cover about 38×10^6 sq km (15×10^6 square mi.) of ocean bottom.

3. Red clay—mostly inorganic material deposited anywhere on the sea bottom, but more notable in the great depths where it is not obscured by organic deposits. It consists of ultrafine particles of volcanic ash, both submarine and terrestrial, desert dust, and some meteoritic material; oxides of iron and manganese give it its red color. It constitutes more than half the Pacific floor and about one-third the combined area of all sea floors. There is no red clay on the land that corresponds to the red clay of the ocean floor. See also: *ocean-bottom deposits.*

pelecypod A benthic aquatic mollusk of the class Pelecypoda. Oysters, clams, scallops, and mussels are typical members. The group is characterized by a bilaterally symmetrical bivalved shell that completely encloses the animal; habitats may be marine, brackish, or freshwater. Different modes of life among the pelecypods have modified their shell shape. Their early classification was based on these differences, but more recent classification is based on the number of hinge teeth. They have been useful as zonal indices. Range, Lower Cambrian to Recent.

Peléean-type eruption See *volcanic eruption, types of.*

Pele's hair See *Pele's tears.*

Pele's tears Named after the Hawaiian goddess of volcanoes and fire. Pele's tears are cylindrical, spherical, or pear-shaped drops of glassy basaltic lava from 6 to 13 mm long. They are ejected as small blobs of liquid lava, and solidify during flight. *Pele's hair* are filaments of solid lava, sometimes exceeding 2m in length. When Pele's tears are ejected, and are still in the liquid state, they generally have these lava threads attached to them. The threads solidify, break off from the parent blobs, and may drift with the wind for great distances. They are frequently found during eruptions of the Hawaiian volcanoes Kilauea and Mauna Loa. See *pyroclastic material.*

pelite Etymol: Greek, *pelos*, "clay" or "mud." A mudstone or lutite, or the metamorphic derivative of a lutite. See also: *psammite; psephite.*

pelitic Pertaining to or derived from pelite; esp. said of a sedimentary rock composed of clay. Also, a descriptive term to further define certain metamorphic or volcanic rocks. Cf: *argillaceous; lutaceous.*

pelmatozoan Any echinoderm living attached to a substrate, either with or without a stem. The term refers to life habit and is not a true taxonomic identification.

pelmicrite Limestone composed of a variable proportion of carbonite mud (micrite and small rounded aggregates of sedimentary material).

penecontemporaneous Formed during or shortly after the deposition of the containing rock stratum.

peneplain Etymol: Latin, *pene-*,

"almost" + plain. A hypothetical surface to which landscape features are reduced through long-continued mass wasting, stream erosion, and sheet wash (*peneplanation*). The concept of such a surface is based on deductions from stream behavior that suggest a reduction to base level of even resistant rocks by continuous stream erosion. In the Appalachians of Pennsylvania, many ridges show the same elevation for long distances, and are widely considered to be remnants of peneplains that were formed at lower elevations.

penetration twin See *twinning*.

peninsula A body of land nearly surrounded by water and joined with a larger body by a neck; e.g., Florida.

Pennsylvanian A name from the state of Pennsylvania for the Upper Carboniferous in North America, coined in 1891 by Henry Williams as an adjunct to the Lower Carboniferous *Mississippian* strata. The term has been formally recognized by the U.S. Geological Survey since 1953. It is used extensively in North America but not elsewhere. See *Mississippian-Pennsylvanian boundary; stratigraphic correlation.*

pentane Any of three paraffin hydrocarbons, formula C_5H_{12}, found in natural gas and petroleum.

pentlandite A bronze to yellow isometric mineral, $(Fe,Ni)_9S_8$, an important ore of nickel. It occurs in massive form, usually intergrown with pyrrhotite.

peocilitic *Poikilitic.*

perched boulder A large detached rock lying in stable or unstable position on a hillside.

percolation The slow, laminar movement of water through any small openings within a porous material. Used as a syn. of *infiltration*.

percussion mark A lunate scar formed on a hard, compact pebble by a sharp blow, possibly indicating high-velocity flow.

perenially frozen ground *Permafrost.*

perennial stream A stream that flows throughout the year.

pergelation The formation of permanently frozen ground, either in the past or present.

pergelisol *Permafrost.*

peri- A prefix meaning "near" or "around."

peridot An olive-green mineral, the gem variety of olivine.

peridotite A coarse-grained intrusive igneous rock; its essential minerals are olivine, clinopyroxene, and orthopyroxene, with little or no feldspar. Limited masses are found inside ophiolitic complexes. Peridotite is often associated with deposits of nickel-bearing minerals.

periglacial Referring to locations, conditions, processes, and topographic features of areas that are adjacent to the borders of a glacier. A periglacial climate is characterized by low temperatures, many fluctuations about the freezing point, and strong wind action, at least during certain times.

perlite Lustrous, pearly gray glass with perlitic texture.

perlitic A texture found in glassy and devitrified rocks, consisting of clusters of concentric, curved fractures. A small kernel of black, uncracked obsidian is sometimes found in the core of a cluster of cracks. Perlitic glasses contain up to 10% by weight of water, while obsidian without cracks usually has less than 1% water. Contraction during cooling is one theory for the development of perlitic texture. Another is that hydration of a thin outer layer of the original obsidian causes this part to expand and crack away from the non-hydrated section. As hydration continues, perlitic cracks develop inward.

permafrost Soil or subsoil that is permanently frozen. It occurs in those regions (arctic, subarctic, alpine) where the mean annual temperature of the soil remains below freezing. Permafrost underlies almost 20% of the earth's land area, where its thickness ranges from 30 cm to more than 1000 m. If permafrost is able to return to an enduring frozen state under existing climatic conditions after it has been thawed naturally or artificially, it is called *active permafrost*. Exploration during the past decade has revealed the existence of subsea permafrost off the Alaskan shore. Syn: *permanently frozen ground; frozen ground; pergelisol.* See also: *pingo.*

permanently frozen ground *Permafrost.*

permeability 1. The measure of the ability of earth materials to transmit a fluid. It depends largely upon the size of pore spaces and their connectedness; it is less dependent on the actual porosity. For example, a clay or shale with a porosity of 30 to 50% may be far less permeable to water than gravel with a porosity of 20 to 40%, because the pore interstices of the clay are so small that flow is inhibited. The customary unit of permeability measurement is the *millidarcy.* adj. Permeable. Cf: absolute permeability; effective permeability; relative permeability. 2. The ratio of magnetic flux density B to the inducing field intensity, expressed as $\mu = B/H$.

permeable Said of a rock or sediment that allows a gas or fluid to move through it at an appreciable rate via large capillary openings.

Permian End of the Paleozoic Era, from 280 to 225 million y.b.p.; named for the district of Perm, U.S.S.R., the site where this sequence was first recognized. Permian rocks are generally subdivided into the Lower and Upper Permian Series, and the boundary placed at the base or middle of the Guadalupan stage of marine deposits in western Texas. *Fusulinids* and *ammonoids* are the most useful index fossils of Permian strata. Where fossils are absent, division is difficult and stratigraphic names and boundaries vary with location. The Permian System is represented by marine and continental sediments. Limestone occurs on the shallow shelves of the Northern Hemisphere, and clastic sediments prevail. Rocks of continental origin are mostly sandstone, but include shale and conglomerate. The Permian sequences are distinct in western North America, particularly Nevada, Texas and Utah. Oil and gas are produced from Permian rocks in Oklahoma and Texas. Permian coals occur in Asia, Australia, and Africa, but are not important in the Western Hemisphere.

Although no clear break exists between Carboniferous and Permian life in the fossil record, the Permian saw the extinction of many fossil groups; trilobites, tabulate and rugose corals, productid brachiopods, and goniatite cephalopods are almost extinct at the close of the Permian. Amphibians and some fish decline, although labyrinthodonts still thrive. Freshwater fish are known in much greater variety than marine forms. The first major reptile evolution took place with the emergence of three principal groups: cotylosaurus; pelycosaurus, and therapsids. Many new insects appear, including beetles and true dragonflies. In the Lower Permian, the descendants of Carboniferous ferns (pteridophytes) and seed ferns (pteridosperms) predominate. During the transition from Early to Late

Permian these primitive forms are replaced by *Gigantoperis* in eastern Asia and western North America, and by *Glossopteris* in the southern continents (*Gondwanaland*). Most horsetails (sphenopsids) become extinct. The great scale trees *Lepidodendron* and *Sigillaria* diminished greatly during the Late Carboniferous-Permian glaciation, but reappeared to a limited extent during the Late Permian.

In general, the Late Permian was a time of widespread mountain building and volcanism. The Variscan orogeny of Europe and the Appalachian orogeny of North America, both episodic during the Permian, were diminishing in intensity by the end of this period. According to the theory of *continental drift*, by the end of the Permian Period all the major continents were nearly or had already locked together to form Pangaea. The Permian Period began with glaciation and ended with aridity. Sedimentary rocks of glacial origin vary in the time and environment of their deposition. Glacial beds may be found from the Mid-Pennsylvanian to Early Carboniferous. Some deposits may be the result of highland glaciation; others were deposited in water. A single Permo-Triassic system has been widely adopted, at least in Britain. It is referred to as *New Red Sandstone*.

permineralization Also called *petrification*. Fossils produced in this manner have literally turned to stone. The hard parts of many organisms may be preserved by mineral-bearing solutions after burial in sediment. Percolating groundwater may infiltrate porous shells and bones, and deposit its mineral content in pores and the open spaces of skeletal parts. These added minerals tend to increase the hardness of the original hard parts of the organisms. Silica, calcite, and various iron compounds are the usual permineralizers, but others are known. Petrified wood is the most common example of this type of preservation. The term *petrified* should be applied only to remains that have been permineralized.

See also: *fossil; replacement; mineralization.*

permissive intrusion Magmatic emplacement wherein the spaces occupied by the magma have been created by forces other than its own; e.g., orogenic forces. Also, the magma or rock body emplaced in this manner. Cf: *forcible intrusion.*

perthite A crystal habit of some alkali feldspars, in which crystals of sodium-rich feldspar (albite) are closely intergrown but distinct from crystals of potassium-rich feldspar (orthoclase or, less commonly, microcline). The sodium-rich phase appears to have exsolved from the potassium-rich phase. Cf: *antiperthite.*

Petoskey stone A water-abraded fragment of Devonian coral from the shore of Lake Michigan at Petoskey, Mich. In that area, much of the stone is cut, polished, and mounted in rings, pendants, etc.

petrification See *permineralization.*

petrified rose *Barite rose.*

petrified wood *Silicified wood.*

petro- A prefix meaning "rock."

petrochemistry An aspect of geochemistry in which the chemical composition of rocks is studied; it is not equivalent to petroleum chemistry.

petrofabric analysis *Structural petrology.*

petrofabric diagram *Fabric diagram.*

petrofacies *Petrographic facies.*

petroglyph Carving or writing incised on a rock by prehistoric or primitive people.

petrographic facies Facies distinguished and grouped on the basis of their composition or appearance, without regard to boundaries or natural relations; e.g., "red-bed facies."

petrographic province A wide area in which most or all of the igneous rocks are believed to have been formed during the same interval of magmatic activity. See also: *comagmatic*.

petrography That branch of geology involved with the description and systematic classification of rocks, esp. by means of microscopic examination. Petrography is more restricted in its scope than is *petrology*.

petroleum A naturally occurring, complex liquid hydrocarbon which, after distillation and the removal of impurities such as nitrogen, oxygen, and sulfur, yields a variety of combustible fuels. See also: *oil sources*.

petroleum geology The branch of geology concerned with the origin, migration, and accumulation of oil and gas, and the search for and location of commercial deposits of these materials.

petrology A branch of geology that is broader in scope that petrography. It deals with the origin, distribution, structure, and history of rocks.

pH A symbol for hydrogen-ion concentration. It is used to express the degree of alkalininty or acidity of a substance in solution.

phacolith A concordant, lenticular igneous intrusion. It is generally assumed that the igneous material was intruded during or before folding of the host rock, and was deformed along with it. Phacoliths sometimes occur in vertical groups in the domes of anticlinal folds. Cf: *laccolith*.

phaneritic The texture of an igneous rock whose individual components are discernible with the naked eye. Syn: *coarse-grained*. Cf: *aphanitic*.

phanerozoic Literally, "evident" or "visible life." That period during which sediments containing obvious plant and animal remains accumulated. It includes the stratigraphic systems from the Cambrian Period to the present. See also: *Precambrian; Cryptozoic*.

phase diagram A graphic method for showing the boundaries of equilibrium between different phases, the parameters usually being temperature, pressure, and composition. Solubilities and sequences of precipitation can be determined from such diagrams; the point whose coordinates represent the chemical composition of a phase is the *composition point*. A phase is a physically and chemically homogeneous part of a system that is separated by an interface from contiguous phases. Several phases can coexist in equilibrium in a system. Interpretative phase diagrams and phase equilibria are facilitated by the use of Gibbs' *phase rule*, which states that the number of components comprising a system is the minimum number of chemical constituents required to define completely the composition of *every* phase of the system. A system is defined by its number of components; a two-component system is a *binary system*; the albite-anorthite-potassium feldspar system is *ternary*; if water is involved in the ternary, the system becomes *quaternary*.

phase rule See *phase diagram*.

phase velocity See *phase velocity and group velocity*.

phase velocity and group velocity Phase velocity is the velocity with which an individual wave or wave crest is propagated through a medium; i.e., the velocity of constant phase. If a line is drawn connecting a point of constant phase in each of several seismograms, such as the crest of a certain wave, its slope gives the *"phase velocity"* of waves of a certain period. If the centers of groups of waves of about the same frequency are connected, starting from the same point as the phase-velocity line, the slope of the second line drawn is the *"group velocity"* for waves of the same period. Both kinds of velocity are greater for longer waves than for shorter waves. See also: *dispersion*.

phenocryst A crystal in an igneous rock, larger than other crystals of the matrix, and formed during the early stages of crystal development. It may vary in size from 1 to 2 mm in width to 10 cm in length; its size is a function of both cooling rate and chemical composition. Phenocrysts will be euhedral if the magma system in which they formed was near equilibrium when solidification took place. Differing rates of nucleation and crystal growth of various mineral families in a cooling magma may produce phenocrysts of mineral composition different from the groundmass minerals. An igneous rock characterized by the presence of these large crystals is said to have *porphyritic* texture. See *porphyry; porphyritic*.

phlogopite A magnesium mica mineral, $K(Mg,Fe)_3AlSi_3O_{10}(OH,F)_2$, light brown or yellowish, often in foliated aggregates. Found in medium- to high-grade metamorphic rocks.

phonolite A fine-grained extrusive rock composed primarily of potassic feldspar (esp. sanidine), nepheline, clinopyroxene, or sodic amphibole; it is the extrusive equivalent of nepheline syenite.

phosphate A chemical compound containing the phosphate ion radical, $(PO_4)^{-3}$. Phosphates comprise a very large group of minerals; however, all but a few are rare. The most abundant phosphate mineral is apatite.

phosphatic nodule A rounded mass, consisting of corals, shells, coprolites, mica flakes, and sand grains, ranging in size from a few millimeters to greater than 30 cm. Phosphatic nodules occur in marine strata; e.g., the Permian beds of the western U.S. and the Cretaceous chalk of England.

phosphorescence A luminescence in which the emission of visible light from a substance continues after termination of the radiation that activated the luminescence. Cf: *fluorescence*.

phosphorite A sedimentary rock that contains at least 20% of phosphate minerals. In ancient phosphorites, the phosphorus is present in fluorapatite in the form of pellets, laminae, oölites, and bone fragments.

photic, photic zone An environmental term meaning "allowing the penetration of light," specifically sunlight. The photic zone in the sea extends down to 200 m; the region below this is called *aphotic* ("without light").

photogeology Although the term may be used in the sense of the geological interpretation of any photograph, in practice it is used almost exclusively for that of aerial photographs. The study of satellite photographs is likely to be very useful in the investigation of large-scale geological phenomena.

phreatic cycle A daily, annual, or other time period during which the water table rises and falls.

phreatic explosion Volcanic explosion or eruption of steam, mud or other material that is not incandescent; i.e., does not contain the usual products of a volcanic eruption. Such eruptions are caused by the rapid conversion of groundwater to steam, as a result of contact between the water and an underlying igneous heat source; e.g., magma. Major phreatic events often produce *tsunamis.* Depending on the location, a subglacial explosion may cause a "glacier run." See also: *jökulhlaup.*

phreatic water Originally applied only to water that occurs in the upper *zone of saturation* under water-table conditions, but now applied to all water in the zone of saturation. Thus, it is synonymous with groundwater.

phreatic zone *Zone of Saturation.*

phyllite A regional metamorphic rock, light silvery-gray in color, with a texture grading from granoblastic (quartzose phyllites) to porphyroblastic (albitic phyllite). Phyllites are derived from clayey sedimentary rocks containing a residue of organic material.

phyllosilicate A silicate mineral whose tetrahedral silicate groups are linked in sheets, each group containing four oxygen atoms, three of which are shared with other groups, so that the ratio is two of silicon to five of oxygen. Ex: mica. See *silicates.*

phylogeny The development or evolution of a given group of organisms, i.e., a kind or type of animal or plant.

phylum In zoological classification, the category between kingdom and class. Pl: phyla.

-phyro, -phyric Combining form denoting *porphyritic.*

physical geography A branch of geography that is the descriptive study of earth's surface in terms of man's physical environment.

physical geology A division of geology in which the processes and forces associated with the evolution and morphology of the earth are studied

along with its constituent minerals and rocks, and the materials of its interior.

physiographic cycle *Cycle of erosion.*

physiographic province A region, all parts of which show similar climate and geological structure, and which has had a geomorphic history such that its relief features are readily differentiated from those of adjacent regions.

physiography Originally a description of natural features, this term later became synonymous with physical geography. Still later, especially in the U.S., the term was limited to the description and origin of landforms. Now outmoded and replaced by *geomorphology.* Its remaining usage is now restricted, as in the differentiation of physiographic provinces.

phyteral Vegetal material in coal whose morphologic forms are still discernible; e.g., cuticle or spore casts. Phyterals are distinguished from *macerals,* the organic compositional units that form the coal mass. See also: *coal.*

phytolith A rock formed of plant remains or by plant activity.

phytoplankton 1. Marine or freshwater microscopic unicellular plant cells that include many algal groups. 2. Drifting and free-floating plants of the sea (such as sargassum weed), the more important microscopic diatoms, dinoflagellates, and many groups of photosynthetic bacteria that are part of the *pelagic* region. See also: *zooplankton.*

piecemeal stoping See *stoping.*

piedmont adj. Situated or formed at the base of a mountain; e.g., a *piedmont terrace.* n. A feature at the base of a mountain; e.g., a *foothill.*

piedmont alluvial plain *Bajada.*

piedmont glacier See *glacier.*

piercement dome *Diapir.*

piezoelectric effect The development, in some crystals, of an electrical potential when pressure is exerted parallel to certain crystallographic directions. Quartz and tourmaline are examples of piezoelectric crystals.

pillow lava Spherical or ellipsoidal structures usually composed of basaltic lava, but which can also be andesitic; generally about 1 m in diameter. These formations are the result of the rapid cooling of hot, fluid magma that comes in contact with water, such as occurs when lava flows into the sea or into water-saturated sediments; e.g., beneath a glacier. Lava pillows are extruded from openings as glassy-skinned lava tongues in much the same way as toothpaste is squeezed from a tube. The margins of pillow lavas are composed of glass, which progressively alters to palagonite. The interiors are composed of more crystalline material. See also: *lava; lava flows.*

pilotaxitic A textural term for groundmasses of holocrystalline volcanic rock in which lath-shaped microlites (usually plagioclase) manifest a feltlike, interwoven pattern, common in flow lines. See also: *felty; hyalopilitic; trachytic.*

pinacoid A crystal form, the faces of which are parallel with two of the axes. Pinacoids occur among crystals of all crystal systems except the isometric system.

pinch-out The end of a stratum or vein that thins progressively until it disappears, and the rocks it had separated are in contact with one another.

pingo Etymol: Eskimo, "conical hill." A large, blister-like mound that is pushed up by the artesian pressure of water within or below the *permafrost* of Arctic regions. These mounds may be up to nearly a kilometer in diameter and exceed 60 m in height. The injected water often freezes to form a tremendous ice mass beneath the soil; when the ice melts, the pingo may collapse into a volcano-like shape.

pinnacle A tall, slender column of rock carved out by water percolating downward through cracks in the upper layer.

pinnacle reef 1. A spire of rock or coral, either submerged or slightly above water level. 2. An isolated stromatoporoid-algal reef mound in Middle Silurian rocks of the subsurface; may be oil productive. The term *pinnacle reef* for such a structure is used especially in the Michigan Basin.

pinnate drainage pattern See *drainage patterns.*

pipe clay A fine white to grayish white, very plastic clay used for making tobacco pipes. Syn: *ball clay; potter's clay.*

piracy *Stream capture.*

pisolite 1. A sedimentary rock, very often limestone, composed mainly of pisoliths; a coarse-grained oölite. 2. A term frequently used for *pisolith*. 3. A single unit in a mass of *accretionary lapilli.*

pisolith An accretionary formation in a sedimentary rock; it resembles a pea in size and shape, and is one of the units making up a *pisolite*. Pisoliths have the same radial and concentric internal structure as *oöliths*, but are larger and less regular. They are often formed of calcium carbonate, and some may be the result of a biochemical algal encrustation process. The term is sometimes used to mean *pisolite.*

pitchblende A brown to black variety of uraninite found in hydrothermal sulfide-bearing veins. It is amorphous or microcrystalline.

pitchstone A volcanic glass with the

luster of resin or pitch. Its color and composition are variable.

piton A term used for the very tall *volcanic spines* found in the West Indies; e.g., Petit Piton (750 m) and Gros Piton (800 m) of Santa Lucia. See *volcanic spine*.

pivotal fault *Hinge fault.*

placer A surficial deposit, such as gravel or sand, containing particles of valuable minerals. Common types are *beach placers* and *alluvial placers*. *Placer mining* is a type of mining in which various methods, e.g., dredging, are used to extract and concentrate heavy minerals from gravel deposits.

placoderm A member of an extinct class of fishes; it was enveloped in armorlike plates and had well-developed, articulating jaws. Range, Early to Late Devonian.

plagioclase A group of triclinic feldspars of general formula $(Na,Ca)Al(Si,Al)Si_2O_8$. It forms a complete solid-solution series, from pure albite $(NaAlSi_3O_8)$ to anorthite $(CaAl_2Si_2O_8)$. The series is divided according to increasing proportions of anorthite; albite (An 0 to 10%); oligoclase (An 10 to 30%); andesine (An 30 to 50%); labradorite (An 50 to 70%); bytownite (An 70 to 90%); anorthite (An 90 to 100%).

planar cross-bedding See *cross-bedding*.

planation The process of leveling a surface by erosional agents.

plane-polarized light Light that has been constrained (e.g., by a Nicol prism) to vibrate in a single plane; ordinary light vibrates in all directions perpendicular to its direction of propagation.

plane surveying A method of surveying in which the curvature of the earth is dis-

regarded; e.g., ordinary field and topographic surveying.

plane table An instrument for obtaining and plotting survey data directly from field observations. It consists of a drawing board mounted on a tripod; the board is fitted with a sighting device, e.g., an *alidade*.

planetology Study of the condensed matter of the solar system, including planets and their satellites, asteroids, meteors, and interplanetary material. The term is often used inaccurately as a synonym for *astrogeology*.

planimetry Determination of angles, areas and horizontal distances by measurements on a map.

plankton Free-floating and drifting organisms that have little or no effective swimming ability. Cf: *nekton*. See also: *pelagic; phytoplankton; zooplankton*.

plankton bloom The sudden development or appearance in a body of water of minute plankton in such concentrations as to color the water. See *red tide*.

planktonic See *pelagic*.

Planosol An intrazonal group of soils having a leached surface layer above a hardpan; these soils develop on flat, upland surfaces, beneath grass or trees and in a humid climate.

plastic Capable of permanent deformation without rupture.

plastic deformation A deformation that is not recoverable or is only partly recoverable, but does not involve failure by rupture. Syn: *plastic flow; plastic strain*.

plastic flow *Plastic deformation.*

plasticity index The range of moisture-

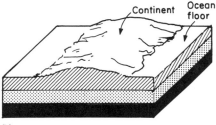

Plate

Plate

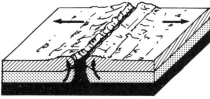

Ridge

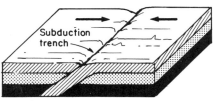

Trench

Basic dynamics of plate tectonics

content values in which a sediment is plastic. It is the difference between the *liquid* and *plastic limits* of the sediment.

plastic limit The moisture-content boundary between the plastic and semisolid states of a sediment; e.g., soil. It is determined by the lowest moisture con-

tent (percent by weight of oven-dried soil) at which the soil cannot be rolled into threads one-eighth inch in diameter without breaking. One of the *Atterberg limits*. Cf: *liquid limit*.

plastic strain *Plastic deformation.*

plate See *plate tectonics.*

plateau basalt Syn: *flood basalt.*

plateau, submarine A comparatively flat-topped sea-floor elevation, usually rising at least 200 m above its surroundings. Some plateaus drop off on all sides, but others are asymmetric and drop off only at one or more sides. 1. *Marginal plateaus* are closely related to continental borderlands. They appear to differ from continental shelves only in depth, which ranges from 2000 to 3000 m. 2. *Mid-ocean plateaus* are preferably denoted as "rises," since depth recording shows their surface to be very irregular. Essentially flat-topped platforms may be of continental origin.

plate boundary See *plate tectonics.*

plate collision See *plate tectonics.*

plate separation See *plate tectonics.*

plate tectonics A synthesis of geological and geophysical observations in which the earth's crust is conceived to be divided into six or more large rigid plates, each of which rotates about its own axis. The movements of these crustal plates are thought to produce regions of tectonic activity along their margins; these regions are believed to be the site of most large earthquakes, volcanism, the formation of island arcs, and mid-oceanic ridges. The theory of plate tectonics, formulated during the late 1960s, unified and expanded the earlier hypothesis of *continental drift* and *sea-floor spreading*. These terms are now considered somewhat inappropriate because individual plates are typically composed of both continental and oceanic crust. (See *crust*.) In determining the cause of sea-floor spreading, a model of three structural elements was used: ridge crests, trench systems, and transform faults. When earthquake epicenters were plotted on a world map, it was found that the seismicity of the earth precisely defines those three structural elements. The simplest type of *transform fault* is that which traverses mid-ocean ridges at right angles to the ridge crest. It appears to be seismically active only between the offset points on the ridge crest. Where plates separate, magma wells up to form an *oceanic ridge* or a *rift valley*; the process is accompanied by seismic and volcanic activity. Where continental plates collide, the heavier of the two will plunge into the asthenosphere, thus forming an oceanic trench. Along this boundary, the *subduction zone*, occurs the most profound seismicity. When plates converge, and one is not subducted beneath the other, the effect is to produce great mountain ranges such as the Himalayas and the Alps. See also: *convection; convection cell*.

platform 1. Any level surface ranging in area from a bench or terrace to a plateau or peneplain. 2. *Wave-cut platform*. 3. The part of a continent that consists of basement rocks covered by essentially flat-lying beds of sedimentary rock; it is part of the *craton*.

platy flow texture A structure found in igneous rock, in which the disposition of tabular sheets resembles stratification. It results from contraction during cooling; the structure parallels the surface of cooling and is often emphasized by weathering. Syn: *planar flow structure*.

playa 1. A flat, dry, barren plain at the bottom of a desert basin, underlain by silt, clay, and evaporites. It is often the bed of an ephemeral lake, and may be covered with white salts. Playas are common in the southwestern U.S. See also: *alkali flat; dry lake*. 2. *Playa lake*. 3. A small sandy area along a bay shore or at the mouth of a stream.

playa lake A shallow, recurring lake that covers a playa after rains, but which disappears during a dry period. Syn: *playa*.

Playfair's law A generalization, proposed in 1802 by John Playfair, concerning accordant stream junctions. It was based on the observation that stream valleys and glaciated valleys differ in the junction of their tributaries, and that all tributaries in stream valleys join at grade, which Playfair felt was compelling evidence that streams form their own valleys.

play of color An optical phenomenon observed in certain minerals, e.g., opal, in which prismatic colors flash in rapid succession. It is caused by a dispersion of light that varies according to the angle of incidence, and the size of the oriented spherical particles of which the mineral is composed and which function as a diffraction grating.

Pleistocene Lower division of the Quaternary Period. See *Cenozoic Era*.

pleochroism The ability of some crystals to exhibit different colors when viewed from different directions under transmitted light. It is a property only of anisotropic crystals, and results from the selective absorption of certain wavelengths of light along different crystallographic directions. It is particularly apparent under polarized light. Tetragonal and hexagonal crystals exhibit *dichroism*, since there are only two directions in which there is differential absorption. In orthorhombic, monoclinic, and triclinic systems there are three directions along which light may be transmitted to yield three different colors; this is called *trichoism*.

pleosponge *Archaeocyathid*.

plication Intense folding on a small scale. adj. *Plicated*. Cf: *crenulation*.

Plinian-type eruption See *volcanic eruption, types of*.

Pliocene Uppermost division or epoch of the Tertiary Period. It is followed by the Quaternary Period. See *Cenezoic Era.*

plucking See *glacial plucking.*

plug See *volcanic plug.*

plug dome A type of *volcanic dome.*

plumbago *Graphite.*

plume A persistent column of magma rising upward from the earth's mantle to the crust; it appears to be the main contributing factor to *hot spots.*

plunge In the geometry of folds, a plunge is the inclination of a fold axis, measured in the vertical plain. Together with the strike of the horizontal projection, the plunge defines the attitude of the hinge of a fold.

plunging fold A fold in which the hinge line is inclined to the horizontal.

pluton A general term for any massive body of igneous rock formed beneath the surface of the earth by the consolidation of magma or by the *metasomatic* replacement of an older rock. See *batholith; dike; laccolith; sill.*

plutonic A term applied generally to rocks, usually igneous, of deep-seated origin. It also describes a rock formed by any process at great depth. Cf: *volcanic.*

plutonic water Juvenile water derived from or in magma, at a considerable depth. Cf: *magmatic water; volcanic water.*

pluvial Pertaining to rain or precipitation; e.g., a *pluvial climate.*

pneumatolysis The alteration of rock or crystallization of minerals by gaseous emissions from solidifying magma. adj. *Pneumatolytic.*

pocket transit *Brunton compass.*

pod An orebody or intrusion of elongate or lenticular form.

Podzol Etymol: Russ., "under ash." A group of zonal soils of which the surface material is organic matter; beneath this there is an ashen-gray layer, and then a zone in which iron and aluminum minerals accumulate. Podzol develops in a cool to temperate and moist climate.

poikilitic 1. Term applied to the speckled appearance of some minerals within igneous rock. Usually, the host completely encloses the inclusions. 2. Textural term describing a condition in which small granular crystals are variously oriented within larger host crystals (*oikocrysts*) of a different mineral. The normal crystallization of magma, in which the inclusions are earlier-formed minerals, can produce poikilitic texture. It can also result from mineral replacement during postmagmatic processes; in this case, the crystal grains indicate remnants of the replaced mineral. Cf: *poikiloblastic.*

poikilo- Prefix meaning "spotted."

poikiloblastic A texture of metamorphic rock, caused by the development during recrystallization of a new mineral around remnants of the original minerals, simulating the *poikilitic* texture of igneous rocks. Cf: *poikilitic.*

point group One of the 32 *crystal classes.*

Poisson's ratio The ratio of unit elongation, c, to unit lateral contraction, s, in a body stressed longitudinally within its elastic limits. The ratio c/s is constant for a given material.

polar glacier See *glacier.*

polarity epoch See *constant polarity epoch.*

polarity interval The fundamental unit of worldwide polarity-chronostratigraphic classification. The term is applied to rock, not time.

polarity reversal *Geomagnetic polarity reversal.*

polarity zone A unit of rock characterized by its particular polarity; it is the basic unit of polarity-lithostratigraphic classification.

polarized light See *plane-polarized light.*

polar migration *Polar wandering.*

polar symmetry A type of crystal symmetry in which the two ends of a central crystallographic axis are not symmetrical. Such a crystal is *hemimorphic.*

polar variation *Polar wandering.*

polar wandering The concept that earth's magnetic poles have migrated over the surface through geological time. The idea is implied in directional changes of the geomagnetic field as determined from the *natural remanent magnetism* of rocks. Pole locations calculated from rocks younger than 20 m.y. show only minor discrepancies from present pole locations. Substantial deviations occur, however, at 30 m.y. and earlier. The fact that polar-wandering curves are different for different locations was one of the first pieces of supportive evidence for *continental drift.* See also: *geomagnetic polarity reversal; paleomagnetism.*

pollen analysis See *palynology.*

poly- A prefix meaning "many."

polyconic projection See *map projection.*

polygenetic 1. Derived from more than one source, or resulting from more than one process of formation; e.g., a mountain range formed by many orogenic incidents. 2. Consisting of more than one type of material; e.g., a conglomerate of materials from several sources. Cf: *monogenetic.*

polygonal ground See *patterned ground.*

polymetamorphism Polyphase or multiple metamorphism, whereby evidence of two or more successive metamorphic events occurs in the same rocks.

polymictic 1. Term applied to a clastic sedimentary rock composed of many rock types. 2. Said of a conglomerate composed of a variety of fragment types. 3. Said of a lake whose water is continuously circulating and shows no persistent thermal statification. Cf: *monomictic; oligomictic.*

polymorphism The characteristic of having two or more distinct forms; e.g., a crystal that has different crystal structures. Iron sulfide crystallizes as two distinct forms: marcasite and pyrite.

polyzoan Syn: *bryozoan.*

pontic Applying to sediments deposited in comparatively deep and still water.

porcellanite A compact, dense siliceous rock having the general appearance of unglazed porcelain; the term is applied to various siliceous rocks of such description.

poriferan *Sponge.*

porosity The ratio of the collected volume of interstices in a soil or rock to the total volume; usually stated as a percentage. Syn: *total porosity.* Cf: *effective porosity.*

porphyritic (adj.) Any igneous rock, regardless of composition, that contains large crystals (phenocrysts) in a groundmass (matrix) of smaller crystals or glass is said to have *porphyritic* texture.

porphyry (n.) A noun often indiscriminately applied to any porphyritic rock. Its use should be restricted to distinctly porphyritic-textured, intrusive igneous rock in which phenocrysts constitute 25% or more of the volume; the groundmass should be fine grained or even *aphanitic*. "Porphyry" is often added to the name of the rock whose texture and composition fit the groundmass part of the rock; e.g., *diorite porphyry* has a fine granular groundmass containing numerous phenocrysts of plagioclase. *Andesite porphyry* is similar, except that the matrix is aphantic. The term "porphyry" should not be applied to porphyritic rocks having a coarse granular groundmass, or to highly glassy porphyritic lava. In such cases, the respective terms *porphyritic diorite* and *porphyritic andesite* should be used, if the rocks so described have the same composition as diorite and andesite. Porphyritic texture is generally interpreted in terms of a two-stage cooling procedure: an initial period of slow cooling, during which large crystals are formed, followed by a period of rapid cooling and the subsequent formation of a finer-grained groundmass. An explanation of porphyritic texture that does not entail two-stage cooling is that crystallization occurs at a fixed degree of undercooling, but different minerals have different rates of nucleation and growth.

porphyroblast *Metacryst.*

porphyroblastic A term used to describe the texture of metamorphic rock in which crystals larger than those found in porphyritic rock are embedded in a finer-grained matrix. Such crystals have grown during metamorphism. Syn: *metacrystic.*

portland cement A cement consisting of a proportional mixture of ground limestone and shale that has been heated until almost fused, and then finely ground.

positive element A sizeable structural feature or area that has had an extended history of continuing uplift.

positive shoreline *Shoreline of submergence.*

possible ore A mineral deposit whose existence and extent are postulated on the grounds of past geological and mining experience.

potash 1. Potassium carbonate. 2. A term rather indiscriminately used for potassium oxide, potassium hydroxide, or even for potassium in such expressions as *potash feldspar.*

potassium-argon age method See *dating methods.*

potassium bentonite See *bentonite.*

potassium feldspar An alkalic feldspar containing the molecule $KAlSi_3O_8$; e.g., orthoclase, microcline, adularia, and sanidine. Syn: *K-feldspar.*

potential ore A mineral deposit not yet located, or for which extraction is not yet economically feasible.

potentiometric surface An imaginary surface that represents the total head of groundwater; it is defined by the level to which water will rise in a well.

potter's clay An iron-free plastic clay, especially suitable for pottery making, modeling, or use on a potter's wheel. Cf: pipe clay.

Pottsvillian Lower Pennsylvanian of eastern North America.

pozzolan A siliceous material that can be finely ground and combined with *portland cement.*

prairie A broad tract of level land or rolling grassland in the interior of North America. It may also refer to one of the grassy plains into which the true prairie of the Mississippi Valley region merges on the west.

Pratt hypothesis A concept of isostasy that assumes a uniform thickness of the earth's crust below sea level, and equilibrium of crustal blocks of varying density. Thus, topographically higher areas, e.g., mountains, would be less dense than topographically lower units. The roots of mountain systems would extend into high-density crustal material called an *antiroot*. Such density variation was explained by the formation of the mountains from the upward expansion of locally heated crustal material, which had a greater volume but lower density after cooling. See also: *Airy hypothesis; isostasy*.

Precambrian That period of time during which the earth's crust was formed and the first life appeared. The terms Archaeozoic, Archean, Azoic, Cryptozoic, and Proterozoic have been used as synonyms or partial synonyms for this period. The duration of the Precambrian is probably not less than 4000 m.y., and covers 90% of geologic time. It is generally divided into the Archeozoic and Proterozoic eras, with subdivisions related to recognizable and datable orogenic episodes. Although presumably worldwide in extent, Precambrian rocks are covered by younger formations in approximately four-fifths of the world's land areas. Exposures are thus limited to: 1. *cores of mountain ranges* bared by uplift and erosion; 2. *gorges*, such as the Grand Canyon; 3. *shields*, or areas that since Precambrian time have never been deeply covered, such as the Canadian Shield. Precambrian rocks of the Canadian Shield contain rich deposits of iron, copper, nickel, silver, and gold. A number of orogenies are known to have occurred during this time, and most Precambrian rocks are therefore strongly metamorphosed. However, some relatively unmetamorphosed Precambrian sediments are known. Fossils of very primitive life forms, as well as trail marks of some wormlike animals, have been found in Precambrian rock. The most abundant fossils are deposits of calcareous algae.

precious metal Gold, silver, or any metal of the platinum group.

precious stone A relatively rare gemstone, such as diamond, ruby, emerald, or sapphire.

precision depth recorder An *echo sounder* with a better than 1-in-3000 accuracy.

pressure gradient 1. The rate of pressure variation in a given direction and at a fixed time; e.g., variation of pressure with ocean depth. 2. Informally, the magnitude of the pressure gradient. Cf: *hydraulic gradient*.

pressure wave P *wave*.

primary dip The slight dip of a bedded deposit that is assumed at its time of deposition. Syn: *original dip*. Cf: *initial dip*.

primary geosyncline *Orthogeosyncline*.

primary gneiss A rock that exhibits characteristics of metamorphism; i.e., planar or linear structures, but lacks observable recrystallization and is therefore considered to be igneous in origin.

primary magma The magma generated below the earth's surface. Sometimes used as a syn. of *parental magma*.

primary mineral A mineral deposited or formed at the same time as the rock containing it, and which retains its original composition and form. Cf: *secondary mineral*.

primary phase The only crystalline phase that can exist in equilibrium with a given liquid; it is the first to appear upon chilling from a liquid, and the last to disappear upon heating to the point of fusion.

primary wave P *wave*.

principal planes of stress Three mutually perpendicular planes, upon each of which the resultant stress is normal; i.e., on which shear stress is zero.

principal stress A stress that is normal to one of the three mutually perpendicular planes that cross at a point in a body on which the shearing stress is zero.

prismatic Said of a crystal with one dimension markedly greater than the other two. May also refer to a metamorphic texture characterized by such crystals.

prismatic cleavage See *cleavage*.

profile A graph or chart that shows the variation of one property with respect to another; e.g., gravity, seismic-wave velocity, and geothermal energy may all be plotted against depth below the surface to obtain desired profiles for seismic or geophysical exploration. See also: *soil profile*.

profile of equilibrium 1. The longitudinal (or long) profile of a graded stream or stream whose gradient at every point is just great enough to enable it to transport its load of sediment. Syn: *graded profile*. 2. The *marine profile of equilibrium* is a smooth, sweeping curve, concave upward. It is steep in the breaker zone and flattens sharply seaward. The slope is just sufficient to keep the material deposited by waves and currents in slow seaward transit.

proglacial Immediately in front of or just outside the limits of a glacier or ice sheet.

propane An inflammable gaseous hydrocarbon, C_3H_8, of the methane series. It occurs naturally in natural gas and crude petroleum.

Proterozoic 1. The later of the two major subdivisions of the Precambrian. Syn: *Algonkian*. Cf: *Archeozoic*. 2. The entire Precambrian.

proto- A prefix meaning "first" or

"foremost," used in the formation of compound words to convey the idea of "essential but not complete." For example, a *protodolomite* is a crystalline calcium magnesium carbonate in which the metallic ions occur in the same crystallographic layers, rather than in alternate layers as in the mineral dolomite.

protore An ore having material too low in concentration or too poor in grade to be mined profitably. See also: *secondary enrichment*.

Protozoa Minute aquatic or parasitic organisms that consist of a single cell or a colonial aggregate of cells without differentiation of function. Protozoa may or may not secrete a skeleton. The typical protozoan is microscopic, but some are as much as 6 cm long. They are divided into a number of groups, but the only two having geological importance are the *foraminifera* and the *radiolarians*.

Foraminifera, or *forams*, are predominantly marine benthic or planktonic protozoans. The test, which may be simple or multi-chambered, is generally composed of calcium carbonate, but may consist of silica or chitin. Certain species build arenaceous tests. Forams are important as age indicators, as rock-building agents, and in sea-floor deposits. Range, Cambrian to Recent. (See *globigerina*).

Radiolarians are marine planktonic protozoans. The test is siliceous, unchambered, and often netlike in appearance. *Radiolarian ooze* is the accumulation on deep sea bottom of the siliceous deposits composed of vast numbers of these tests. These siliceous exoskeletons may be partially responsible for some flint and chert deposits. (Certain cherts are referred to as *radiolarite*.) Their fragility and minute size somewhat limit the radiolarian test in fossil use; however, they have been recorded in rocks from Cambrian to Recent time. See *pelagic deposits*.

provenance A place of origin; in

particular, the area or region from which the constituents of a sedimentary rock or facies were derived. Cf: *distributive province.*

province Part of a region defined by certain peculiarities of climate, geology, flora, fauna, etc.

psammite Etymol: Greek, *psammes,* "sand." A term applied to a sandstone or arenite, or to the metamorphic derivative of an arenite. See also: *pelite; psephite.*

psephite Etymol: Greek, *psephos,* "pebble." 1. A coarse sediment such as gravel, breccia, or conglomerate; equivalent to the Latin-derived *rudite.* 2. The metamorphic derivative of a rudite. See also: *pelite; psammite.*

pseudo- A prefix meaning false. In most scientific terms it denotes a deceptive resemblance to the object or substance to whose name it is prefixed.

pseudofossil An object of inorganic origin that resembles one of organic origin; often found in sedimentary or metamorphic rock. Common false fossils are *dendrites, concretions, cone-in-cone* structures, *septarian nodules,* and *slickensides.* Cf: *fossils.*

pseudotachylyte 1. A dense rock associated with intense fault movement involving extreme mylonitization and/or partial melting. 2. A dark-gray or black rock bearing an external resemblance to *tachylyte* and typically occurring in erratically branching veins.

pteridophyte A spore-bearing vascular plant, belonging to a division that appeared in the Devonian, and including horsetails, club mosses, and ferns. Cf: *bryophyte; spermatophyte; thallophyte.*

pteropod A marine gastropod of the order *Pteropoda.* Pteropods have thin, calcareous shells which, in different species, may be snail-like, slender, conical, globular and spinous, straight, or curved. Their skeletal remains accumulate in the ocean bottom to form *pteropod ooze,* one of the constituents of *pelagic deposits.* Range, Cretaceous to present.

puddingstone Syn: *conglomerate.*

pumice A white or gray, highly vesiculated pyroclastic material that was almost completely liquid at the moment of effusion. It is a form of frothlike volcanic glass, the appearance of which is due to the sudden release of dissolved vapors when it solidified. Its specific gravity is usually less than 1 gm/cm^3. It would have formed obsidian had it cooled under greater pressure. If conditions permit, any type of lava may assume the pumiceous form, though basalts and andesites are less common than rhyolites and trachytes. However, basaltic pumice is found in Hawaii both as pyroclastic material and as froth on lava flows. Cf: *scoria.* See *pyroclastic material.*

pure shear A strain in which a body is lengthened in one direction and shortened at right angles to that direction.

P wave See *seismic waves.*

pyrargyrite A sulfide mineral, Ag_3SbS_3, an important ore of silver.

pyrite A common yellow sulfide mineral, FeS_2, dimorphous with marcasite. Pyrite forms under a wide range of pressure-temperature conditions, and so is found in many geological environments. It occurs in masses, associated with chalcopyrite, among magmatic segregation deposits in mafic rocks. Pyritized concretions are formed by chemical deposition under water. It is the most abundant and widespread of the sulfide minerals and is an important ore of sulfur. Its yellow-gold color has often led to its being mistaken for gold. Syn: *fool's gold.* Cf: *pyrites.* See also: *pyritization.*

pyrites Any of various metallic-looking sulfides, of which pyrite is the most common. The term is used with some qualifying term that designates the component metal; e.g., "copper pyrites" (chalcopyrite).

pyritization 1. The introduction of iron disulfide, FeS_2, into any type of rock. 2. A type of fossilization wherein the original hard parts of an organism are replaced by pyrite, FeS_2. The iron disulfide is probably formed from the sulfur of the decomposing organisms, and the iron from the encasing sediments. Remains of brachiopods and some mollusks are often found in a pyritized state. See *fossil; replacement.*

pyroclast An individual fragment of any size ejected during a volcanic eruption.

pyroclastic 1. Pertaining to fragmental rock materials formed by volcanic explosion or aerial ejection from a volcanic vent. Not synonymous with *volcaniclastic* (fragmental volcanic), insofar as the latter would include fragmental lava. (See *pyroclastic material.*) 2. In the plural, the term is used as a noun, and is a synonym for *pyroclastic material.* 3. A textural term describing the fabric of a rock whose fragmented appearance is the result of explosive volcanic processes; e.g., *volcanic tuff* and *volcanic breccia.* Cf: *volcaniclastic.*

pyroclastic flow A synonym for *ash flow.* See also: *nuée ardente.*

pyroclastic material Pyroclastic material includes those clastic rock materials formed by volcanic explosion or aerial ejection from a volcanic vent. Although in the full geologic record of the earth, the weight of material from lava flows exceeds the weight of pyroclastic products, the latter are more extensive during most volcanic eruptions that are observed. Pyroclastic material may be classified according to size, origin or petrographic composition.
1. *Bombs* and *blocks*—fragments between 64 mm and a few meters.
2. *Lapilli* (It: "little stones")—fragments between 64 and 2 mm.

3. *Volcanic ash*—fragments between 2 and 0.25 mm.
4. *Volcanic dust*—fragments less than 0.25 mm.
5. *Pumice*—a light-colored, highly vesicular glassy material; commonly rhyolitic in composition.
6. *Scoria*—a vesicular material that is usually heavier, darker, and more crystalline than pumice. Both pumice and scoria may exist as pyroclastic material or as froth or lava flows. Ashes, pumice, scoriae, and bombs are formed from ejected fragments of fluid lava. Other ejecta, including lapilli and blocks, consist of fragments of all sizes thrown out in a completely solid or semisolid state.

Pyroclastic material having a rhyolitic composition is generally white or light gray in color; basaltic material is commonly black or red, and andesitic material, intermediate in color. See also: *pyroclastic rocks.*

pyroclastic rocks Rocks consisting of consolidated volcanic ejecta. Such rocks are generally classified according to the size of the particle.

Particle	Size in Millimeters	Aggregate rock
ash	‹ 2 mm	Tuff
lapilli	2-64 mm	Lapilli stone
bombs, blocks	› 64 mm	Agglomerate, volcanic breccia
consolidated lapilli and ash		Lapilli tuff
consolidated volcanic breccia and ash		Tuff breccia

Rocks composed entirely of lapilli or of volcanic blocks are rare, since finer particles usually fill in the spaces. Therefore, consolidated mixtures of ash and lapilli (lapilli tuff) and of ash and blocks (tuff breccia) are common. See *ignimbrite; pyroclastic material.*

pyrogenesis A term encompassing the intrusion and extrusion of magma, and products derived from it. adj. *Pyrogenetic.*

pyrolusite A soft, black tetragonal mineral, MnO_2, the most important ore of manganese. It usually occurs as fibrous, dendritic, or concretionary aggregates.

pyrometasomatism See *metasomatism.*

pyrope The magnesium-aluminum end-member of the garnet series, $(Mg,Fe)_3Al_2(SiO_4)_3$. Its color varies from deep red to almost black. It is found in detrital deposits as fragments, and in kimberlite, an ultrabasic igneous rock in which pyrope crystals occur together with large crystals of olivine within a matrix of serpentine and other minerals. See also: *garnet.*

pyrophyllite A yellowish white, gray, or pale-green mineral, $AlSi_2O_5(OH)$, resembling talc. It usually occurs in lamellar or radiating foliated aggregates.

pyroxene A group of common rock-forming minerals, with general formula $ABSi_2O_6$, where A is Mg, Fe^{+2}, Ca, or Na, and B is Mg, Fe^{+2}, or Al. The minerals crystallize in both orthorhombic and monoclinic systems, and all species have similar cleavage. The principal orthorhombic members of the group are enstatite and hypersthene; the monoclinic members are diopside, augite, and jadeite. Cf: *clinopyroxene.*

pyroxene-hornfels facies A metamorphic rock facies characteristically formed near larger *granitic* or *gabbroic* bodies at depths of a few kilometers; the mineral assemblages reflect lower temperatures. In pelitic rocks, minerals such as quartz, orthoclase, sillimanite, hypersthene, and plagioclase occur. Minerals found in calcareous rocks include plagioclase, diopside, and grossularite.

pyroxenite An ultramafic, intrusive igneous rock composed essentially of pyroxene with accessory olivine, hornblende, or chromite.

pyroxenoid A mineral, such as rhodonite or wollastonite, that is chemically analogous to pyroxene, but whose SiO_4 tetrahedra are joined in chains with a repeat unit of 3, 5, 7, or 9.

pyrrhotite A yellowish-brown, pseudohexagonal mineral, $Fe_{1-x}S$. Its composition is close to that of iron sulfide, but it is deficient in iron. Some pyrrhotite is magnetic; nickel-bearing deposits are among the main sources of nickel.

Q

Q A measure of dissipation in energy-storing systems; it is proportional to the energy stored in the system divided by the energy dissipated during a cycle. Among the many ways in which Q can be characterized is as the degree of perfection of elasticity; one measure of it in the earth is the width of spectral peaks in free oscillations. The decay of *Love* and *Rayleigh waves* after repeated circling of the earth is an alternative method. Using both methods, it as found that the Q of rocks increases with pressure, decreases with temperature, and depends upon the rock's composition and physical state.

quaking bog See *bog.*

quartz Crystalline silica, SiO_2; it occurs as an important constituent of many igneous, sedimentary, and metamorphic rocks. It is the principal mineral in sandstones and quartzites, as well as in unconsolidated sand and gravel. Next to feldspar, it is the commonest mineral, occurring either in coarsely crystalline or cryptocrystalline masses. Colorless and transparent quartz, if found in good crystals, is known as *rock crystal*. Other coarsely crystalline colored varieties are used as gemstones: *amethyst*, purple to blue-violet; *rose quartz*, pink; *citrine*, orange-brown; *smoky quartz*, pale yellow to deep brown. See also: *alpha quartz; beta quartz.*

quartzarenite A sandstone composed primarily of quartz; the term is essentially equivalent to *orthoquartzite.*

quartz diorite A coarse-grained plutonic rock composed essentially of plagioclase, usually with a small amount of orthoclase; the approximate extrusive equivalent of *dacite*. Common dark minerals found in quartz diorite are biotite and hornblende. Its composition is the same as that of *diorite*, but with more quartz (between 5 and 20% of the light-colored constituents). As the alkali feldspar content increases, quartz diorite grades into *granodiorite*. Syn: *tonalite.*

quartz index 1. A quantity in a particular system of rock classification that is an indicator of the degree of silica saturation in a rock. 2. The mineralogical maturity of a sandstone, expressed as the ratio of quartz plus chert to the combined percentage of feldspar, rock fragments and, clay matrix.

quartzite 1. A metamorphic rock consisting primarily of quartz grains, formed by the recrystallization of sandstone by thermal or regional metamorphism; a *metaquartzite*. 2. A sandstone composed of quartz grains cemented by silica; an *orthoquartzite.*

quartz monzonite A granitic rock in which 10 to 50% of the felsic constituents comprise quartz whose ratio of alkali feldspar to total feldspar is between 35 and 60%. It is the approximate plutonic equivalent of *rhyodacite*, grading into *granodiorite* with an increase in plagioclase and femic minerals, and into

granite with more potassium-rich alkali feldspar. Syn: *adamellite.*

quartzose A sediment or rock (esp. sands and sandstones) containing quartz as a principal constituent.

Quaternary The most recent period of geologic time. A division of the *Cenozoic*. See *Cenezoic Era.*

quaternary system See *phase diagram.*

quicksand Quicksand is not a particular kind of sand as far as its chemical composition is concerned; it is a mass or bed of fine sand comprising smooth grains that do not adhere to each other. Water usually flows up through the pores, resulting in the formation of an extremely unstable mass that yields readily to pressure.

Q wave *Love wave.*

R

radial drainage pattern See *drainage patterns.*

radial symmetry A pattern of symmetry in which similar parts of an organism are regularly arranged about an axis or central point; e.g., a starfish or flower. Cf: *bilateral symmetry.*

radioactive age determination *Radiometric dating.*

radioactive decay The spontaneous transformation of a nucleus into one or more different nuclei by the emission of a particle or photon, by fissioning, or by the capture of an orbital electron. See also: *alpha decay; beta decay; gamma decay.*

radioactive series A series of isotopes, each of which becomes the next in sequence by some type of decay until a stable element is reached. The major radioactive series are the actinium, thorium, and uranium series. See also: *parent element; daughter element; end product.*

radioactivity A property of certain elements that spontaneously and constantly emit ionizing and penetrating radiation. See also: *radioactive decay.*

radioactivity log See *well logging.*

radiocarbon Any of the radioactive isotopes of carbon: ^{10}C, ^{11}C, ^{14}C, ^{8}C.

radiocarbon dating *Carbon-14 dating.*

radiogenic isotopes An isotope produced by radioactive decay; it is not necessarily radioactive itself. Cf: *radioisotope.*

radioisotope A radioactive isotope of an element. Cf: *radiogenic isotope.*

radiolaria See *Protozoa.*

radiolarian ooze See *pelagic deposits.*

radiometric dating See *dating methods.*

rain shadow A dry region on the lee side of a topographical obstacle, generally a mountain range, where rainfall is significantly less than on the windward side. Parts of western North America lie in rain shadows behind high mountains (e.g., the Sierra Nevada) that force the prevailing winds to rise, cool, and release most of their moisture in crossing.

range finder An instrument for determining the distance from a single point of observation to other points at which no instruments are located.

range-zone A body of strata that represents the total extent of occurrence of any particular fossil form selected from the collection of fossil forms in a stratigraphic sequence. Both horizontal and vertical directions are included in the word "range." Cf: *biozone; acme-zone.*

rank 1. The rank of coal is largely determined by the amount of fixed carbon, volatiles, and water it contains. In general, it reflects the position of a coal in the transition series from peat to anthracite. Cf: *grade; coal type.* 2. Metamorphic grade.

rapids A part of a waterway where the current is moving swiftly and over obstacles.

rare earths A series of metallic elements having closely similar chemical properties. The series consists of the lanthanide elements (atomic numbers 57 to 71), scandium (21), and yttrium (39). Abbrev: REO.

rational face A crystal face peculiar to the internal atomic structure of the mineral type to which the crystal belongs.

ravinement See *recession.*

ray A vector normal to a wave surface, indicating direction and sometimes velocity of propagation.

Rayleigh number A parameter used in the mathematical description of fluid convection. It is proportional to the ratio of the time required to heat a fluid layer by conduction and the time required for fluid particles to circulate once around the *convection cell*. Rather large Rayleigh numbers occur in the mantle, and have a lower limit of about 10^{6}. For small values, thermal convection cannot occur; for large values, the system becomes unstable and develops disturbances that grow into convection cells.

Rayleigh wave See *seismic waves.*

raypath The imaginary line along which wave energy travels. It is always normal to the wave front in isotropic media.

reaction pair In geology, two minerals that exhibit the reaction principle; i.e., one mineral species is converted into a different mineral by the reaction of the crystal phase with the liquid magma containing it. Thus, forsterite, which is stable at high temperatures, is converted into enstatite at a lower temperature by the addition of silica from the magma containing it. See also: *reaction series*.

reaction point A point on a melting relations diagram in which it is impossible to state the composition of the liquid in terms of all the solid phases in equilibrium at this point.

reaction rim A peripheral zone around mineral grains that are typically anhedral corroded grains of the unstable high-temperature phase of a mineral. The rim is composed of another mineral species, and represents the reaction of the earlier solidified mineral with the surrounding magma. Cf: *corrosion border*.

reaction series A sequence in which early-formed minerals react with the melt to form new minerals that are further down in the series. This concept, originally proposed by N.L. Bowen, suggests that the ferromagnesian minerals form a *discontinuous reaction series*, and the feldspars a *continuous series*. A continuous reaction series is one in which there are no abrupt phase changes during the reaction of early-formed crystals with later melts. A discontinuous reaction series is one in which reactions of early-formed crystals with later melts represent a sharp phase change. In the *discontinuous reaction series*, early-formed crystals of olivine react with the melt to form pyroxene, which later reacts to form amphibole. Simultaneously, a continuous reaction in the feldspar group is in progress. Anorthite-rich plagioclase is the first mineral to crystallize; unless removed, the anorthite plagioclase reacts with the melt to form plagioclase that is continually enriched in albite. See also: *reaction pair*.

realgar A monoclinic mineral, arsenic sulfide, AsS, usually found in compact aggregates and brilliant red-orange films. It is deposited in hot springs and low-temperature hydrothermal veins. If exposed to light and air, it gradually converts to orpiment.

Recent *Holocene.*

recession 1. Glacial recession. 2. Continuing landward progression of a shoreline as a result of erosion; also, the net landward movement during a specific time interval. Ant: *advance*. 3. Backward movement of an eroded escarpment, or the retreat of a slope from a former position without a change in its angle. When considering coastal recession and barrier islands, coastal specialists may use the term *ravinement*. This is a process whereby material from a beach is displaced to the opposite side of the island without significant loss, and is thus not a true erosion process.

recessional moraine See *moraine*.

recharge The sum of the processes involved in the addition of water to the zone of saturation, or the amount of water added. Syn: *intake*.

recrystallization The formation, essentially while in the solid state, of new crystalline mineral grains in a rock. The new grains are usually larger than the primary grains, and their mineral composition may differ or be the same.

rectangular drainage pattern See *drainage patterns*.

recumbent fold See *fold*.

red beds Predominantly red sedimentary strata composed primarily of sandstone, siltstone, and shale; their color is due to the presence of hematite (ferric oxide). They are generally considered to indicate sedimentation in an arid, continental environment, and are commonly associated with evaporate deposits. Examples of red beds include the Old Red Sandstone facies of the European Devonian and the Permian and Triassic sedimentary rocks of the western U.S.

red clay A fine-grained pelagic deposit, reddish-brown or chocolate colored, formed by the accumulation of material at depths generally greater than 3500 m, and at a far distance from the continents. Red clay contains rather large amounts of meteoric and volcanic dust, pumice, shark teeth, manganese concretions, and ice-transported debris. Cf: *red mud.*

red mud See *terrigenous deposits.*

red ocher A red, earthy hematite used as a pigment. See also: *ocher.*

redox potential 1. Oxidation-reduction potential. Symbol E or Eh. 2. The oxidation-reduction potential of an environment; i.e., the voltage obtainable between a normal hydrogen electrode and an inert electrode placed in the environment.

red tide A discoloration of the water caused by the rapid multiplication (bloom) of one or several species of unicellular algae or protozoa. Red tides may be harmless or lethal among invertebrates and fish. Virulent poisons have been isolated from blooms of blue-green algae and of *dinoflagellates.*

reduzates Sediments that have accumulated under reducing conditions, and which are therefore typically rich in organic carbon and iron sulfide; black shale and coal are examples. Cf: *resistates; evaporites; hydrolyzates; oxidates.*

reef 1. A moundlike or ridgelike formation built by calcareous algae and sessile calcareous animals, esp. corals, and composed largely of their remains. It may also be such a structure built in the geological past and now enclosed in rock, usually of dissimilar lithology. Syn: *organic reef.* See also: *barrier reef; coral reef; atoll;*

bioherm. 2. A mass of coral, sand, and gravel rising above a sea or lake bottom, and dangerous to navigation. See also: *shoal.* 3. *Saddle reef.*

reef rock A resistant, massive, unstratified rock composed of the calcareous remains of reef-building organisms, often interspersed with carbonate sand; the entire substance is cemented by calcium carbonate.

reference locality A place containing a reference section designated as a supplement of the *type locality.*

reference plane *Datum plane.*

reference section A rock section specified as a supplement of the *type section,* or as a replacement if the latter is no longer exposed; it also provides a correlation standard for a particular section of the geologic column. See also: *standard section.*

reflection The return of a wave that is incident upon a surface to the original medium. A seismic wave that is reflected at an interface between different media is called the *reflection wave.*

reflection shooting See *seismic shooting.*

reflux A process in which heavy, concentrated brines move downward through the floor of an evaporite basin. Because such brines may be rich in magnesium, reflux is thought to contribute to the dolomitization of carbonate rocks in some sequences. The depth of dolomitization in a *sabkha* may depend on the extent to which reflux occurs.

refraction In physics, the directional change of a ray of light, heat, or sound, or an energy wave, such as a seismic wave, upon passing obliquely from one medium into another in which its speed is different.

refraction shooting See *seismic shooting.*

refractive index *Index of refraction.*

refractory ore Ore from which it is difficult to recover valuable substances.

reg Arabic term meaning "stony desert." A *reg*, as it is called in Algeria, is a stone-littered desert surface from which fines have been removed by deflation and other erosion processes. Similar desert surface is called *serir* in Libya. Cf: *erg; hammada.*

regelation A twofold process involving the melting of ice that is subjected to extreme pressure, and the refreezing of the resultant meltwater upon removal of that pressure. Regelation is one of the factors involved in the *basal sliding* of glaciers.

regime 1. In a stream channel, the existence of a balance between deposition and erosion over several years. 2. The condition of a stream with respect to the rate of its average flow, as determined by the volume of water flowing past different cross-sections within a defined time period. 3. A regular pattern of occurrence, or a condition showing widespread influence; e.g., a sedimentary regime. 4. In glaciology, syn. of *balance.*

regimen 1. The flow characteristics of a stream, such as volume, velocity, and sediment-transport capacity. Cf: *regime.* 2. The total quantity of water associated with a drainage basin, and its behavior, as determined from such quantities as rainfall, surface and subsurface flow, and storage. 3. An analysis of the total quantity of water associated with a lake over a specified time interval. 4. Glacial *balance.*

regional dip A nearly uniform inclination, usually at a small angle, of strata extending over a wide region. Syn: *normal dip.*

regional metamorphism See *metamorphism*

regolith Unconsolidated rock material resting on bedrock, found at and near earth's surface. Residual regolith is formed by the mechanical and chemical weathering of bedrock; transported regolith is moved and deposited by processes acting at or near earth's surface. Regolith material includes alluvium, till, loess, dune sand, and volcanic dust.

regression Retreat of the sea from land areas; also, any change that moves the boundary between marine and non-marine deposition toward the center of a marine basin or converts deep-water conditions to near-shore, shallow-water conditions.

regressive A descriptive term applied to some feature that was formed or deposited during a time of sea withdrawal or land emergence.

rejuvenation The reversal or renewal of some landscape feature or geological form to a former degree of effectiveness, or to a condition leading to a new cycle of erosion. A stream is said to be rejuvenated when, after having developed to maturity, it has its erosive ability renewed as a result of regional uplift. Other factors commonly related to rejuvenation are the lowering of sea level and climate change.

relative age The geological age of a rock, fossil, or event, as defined relative to other rocks, fossils, or events, rather than in terms of years. Cf: *absolute age.*

relative dating The chronological arrangement of fossils, features, or events with respect to the geologic time scale, without reference to their absolute ages.

relative humidity The ratio, expressed as a percentage, of the actual amount of water vapor in a given volume of air to the amount that the same volume would hold if the air were saturated at the same temperature.

relative permeability The ratio

between *effective* and *absolute* permeability in a rock, where effective permeability is the permeability to a given fluid at partial saturation, and absolute permeability is the permeability at 100% saturation.

relic 1. A landform, such as an erosion remnant that has survived disintegration or decay. 2. A metamorphic relict.

relict A feature or fabric structure of a previous rock that still persists in a later rock, despite metamorphism, is said to be *relict*. The term is also applied to topographic features that remain after other parts have disappeared; e.g., a *volcanic neck*.

remanant magnetism See *natural remanent magnetism.*

remote sensing The collection of information or data without the direct contact of a recording instrument. Devices such as cameras, infrared detectors, microwave frequency receivers, and radar systems are used in remote sensing.

reniform A crystal structure in which radiating individuals terminate in rounded, kidney-shaped masses; e.g., *reniform hematite.* Cf: *botryoidal; colloform.*

REO *Rare-earth oxides.*

replacement 1. *Metasomatism.* 2. A process of fossilization also referred to as *mineralization.* (Cf: *permineralization.*) A type of preservation by which the original hard parts of an organism are removed by solution, and the resulting voids filled by deposits or minerals; e.g., calcite, silica, pyrite, limonite. See also: *fossil.*

resequent fault-line scarp See *fault-line scarp.*

resequent stream A river or stream that flows along the dip of underlying strata in the same direction as a consequent stream, but at a lower level and on a surface lower than the original surface.

reservoir rock Any permeable and porous rock that yields gas or oil. Sandstone, dolomite, and limestone are the most prevalent types.

residual clay Clay material formed in place by the weathering of rock; it is derived either from chemical decomposition of feldspar, or by the removal of non-clay mineral constituents from clay-bearing rock.

residual liquid The still-molten portion of a magma that remains in the chamber after some crystallization has occurred through differentiation. Syn: *residual magma; rest magma; ichor.*

resistates Sediments composed of chemically resistant minerals with abundant weathering residues; e.g., highly quartzose sediments. Cf: *hydrolyzates; oxidates; reduzates; evaporites.*

resistivity 1. (Electrical) The resistance per unit length of a unit cross-sectional area of a material. It is the reciprocal of the electrical conductivity and is measured in ohm-cm. $\rho = RA/l = $ ohm-cm, where ρ is the resistivity, R the conductor resistance, A and l the cross-sectional area and length of the conductor, respectively. Also called *specific resistance.* 2. (Thermal) Reciprocal of *thermal conductivity.*

resistivity log See *well logging.*

resistivity method An electrical method of geophysical exploration in which a voltage is applied between electrodes implanted in the ground, and the resulting current is measured. Measurements are usually made between several electrodes crossing the area under study. A change in resistivity, as noted by a change in current, usually constitutes a resistivity anomaly, and may indicate some obscured feature, such as an orebody or fault.

reticulate Said of something having a netlike structure, such as a vein or a lode. A rock texture in which partially altered crystals form a network is said to be *reticulate*.

retrograde metamorphism The response of mineral assemblages to decreasing temperature and pressure. The minerals may form new combinations that are capable of stability within the adjusted conditions. See *metamorphism*.

reverse fault A fault along which the *hanging wall* is displaced upward with respect to the *footwall*. Partial syn: *thrust fault*. Cf: *normal fault*.

reversed polarity A natural remanent magnetism opposite to the present geomagnetic field direction. Cf: *normal polarity*. See also: *geomagnetic polarity reversal; constant polarity epoch*.

reversed zoning *Inverse zoning*.

Reynolds number The parameter that determines whether laminar or turbulent flow will occur in a given condition. For flow in a circular pipe, the Reynolds number $R = 2\rho a v/\eta$, where a is the tube radius, ρ the density of the fluid, v its velocity, and η the viscosity coefficient. For Reynolds numbers below 2000, pipeflow is laminar, and for values above 2000, it is turbulent. See also: *laminar flow; turbulent flow*.

rheomorphism The process by means of which a rock becomes mobile as a result of some degree of fusion; the change is often accompanied by the addition of new material by diffusion. Cf: *mobilization*.

rhodocrosite A banded pink rhombohedral mineral, $MnCO_3$, commonly occurring in granular, concretionary, or stalactitic masses. It forms in medium-temperature hydrothermal veins in association with copper, silver, and lead sulfides, and other manganese minerals. It is used as an ore of manganese when quantities are sufficient, but its principal use is as a semiprecious stone.

rhodonite A deep-pink to brown silicate mineral of the triclinic system, $(Mn,Fe,Mg)SiO_3$; it occurs as compact granular masses with black veins of manganese dioxide. Rhodonite is a mineral typical of the metamorphism of impure limestone.

rhombic system *Orthorhombic system*.

rhombohedral cleavage See *cleavage*.

rhombohedral packing See *packing*.

rhyodacite An extrusive rock, usually porphyritic, with phenocrysts of plagioclase, sanidine (an alkali feldspar), and quartz in a glassy to microcrystalline groundmass. Mafic phenocrysts of any type may be present, but are usually sparse to absent. See also: *rhyolite; dacite*.

rhyolite One of a group of extrusive rocks commonly showing flow texture, and typically porphyritic, with phenocrysts of potassium feldspar in a glassy to microcrystalline groundmass. Rhyolite is the extrusive equivalent of *granite*. See also: *rhyodacite; dacite*.

rhythmites Regularly banded deposits developed by cyclic sedimentation. Fine-textured, dark laminae alternate with coarser-grained lighter layers. An individual pair of such laminae is a "couplet." Rhythmites readily develop on the floors of cold freshwater lakes into which meltwater streams intermittently flow. They are rarely found in salt or brackish water. Nonglacial rhythmites are also known; e.g., some Pleistocene sequences. Many couplets are annual deposits; i.e., a pair of laminae, one fine and dark, the other coarser and lighter, that have been deposited within a 1-year period. Such couplets are called *varves* (Swedish, *varv*, "layer"), and can be used as a means of dating. However, great care must be exercised to ascertain the annularity of a particular rhythmite before it is used for this purpose. In the case of glacial lake laminations that are true varves, the fine and coarse layers are deposited in winter and summer, respectively. There are many conditions that may produce *non-annual rhythmites*; e.g., warm and cold spells of a non-annual nature. In areas where non-annual rhythmites are rare, varves have provided a time scale for the Late Pleistocene.

ria Etymol: Span., *ria*, from *rio*, "river." 1. A long, narrow arm of the sea, produced by the partial submergence of an area dissected by subaerial erosion; it is shallower and shorter than a *fjord*. 2. A long, narrow river inlet that gradually decreases in depth from mouth to head.

Richter scale See *earthquake measurement*.

ridge See *mid-oceanic ridge*.

ridge-and-valley topography A land surface distinguished by parallel valleys and ridges in close succession. Such topography develops as a result of the differential erosion of strata having various degrees of erodibility. In the type region, the Ridge and Valley province of the Appalachian Mountains, the rocks are intensely folded, and show many low-angle thrust faults.

riebeckite A dark-blue or black monoclinic mineral of the amphibole group. It occurs as bluish fibrous crystals in asbestiform aggregates (crocidolite), and has a characteristic chatoyancy.

rift See *rift valley*.

rift trough *Rift valley*.

rift valley A valley that has developed along a tectonic rift resulting from plate separation. Where the plates are separating rapidly, as along the East Pacific Rise, the rift is filled by the magma that wells up to form new sea floor that migrates along the sloping flanks on either side of the

original fissure. Where separation of the plates is slower, as along the Atlantic and Indian ridges, the upwelling magma does not cover over the fissure; instead, it adheres to the trailing edge of the plate on either side of the rift, creating precipitous scarps on each side. The original rift has now become a rift valley, which marks the center of the ridge. The deep cleft along the crest of the *mid-oceanic ridge* is called the *mid-ocean rift*. See also: *plate tectonics*.

right-handed separation *Right-lateral separation*.

right-lateral fault See *strike-slip fault*.

right-lateral separation Displacement along a fault such that the side opposite the observer appears shifted to the right. Syn: *right-handed separation*. Cf: *left-lateral separation*.

rigidity modulus *Modulus of rigidity*.

rille One of many variously shaped, trenchlike valleys on the lunar surface. Rilles are thought to represent fracture systems that originated in brittle material.

ring complex A composite igneous intrusion, the individual components of which appear as circular or arcuate outcrops. *Ring dikes* and *cone sheets* are the two main members of ring complexes. Cf: *cauldron subsidence*. Both components can be created from the upsurge of magma and its migration into subsequent tensile fractures in the host rock. Ring complexes are well-known along the western coast of Scotland. See *dike*.

ring dike See *ring dikes and cone sheets*.

ring dikes and cone sheets Curved dike forms, commonly found together. *Ring dikes* are annular or arcuate in plan, with a more or less vertical axis. The diameter varies considerably; some measure a few hundred meters, others reach several kilometers. The curvature

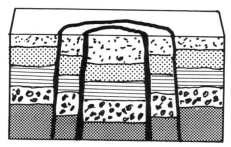

Ring dike

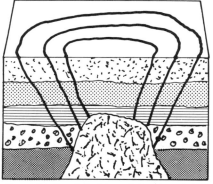

Cone sheets related to an intrusion at depth

seldom encompasses as much as 360°. In the U.S., New Hampshire is known for its many ring dikes; esp. the Ossippee Mountains. *Cone sheets* have the form of a conical surface converging to a buried focus (presumably a magmatic center). These dike forms, often occurring in concentric sets, are generally thin and interrupted, but some do circumscribe the larger part of a circle. The formation of cone sheets is ascribed to the injection of a body of magma into the earth's crust; magmatic pressure produces cracks in the overlying crust, into which magma from the primary body is injected. As the magma subsides, relaxation of pressure allows the structure to slump (*cauldron subsidence*) into the primary magma

body, and causes magma to well up along vertical fractures to form ring dikes. Cone sheets may be likened to cracks made in a window pane by BB shot, i.e., conical, with the apex toward the impact point; ring dikes resemble a collapsing arch. See also: *ring complex*.

Ring of Fire The volcanic chain that describes a rough circle around the perimeter of the Pacific. See *volcanoes*.

ring structure See *ring complex*.

riparian Pertaining to or situated on a riverbank.

ripple mark Small-scale troughs and ridges formed in loose sand by wind. Rippling occurs during gentle winds; when wind velocity rises, the ripples are destroyed; if the wind subsides gradually, the hollows tend to fill. A *crescent ripple mark*, which is asymmetric in form, is a result of the movement of air or wind more or less continuously in one direction. *Oscillation ripple marks* are symmetrical in outline and are produced by the oscillating or "lapping" movement of water. Two sets of differently oriented ripple marks on the same surface sometimes create a crossed or cell-like pattern called an *inteference ripple mark*.

river bar An accumulation of alluvium in the channel, at the mouth, or along the banks of a river or stream. At low water, it is commonly exposed and constitutes a navigational obstruction. *Channel bars* are located within the course of a stream, and seem to be most characteristic of braided streams, although not restricted to them.

roche moutonnée Etymol: Fr., "fleecy rock." A glaciated bedrock surface in the form of asymmetrical mounds of varying shape. The up-ice side of a roche moutonnée has been glacially scoured and is smoothly abraded; the down-ice side has a steeper, jagged slope as a result of glacial plucking. Thus, the ridges dividing up- and down-ice slopes are perpendicular to the general flow of the former ice masses. Roche moutonnée landscapes are characteristic of glaciated crystalline shield areas. Also called *sheepback rock*. Groups of roches moutonnées were probably named in the 1700s after the curls on sheepskin wigs (*peruques moutonnées*). Much later, the term was applied to individual glaciated outcrops because they resembled sheeps' backs in form and texture. A *crag* and *tail* differs from a roche moutonnée by the presence of a long, tapered sediment ridge extending down-ice. See also: *glacial plucking; glacial scouring.*

rock association A group of igneous rocks within a petrographic province that are related both chemically and petrographically.

rock crystal See *quartz.*

rock cycle See *geochemical cycle.*

rock flour Finely ground rock particles resulting from glacial erosion.

rock-forming Referring to those minerals that are part of the composition of rocks and determine their classification. Among the more important rock-forming minerals are quartz, feldspars, micas, amphiboles, and pyroxenes.

rock glacier See *talus.*

rock mantle *Regolith.*

rock pedestal One of variously formed land features (most common on desert surfaces) sculpted by wind erosion. The more resistant parts of a rocky mass, formed of alternate layers of hard and soft rock, are worn away more slowly; because maximum abrasion occurs at ground level, the mass may eventually resemble an object atop a pedestal.

rock pediment A pediment formed on a bedrock surface.

rock salt The term used for *halite* when

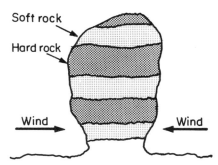

Rocky mass formed of alternate layers of hard and soft rock

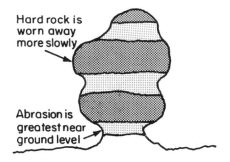

Formation of rock pedestal

it occurs as a massive or granular aggregate.

rock-stratigraphic unit *Lithostratigraphic unit.*

rock stream *Block stream.*

rock unit *Lithostratigraphic unit.*

rodding structure See *mullion structure.*

roof The rock lying above a coal bed; also, the country rock bordering the upper surface of an igneous intrusion.

room-and-pillar A system of mining in which ore is mined in chambers supported by columns of undisturbed rock that have been left intact for the purpose of roof support.

root zone That zone in the earth's crust from which thrust faults emerge. *Root zone* also refers to the source of original attachment of the basal part of a fold nappe. Also see: *isostasy.*

ropy lava *Pahoehoe.*

rose quartz See *quartz.*

Rossi-Forel scale See *earthquake measurement.*

rotary drilling The principal method of drilling deep wells, especially for gas and oil. Rotary drilling may be adapted for any angle, and is suitable for underground mining. In this method the drill bit rotates while bearing down. A great advantage of rotary drilling over *cable-tool drilling* is that the borehole is kept full of drilling mud during the process. *Drilling mud* is a weighted fluid that is pumped down the drill pipe, out through openings in the drill bit, and back up to a surface pit, where it re-enters the mud pump. In addition to cooling and lubricating the drill bit, this fluid, by its hydrostatic pressure, inhibits the entry of formation materials into the well, thus preventing "blowouts" and "gushers." The drilling mud also carries the crushed rock to the surface. Cf: *cable-tool drilling.* See also: *air drilling.*

rotary fault *Rotational fault (hinge fault).*

rotation axis *Axis of symmetry.*

rotational fault *Hinge fault.*

rotational movement An apparent fault-block displacement in which the blocks have rotated in relation to each other; the alignment of previously parallel features is disturbed.

rotational wave *S wave.*

rounded Said of a *sedimentary* particle whose original edges and faces have been almost removed by abrasion, although the original shape is still clearly seen. The degree of abrasion of a *clastic* particle, as shown by the sharpness of its corners and edges, is called *roundness.*

rubellite A pale, rose-red to deep-red lithian variety of tourmaline that is used as a gemstone.

rubidium-strontium age method See *dating methods.*

ruby The red gem-variety of *corundum*; its color is attributed to the presence of a small amount of chromium.

ruby silver The red silver-sulfide minerals *proustite* (light ruby silver) and *pyrargyrite* (dark ruby silver).

rudaceous Said of a sedimentary rock composed of a significant proportion of fragments larger than sand grains (pebbles, gravel), or said of the texture of such a rock.

rudite Etymol: Latin, *rudus,* "rubble." Any sedimentary rock composed of coarse fragments larger than sand grains (2 mm) in diameter. Included are both conglomerate (rounded pebbles and grains) and breccia (angular pieces). A sedimentary rock consisting of more than 50% detrital calcite particles that are greater than sand size is referred to as *calcirudite*; it may be a consolidated calcareous gravel, or a limestone breccia or conglomerate. The term *rudite* is equivalent to the Greek-derived *psephite.* See also: *lutite; arenite.*

rugose coral Any zoantharian of the order Rugosa. Rugose corals may range in form from simple solitary to complex colonial types; solitary species are called horn corals. The corallite wall is massive, and various columella structures are found in the center of each corallite. Some serve as guide fossils and, because of their wide growth, are used as biological clocks to determine day and year length in the geological past. Range, Lower Ordovician to Permian.

runoff That part of precipitation that appears in or "makes its way to" surface streams. The runoff that reaches stream channels immediately after a rainfall or the melting of snow is *direct runoff*. Soil and rock permeability, vegetation, temperature, and ground slope all affect runoff. Ground that is water-saturated or filled with ice favors runoff.

rutile A yellow, red, brown, or black tetragonal mineral, TiO_2, an important ore of titanium. Rutile occurs as a very common accessory mineral in intrusive igneous rocks, or dispersed through quartz veins.

rutilated quartz Quartz in which needlelike crystals of *rutite* are enclosed. See also: *sagenite*.

R wave *Rayleigh wave.*

S

sabkha A halite-encrusted surface of salt flats. The sabkha zone parallels the coast; during times of strong offshore winds, together with spring tides, extensive areas of sabkha are flooded. Subsequent intense evaporation causes the upward movement of concentrated brine, with the resultant precipitation of halite, gypsum, aragonite, and celestite. Underlying dominantly aragonitic sediment is dolomotized by calcium carbonate. (See also: *reflux.*) Sabkhas are found on many modern coastlines; e.g., Persian Gulf, Gulf of California. The facies may be indicated by evaporites, the absence of fossils, pebble conglomerates, algal mats, and dolomitization. This environment may have been significant in the formation of petroleum deposits.

saccharoidal Having a granular texture resembling that of a loaf of sugar. The term is used to describe some marbles and sandstones. Syn: *sucrose.*

saddle reef A mineral deposit found in the crest of an anticlinal fold following the bedding planes. These deposits are usually found in vertically stacked succession; e.g., in Bednigo, Australia, gold-bearing deposits occur in the domes of anticlinal folds of Lower Ordovician quartzites and slates. In Nova Scotia, gold-bearing saddle reefs are found in association with folded beds of slate.

sagenite A variety of rutile that occurs in crossed needlelike crystals, often enclosed in quartz. Similar crystals of tourmaline, goethite, and other minerals that penetrate quartz are also referred to as sagenite.

salic A mnemonic term derived from silica and aluminum and applied to the group of standard normative minerals (CIPW classification) in which one or both of these elements occur in large amounts. These include quartz, feldspars, and feldspathoids. The corresponding term for the silicic and aluminous minerals actually present in a rock is *felsic*. Cf: *femic; mafic.*

salina Etymol: Span., "salt mine." A salt-encrusted playa or a place where crystalline deposits are found.

saltation A type of sediment transport intermediate between suspension and traction; particles are moved forward in short, abrupt leaps.

salt dome A circular piercement structure (*diapir*) produced by the upward movement of a pipelike plug of salt. Salt domes occur in certain areas underlain by thick layers of sedimentary rock, and are derived from deposits of deeply buried halite. A *cap rock* of gypsum, anhydrate, and varying amounts of sulfur and calcite are found on top of many salt plugs, and some contain petroleum deposits. Oil and gas may also accumulate in traps created by the deformation of the rocks surrounding the salt plug.

salt flat *Alkali flat.*

salt lake A body of water in an arid or semi-arid region, having no outlet to the sea and containing a high concentration of salt (NaCl); e.g., Great Salt Lake in Utah; the Dead Sea in the Near East. See also: *alkali lake.*

salt lick A place to which animals (deer, bison, cattle) go to lick up the salt lying on the ground, such as the area surrounding a salt spring.

salt marsh See *marsh.*

saltpeter 1. A naturally occurring potassium nitrate (niter); 2. A general name for cave deposits of nitrate minerals.

salt tectonics The study of the formation and mechanics of emplacement of salt domes and other salt-related structures.

sand 1. A detrital particle larger than a silt grain and smaller than a granule, having a diameter in the range of 0.062 to 2 mm. 2. A clastic sediment of such particles. 3. *Sandstone.* 4. Driller's term for any productive porous sedimentary rock in a well. See also: *oil sand.*

sand crystal A large crystal with irregular external form, consisting of as much as 60% sand inclusions. Calcite, barite, or gypsum sand crystals develop by growing within a deposit of unconsolidated sand during cementation.

sand dune *Dune.*

sand-shale ratio The ratio in a stratigraphic section of the amount of sandstone and conglomerate to that of shale, without considering the nonclastic material. Cf: *clastic ratio.*

sandstone A sedimentary rock composed of sand-sized grains in a matrix of clay or silt, and bound together by a cement that may consist of calcium carbonate, silica, or iron oxide. Quartz forms about 65% of the detrital fraction of the average sandstone, and feldspars about 10 to 15%. The mineral composition of a sandstone can be used to determine the character of the source rock from which the grains were derived, and the diagenetic events that have affected the sandstone. The color of sandstones, which may be red, white, black, gray, or green, is typically determined by the cementing agents.

Most sandstones originate as underwater deposits, usually marine; many sandstones were originally beach deposits. Non-marine sandstones were deposited in lakes and, infrequently, along stream courses. A few sandstones are thought to have initially been sand dunes rather than water-laid deposits. Although there are various ways of classifying sandstones, they are commonly separated into four major groups, according to mineral composition: *orthoquartzite*—richest in silica, both in the matrix and the grains themselves; *arkose*—richer in feldspar than is quartzite; *graywacke*—which represents a mixture of rounded to angular quartz, feldspar, and mica in various cementing materials; *subgraywacke*—which contains less feldspar than does graywacke, and whose quartz grains are more rounded.

sanidine A colorless or whitish silicate mineral, $KAlSi_3O_8$, a high-temperature potassium feldspar. It is typical in recent volcanic rocks of trachytic or high-potassium type.

sanidinite facies A facies represented by small fragments of aureole materials that have often been fully immersed in silicate liquids. The assemblage of minerals indicates maximum temperatures, often at very low pressures. Pelitic rocks contain minerals such as sillimanite, sanidine, hypersthene, and anorthite. Calcareous rocks tend to lose almost all carbon dioxide, but pure calcite may survive. Typical metamorphic minerals are quartz, wollastonite, anorthite, and diopside.

saponite A trioctahedral, magnesium-rich clay mineral of the montmorillonite group. It represents an end-member in which magnesium almost completely replaces aluminum.

sapphire Any pure, gem-quality corundum other than *ruby*. Varieties other than the well-known *blue sapphire* are green sapphires and yellow sapphires. The color of blue sapphire is attributed to the presence of small amounts of oxides of cobalt, chromium, and titanium.

saprolite A soft, earthy, red or brown, clay-rich and totally decomposed rock, formed in place by chemical weathering of igneous or metamorphic rocks, particularly in humid climates. Structures that were still in the unweathered rock are preserved in saprolite. Cf: *laterite.*

sapropel An unconsolidated deposit composed chiefly of the remains of certain algae, with mineral grains and spores as minor constituents; these remains accumulate and decompose in anaerobic conditions on the shallow bottoms of lakes and seas. When consolidated, sapropel becomes oil shale, bituminous shale, or *boghead coal.* It is distinguished from peat by its high content of fatty and waxy substances (due to algal remains) and small amount of cellulose material.

sardonyx A gem variety of chalcedony that has straight, parallel bands of sard (a reddish-brown chalcedony) alternating with chalcedony of another color, usually white.

sastrugi features From the Russian, meaning "irregularities," these are wave formations caused by persistent winds blowing on a snow surface. Sastrugi features vary in size according to the duration and force of the wind and the condition of the snow surface in which they are formed.

satin spar A white, translucent, fibrous gypsum with a silky luster.

saturated 1. Describing the condition in which pores of a material are filled with water or another liquid. 2. Said of a rock having quartz as part of its norm. (See *silica saturation.*) 3. Said of a mineral that can form in the presence of free silica. See *silica saturation.*

saturated zone *Zone of saturation.*

saturation 1. The degree to which the pores in a rock hold water, oil, or gas; generally expressed in percent of total pore volume. 2. The degree of *silica saturation* in an igneous rock. 3. The maximum possible amount of water vapor in the earth's atmosphere for a specific temperature.

saturation line 1. A line, on a compositional variation diagram for an igneous rock series, that represents the degree of saturation with respect to silica. Rocks to the right of it are oversaturated, those to the left are undersaturated. (See *silica saturation.*) 2. The boundary on a glacier between the zone of partial melting (where the firn layer is not fully soaked) and that where the firn layer is saturated with meltwater.

sausage structure *Boudinage.*

saussurite A compact mineral aggregate consisting of albite and zoisite or epidote; it is an alteration product of calcic plagioclase.

savanna Etymol: Span., *sabana,* "flat country." The common name for tropical or subtropical grassland characterized by scattered shrubs or trees and a pronounced dry season. Savanna areas occur primarily in Africa (veld or bushveld) and South America (llanos), with similar areas in Australia and Madagascar. Also spelled *savannah.*

scabland The type area of this topography is the Columbia lava plateau of eastern Washington. The basaltic plateau is broken up into a maze of mesas, buttes, and canyons that were formed by the erosive action of glacial meltwaters. Among the more striking features are reticulated channels, abandoned canyons, dry waterfalls, and hanging valleys.

scale trees Paleozoic species of Lycopsids are known as "scale trees" because their bark pattern has a scalelike appearance. Some attained heights of 30 m. Two Late Paleozoic species, *Lepidodendron* and *Sigillaria*, are commonly found in Mississippian and Pennsylvanian coal deposits. These are the "coal plants." *Lepidodendron* is tall and branching, with slender leaves and diamond-shaped leaf scars. *Sigillaria* has a stout trunk, bladed leaves, and vertical leaf-scars. See *fossil plants*.

scaphopod A marine univalve of the class Scaphopoda, having an elongated, bilaterally symmetrical tusklike shell that is open at both ends. Maximum length is 13 cm, but usually less. Scaphopods range from the Devonian to present, but are of minor importance as fossils.

scapolite A group of calcium and aluminum silicate metamorphic minerals occurring in schists, gneisses, and crystalline limestone.

scarp A sequence of cliffs produced by faulting or erosion; it is a truncation of the term *escarpment*.

scattering layer See *deep-scattering layer*.

scheelite A yellow, green, or reddish-gray tetragonal mineral, $CaWO_4$, an ore of tungsten. It is found in high-temperature pegmatitic and hydrothermal veins.

schist A foliated metamorphic rock that is not defined by mineral composition, but by the well-developed parallel orientation of more than 50% of the minerals present; in particular, those minerals of lamellar or elongate prismatic habit, such as mica and hornblende. Cf: *gneiss; phyllite*.

schistosity The foliation or layered appearance observed in schists or other coarse-grained crystalline rocks, due to the parallel alignment of platy or ellipsoidal mineral grains. adj. *Schistose*.

schlieren Pencil-like, discoidal, or bladelike mineral aggregates, varying greatly in size, that occur in plutonic rocks. They have the same general mineralogy as the host rock but, because of differences in mineral fractions, the color of schlieren is either lighter or darker. Cf: *flow layer*.

scintillation counter An instrument that measures radiation by counting the individual scintillations emitted by the substance or object being investigated. The counter consists of a phosphor and a photomultiplier. The phosphor emits visible light when struck by high-speed particles, and the photomultiplier tube registers the phosphor flashes.

scissors fault A fault on which there is increasing separation along the strike from an initial point of zero separation, with reverse offset in the opposite direction. (Ex: The great Uinta Fault on the north side of the Uinta Mountains of Utah.) Such a separation may be due to a pivotal movement on the fault, or be the result of uniform strike-slip movement along a fault across an anticlinal or synclinal fold. The term is not used in a precise or strict sense; *hinge fault, pivotal fault, rotary fault,* and *rotational fault* are similarly used.

scoria A substance that may occur as a vesicular *pyroclastic material* or as froth on a lava flow; it is composed of heavier ferrugenous lavas and so is usually heavier and darker than pumice. However, the distinction between the two can be rather arbitrary; the low specific gravity of some scoriae is equivalent to that of some pumice. Certain scoriae, sometimes called *cinders* or volcanic cinder, resemble clinkers or cinders from a coal furnace. adj. *Scoriaceous.* Cf: *pumice.* See *pyroclastic materials.*

scour-and-fill The alternate scouring out and refilling of a channel or depression by water that fluctuates in current speed and volume.

scree A heap of rock debris produced by weathering at the base of a cliff, or a sheet of coarse waste covering a mountain slope. "Scree" is frequently considered to be a synonym of "talus," but is a more inclusive term. Whereas talus is an accumulation of debris at a cliff base, scree also includes loose debris lying on slopes without cliffs. The term "scree" is more commonly used in Great Britain, whereas "talus" is more commonly, but often incorrectly, used in the United States.

sea arch See *natural bridges and arches.*

sea cave A marine erosion feature that forms along a line of weakness at the base of a cliff that has been subjected to prolonged wave action. It is a cylindrical tunnel extending into the cliff. If a *blow hole* is opened on the cliff top, continued wave action ultimately causes the roof of the cave to collapse, and a narrow sea inlet forms.

sea cliff See *cliff.*

sea-floor spreading A hypothesis, formed by the U.S. geophysicist Harry Hess at Princeton University in 1960, that oceanic crust forms along the *mid-oceanic ridge* system and spreads out laterally away from it. The concept was pivotal in the development of *plate tectonics.* Until the 1950s, it was believed that the ocean floor represented the oldest parts of earth's crust. Discovery in the late 1940s of the mid-ocean ridge system and new knowledge that the crust

forming the ocean floor is thinner and younger than the continental crust provided the basis for the Hess hypothesis: Magma that continuously upwells along the mid-ocean ridges flows down the flanks on either side; cool magma that is pushed away by molten material becomes new sea floor.

The sea-floor hypothesis has been supported by a great deal of substantial evidence. Thermal anomalies over the mid-ocean ridges (three to four times normal values) are considered to reflect the intrusion of molten material near the ridge crests. Anomalously low seismic-wave velocities have also been observed at the ridge crests; thermal expansion and fracturing associated with the upwelling magma are thought to be the responsible factors. Sea-floor spreading was further corroborated when investigation of oceanic magnetic anomalies revealed a pattern of magmatic stripes running parallel with the ridges. The geomagnetic field is alternately high and low with increasing distance from the axis of the mid-ocean ridge system. These linear trends seem to be underlain by alternating bands of normally and reversely magnetized basaltic rocks generated at the ridge axis, where they acquired a thermoremanent magnetization as they cooled; their magnetic polarity is determined by the polarity of the geomagnetic field at the time of solidification. (See *geomagnetic polarity reversal.*) The farther distant from the ridge axis a band is located, the older the rock. The oldest marine bottom sediments, as determined from core samples, date only to the Jurassic (190 m.y.a.).

The concept of sea-floor spreading was instrumental in establishing the theory of *continental drift.* It is believed that the spreading out of sea floor and continuous upward flow of molten material are responsible for the migration of the continents. See also: *plate tectonics.*

seamount An isolated submarine elevation of at least 700 m. Seamounts vary in shape, and are found singly as well as in groups or chains. They may be relatively small conical peaks or massive structures nearly equal to the size of Rhode Island. These elevations have been found in all major ocean basins; they are generally formed in rifts in the earth's crust (e.g., Iceland), and in the center of oceanic plates (e.g., Hawaii). Seamounts formed on weak, new oceanic crust tend to have highly localized geoid deflections; those formed on thicker, older crust produce a broader geoid deflection; both types have an equal mass below the crust. Although evidence indicates that they are all of volcanic origin, it is not known if volcanism in general occurs before or after crust formation. See also: *guyot*.

sea stack *Stack.*

secondary 1. A term applied to rocks and minerals formed by the alteration of pre-existing minerals (primary minerals). Secondary minerals may be found at the site as paramorphs or pseudomorphs, or may be deposited from solution in rock interstices through which the solution is percolating; e.g., amygdules. 2. Formed of material derived from the disintegration or erosion of other rocks; e.g., clastic rocks.

secondary consolidation The consolidation of sediment resulting from internal processes such as recrystallization, and occurring at essentially constant pressure.

secondary enrichment A term used especially for a near-surface process of mineral deposition, by which a primary ore body or vein is later enriched to a higher grade. The enrichment process often consists of two phases: (1) oxidation of overlying ore masses above the water table (zone of oxidation)— oxidation may then produce acidic solutions that leach out metals and carry them downward toward the water table; (2) below the zone of oxidation, reaction of the solutions with the primary material by coating or replacement with an ore of more valuable content. The water table thus marks the lower limit of oxidation, near which there may be a zone of secondary enrichment. This process has been significant in the formation of many copper deposits. Syn: *supergene enrichment.* See also: *gossan; oxidized zone.*

secondary mineral A mineral formed after the rock enclosing it has been formed; the process usually involves the alteration of a *primary mineral* by metamorphism, weathering, or solution.

secondary wave *S wave.*

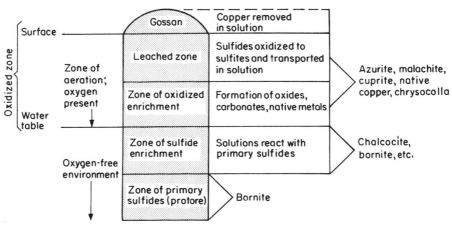

Process of secondary enrichment

second law of thermodynamics See *entropy.*

secular variation A slow, steady, progressive change in part of earth's magnetic field, due to the internal state of the planet. These changes are immediately apparent from yearly averages of the values of magnetic elements as reported by worldwide magnetic observatories. By taking period averages, transient disturbances caused by field interaction with solar-particle emissions are eliminated, and what remains are slow variations of internal origin.

sedifluction The subaerial or subaqueous movement of material in unconsolidated sediments, that occurs in the primary stages of diagenesis.

sediment Solid material, organic or inorganic in origin, that has settled out from a state of suspension in a liquid and has been transported and deposited by wind, water, or ice. It may consist of fragmented rock material, products derived from chemical action, or from the secretions of organisms. Loose sediment such as sand, mud, and till may become consolidated and/or cemented to form coherent sedimentary rock. Cf: *deposit.*

sedimentary Containing sediment or formed by its deposition.

sedimentary cycle *Cycle of sedimentation.*

sedimentary dike A tabular mass of sedimentary material that traverses the structure of pre-existing rock, thus resembling an igneous dike. A sedimentary dike is formed by the forcible injection of sediments under abnormal pressure into a fissure. Such pressure may be due to enclosed gas or to the weight of overlying rocks.

sedimentary facies An areally confined part of an assigned stratigraphic unit that shows characteristics clearly distinct from those of other parts of the unit.

sedimentary rock A rock formed by the consolidation of sediment settled out of water, ice or air and accumulated on the earth's surface, either on dry land or under water. *Clastic* or *detrital* sedimentary rocks originate with the accumulation of discrete mineral or rock particles derived from weathering and the erosion of pre-existing rocks. Nonclastic varieties originate from molecular quantities of minerals of chemical precipitation (e.g., cave travertine), or from the biogenic action of organisms (e.g., coal, diatomite). Sediments are consolidated into a rock mass by *lithification.* Sedimentary rock is typically stratified or bedded; beds can vary greatly in thickness.

Texture and *composition* are the two principal properties used in identifying sedimentary rocks. The texture, in this case, usually refers to particle size; the particle reflecting the nature of weathering and of transportation of particles and conditions at the site of deposition. Texture of sedimentary rock is described as *rudaceous* (coarse), *arenaceous* (medium) or *lutaceous* (fine); rocks characterized by these textures are referred to as *rudite, arenite,* and *lutite,* respectively. These are purely textural terms, independent of composition. Rocks showing the presence or absence of a wide range of particle sizes are distinguished as *poorly sorted* or *well-sorted.* The texture of rocks consisting of chemical or biogenic precipitates is called *massive,* regardless of crystal size.

The composition of sedimentary rocks is determined by conditions prevailing at the source of the sediments and at the site of deposition, as well as by the nature of transport. The rocks may contain fragments of any pre-existing rock or mineral. For example, a *conglomerate* may include fragments of quartz, limestone, chert, etc. Rocks resulting from biogenic or chemical accumulation are usually composed of calcium or magnesium carbonate, silica, or *evaporite* salts. Primary structures exhibited in sedimentary rock, such as bedding (stratification) and fossil inclusion, reflect environmental and climatological conditions at the time of deposition.

The major types of sedimentary rocks are mudrocks (65%), sandstones (20 to 25%), and carbonate rocks (10 to 15%). There are transitional rocks such as *coquina* that are both sandstone and limestone, but are usually included with the limestones.

sedimentation The process of depositing sediments, including the separation of rock particles from their origin, the transportation of these particles, their site of deposition, and all changes and stages leading to the eventual consolidation of the sediment into sedimentary rock.

sedimentology The scientific study of sedimentary rocks and the processes responsible for their formation; also, the origin, description, and classification of sediments.

segregation 1. *Magmatic segregation.* 2. A secondary feature, such as a nodule of iron sulfide, that is formed by the chemical rearrangement of minor components within a sediment subsequent to its deposition.

seiche Periodic oscillation of a moderate-sized, enclosed body of fluid, characteristic of the system only, and independent of the exciting force, except as to initial magnitude. Where water is the medium, the oscillating system is characterized by the shape of the basin and the depth of the water. The restoring force is provided by gravity. Earthquakes commonly effect seiches in ponds and lakes. Sudden barometric pressure changes plus wind velocity and direction are also important. See also: *tsunami.*

seif Etymol: Arabic, "sword." A long chain of dunes or a very large *longitudinal dune* with a curved, knifelike ridge. Seifs grow in width and height largely through the action of cross winds, and increase their length during those intervals in which the prevailing wind parallels the trend of the seif. Dunes of this type may reach 200 m in height. Some individual seif ridges can be 100 km long and, when in groups, extend for more than 300 km. There is some belief that the seif form is a modification of the *barchan* form. See also: *dune.*

seism Earthquake.

seismic Pertaining to a naturally or artificially induced earthquake or earth vibration.

seismic activity Seismicity.

seismic area An *earthquake zone,* or the region affected by a particular earthquake.

seismic detector An instrument, such as a seismometer or geophone, that receives seismic impulses and converts them into readable signals.

seismic discontinuity Discontinuity.

seismic event An earthquake or a similar transient earth motion caused by an explosion.

seismic exploration The use of seismic surveying and explosion seismology in prospecting for oil, gas, or other mineral sources. The use of such methods in seismic prospecting has led to important mathematical findings about the transmission of elastic waves from particular source types and through media of various complexities. The two basic practical methods commonly used are *reflection shooting* and *refraction shooting.* Syn: *seismic prospecting.*

seismic focus In seismology, the point within the earth that is the center of an earthquake. It is the initial rupture point of an earthquake, where strain energy is transformed into elastic wave energy. Cf: *epicenter.*

seismic intensity The average rate of flow of seismic wave energy through a unit cross-section, perpendicular to the direction of wave propagation.

seismicity The phenomenon of earth movements and their geographic distribution. Syn: *seismic activity*.

seismic magnitude See *earthquake measurement*.

seismic map A contour map constructed from seismic data. The data values may be either in units of depth or time, and may be plotted with respect to the observing (receiving) station at the surface or with respect to a series of subsurface reflecting or refracting points.

seismic moment A relatively new method of measuring the strength of an earthquake, and a more physical measure of its size than that provided by earlier methods. It is defined as the product of the rigidity of the rock, the area of faulting, and the average amount of slippage. The seismic moment is determined by the Fourier analysis of seismic waves of such long period that the details of the rupture are smoothed out, with the effect that the entire fault may be considered to be a point-source. The periods at which the seismic moment are determined increase with the size of the fault. If the fault is based on such long-period waves, the slip from unruptured to ruptured state appears to be instantaneous. The actual pattern of the seismic radiation emitted by the instantaneous rupture is mathematically equivalent to the theoretical radiation pattern emitted by two hypothetical torque couples embedded in an unruptured elastic medium. The torque couples, rotating in opposite directions, deform the medium, thus radiating elastic waves in a pattern identical with that in which an earthquake source radiates seismic waves. The moment can be calculated from this model. See also: *earthquake measurement*.

seismic prospecting *Seismic exploration*.

seismic regions A division of the earth's interior into seven regions, (A to G), according to depth based on seismic velocity distributions.

Region	Depth (km)	Features of Region
A		Crustal layers
	33	
B		Steady positive P and S velocity gradients
	413	
C		Transition region
	984	
D		Steady positive P and S velocity gradients
	2898	
E		Steady positive P velocity gradient
	4982	
F		Negative P velocity gradient
	5121	
G		Small positive P velocity gradient
	6371	

seismic sea wave See *tsunami*.

seismic shooting A method of geophysical prospecting, using techniques of explosion seismology, in which elastic waves are produced in the earth by firing explosives. Of several methods used, *reflection shooting* and *refraction shooting* have been the most successful. In reflection shooting, the travel times of seismic waves from small explosions in shallow holes are recorded on portable seismographs, thus permitting identification of various strata by their differing elastic properties. A comparison of travel times to different points indicates the structure of buried strata.

Refraction shooting is designed to determine seismic velocities and configurations of layers in those regions where velocity does not decrease with depth. Recording instruments are set up at distances from an explosion source that are several times the depth of the most remote layer being investigated. This contrasts with reflection shooting, where the recording instruments are close to the source, the object being to record near-vertical reflections from the boundaries of the layer being investigated. Various filtering procedures are used to isolate the desired phases from background effects such as scattered waves and surface waves.

seismic velocity See *seismic waves.*

seismic waves The manifestation of energy that is released when rocks within the earth fracture or slip abruptly along fault planes during an earthquake. Seismic waves are of two main types: *body waves* and *surface waves.* Body waves, appropriately, travel through the body of the earth; *primary (P) waves,* which are compression waves, are the fastest of this type and can travel through both solid and liquid matter; *secondary (S) waves,* the slower body waves, are transverse shear waves that can propagate through solid but not liquid matter. Primary waves, which are similar to sound waves, emerge from the earthquake focus and alternately compress and stretch the medium through which they travel. As secondary waves travel through a medium, individual particles vibrate at right angles to the direction of wave propagation. The velocities of both P and S waves vary in different media. The ratio of their velocities, P_v/S_v, is almost always a constant, but decreases some weeks, months, or even years before an earthquake, and increases to almost normal level just prior to the earthquake. See *earthquake prediction.*

When body waves travel outward through the earth to the surface, some are transformed into surface waves, two types of which predominate: *Rayleigh waves,* which have a rotating motion in the vertical plane, aligned in the direction of travel, and *Love waves,* which have only horizontal shear motion. Rayleigh and Love waves have lower frequencies than

Primary wave

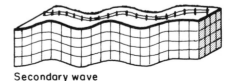
Secondary wave

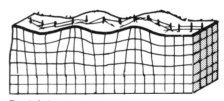
Love wave

Rayleigh wave

Types of seismic waves

the body waves from which they originate, and usually travel more slowly. But they persist for great distances, sometimes circling the earth many times before subsiding.

The use of seismic waves has been invaluable in determining physical properties of the earth and the structure of its interior. *Arrival times* are used to locate reflecting discontinuities within the earth, and the velocities of P and S waves give information about the ratio of any two of the parameters incompressibility, elastic rigidity, and density of the material through which the waves pass. Seismological evidence for a distinct core of the earth, and for a clearly delineated inner core, has been derived from reflected P waves.

seismogram The record made by a *seismograph*.

seismograph An instrument that detects and records seismic vibrations. The basic principle of operation for all seismographs is essentially the same. A frame that is rigidly fastened to a body of rock capable of transmitting earth movements will move in conjunction with the rock support. Another element of the seismograph is a free-swinging pendulum to which is attached some scribing device that records the motion of the pendulum and, thus, of the earth as well. Since seismic waves can cause the earth to move vertically as well as horizontally, seismographs are designed accordingly.

seismology The science of earthquakes, including their origin, propagation, energy manifestations, structure of the earth, and possible techniques and methods of prediction.

seismometer *Seismic detector.*

selenite A variety of gypsum occurring as colorless and transparent monoclinic crystals, or large crystalline masses that yield broad folia upon cleavage.

selenology A division of astronomy that deals with various aspects of lunar science, including lunar geology.

selvage 1. *Fault gouge.* 2. A marginal zone of a rock form, showing some distinguishing peculiarity, such as in composition or fabric.

Senecan Lower Upper Devonian of North America.

sepiolite A phyllosilicate clay mineral, $Mg_4(Si_2O_5)_3(OH)_2 \cdot 6H_2O$, occurring as white or yellowish concretions or porcellaneous masses. Sepiolite is found as a surface alteration product of

magnesite or serpentine, and is used in the manufacture of tobacco pipes. Syn: *meerschaum.*

septarian A term applied to the irregular pattern of internal cracks in a *septarium*, resembling the polygonal shrinkage cracks developed in mud during desiccation. Also applied to the mineral deposits, such as calcite, that may occur as fillings of these cracks.

septarian nodule See *septarium*.

septarium A large, roughly spheroidal concretion, generally of limestone or clay ironstone, characterized by a network of radiating and intersecting mineral-filled cracks; the mineral is usually calcite. Internally, a septarium consists of irregular, polygonal blocks cemented together by the mineral deposits within these cracks. The formation of septaria (pl.) is similar to that of geodes. When flattened oval septaria weather so as to leave the veins in relief, they resemble turtles in form and surface marking; some are mistaken for fossil turtles. Other names for septarium are *septarian nodules* and *turtle stone.*

serac A jagged pinnacle of ice that is formed when several glacial crevasses intersect each other at the terminus of a glacier. The term "seracs" is derived from their similarity to a kind of curdy Alpine cheese of this name.

sericite A fine-grained mica, generally occurring in small flakes. It is particularly common in schists, but is also found as an alteration of wall-rock material. Sericite is used in building work for roof covering and outside facing.

series 1. A *chronostratigraphic unit* above *stage* and below *system*; it includes the rocks formed during an epoch of geological time. Some series are worldwide, whereas others are regional. 2. *Igneous rock series*. 3. A term sometimes used erroneously instead of group for an association of formations. 4. *Radioactive series*.

serpentinite A metamorphic rock, composed almost completely of serpentine-group minerals, e.g., antigorite, magnetite, chrysotile. Its color varies from dark green to black, and its texture may be lamellar or feltlike with zonations. It is found in greenschist facies, where it is commonly derived from the alteration of peridotite. Serpentinite is used for decorative purposes.

serpentine A group of common, rock-forming minerals containing many polytypes, $(Mg,Fe)_3Si_2O_5(OH)_4$, almost always occurring as fibrous yellowish-white or green aggregates. Serpentines are found in low-grade metamorphic environments and are derived from magnesium silicates like olivine, pyroxene, and amphibole.

serpentine asbestos *Chrysotile.*

serpentine marble *Verd antique.*

serrate Said of topographic features that exhibit a notched or saw-toothed profile.

shadow zone 1. A region 100° to 140° of arc from the focus of an earthquake, in which there is no direct penetration of seismic waves. It is a 4350 km belt circling the earth. *Primary waves* are not observed in this region because they are

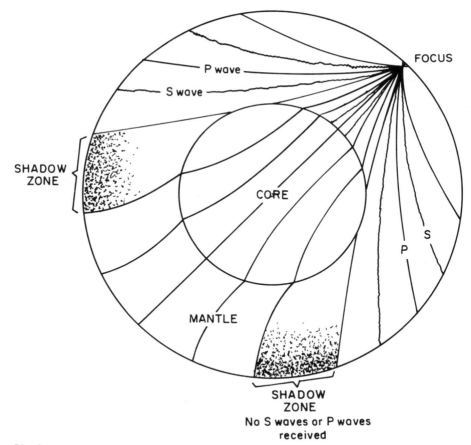

Shadow zone

reflected or refracted below the boundary between the earth's core and mantle. *Secondary waves* are not observed beyond 100° from their course because they cannot travel through a liquid. The non-penetration of secondary waves within this zone was used as evidence that the earth's core is part liquid. See also: *Gutenberg discontinuity*. 2. *Wind shadow*.

shale A fine-grained sedimentary rock formed by the compaction of silt, clay, or sand that accumulates in deltas and on lake and ocean bottoms. It is the most abundant of all sedimentary rocks. Shales may be black, red, gray, or brown, and have excellent fissility. A material that has the appearance or character of a shale, esp. fisillity, is said to be *shaly*. Cf: *argillaceous*. See also: *oil shale*.

shallow-focus earthquake An earthquake with a focus at less than a 70 km depth. Most earthquakes occur at this depth. Cf: *intermediate-focus earthquake; deep-focus earthquake*.

shape-preferred orientation The preferred orientation of flattened or elongated crystal axes, due to crystal gliding, magmatic flow, or dynamic recrystallization. Cf: *lattice-preferred orientation*.

shatter cone A striated, conical rock fragment along which fracturing has occurred; it varies from a centimeter to several meters in length, and is usually found in nested or composite structures that bear a resemblance to cone-in-cone formations in sedimentary rocks. Shatter cones are found in limestone, dolomite, sandstone, shale, and granite; they are believed to have been formed by shock waves created by meteorite impact.

shear A change in shape due to stresses that cause one part of a body to slide past a contiguous part, like members of a deck of cards. Shear stress acts tangentially or parallel to the plane along which a solid body will ultimately rupture or deform by shear.

shear fold A fold that results from minute displacements along closely spaced fractures, each of which is a minute fault. If the fractures are very close, and the beds tend to parallel the fractures, the resulting structure is a major fold with several associated minor folds. Syn: *slip fold*.

shear fracture A fracture due to stresses that tend to slide one part of a rock past the adjacent part.

shear modulus *Modulus of rigidity*.

shear strain The amount by which parallel lines have slid past one another by deformation.

shear strength The internal resistance of a body to shear stress. The total shear strength of an isotropic substance is the sum of the internal friction and the cohesive strength.

shear stress That component of a stress that is tangential to a plane passing through any specified point in a body.

shear wave *S wave*.

shear zone A tabular zone of rock showing evidence of shear stress in the form of crushing and brecciation by many parallel fractures.

sheepback rock *Roche moutonnée*.

sheeted-zone deposit A mineral deposit comprised of veins or lodes that occupy a *shear zone*.

sheet erosion The removal of surface material from a wide area of gently sloping or graded land by broad continuous sheets of running water rather than by streams. Cf: *gully erosion*.

sheetflood A broad sheetlike area of moving water. It is characteristic of desert fans, whose loose silt and sand become water-loaded so quickly that the water is unable to scour into the surface. If it is diverted by cobbles and clumps of vegetation, a sheetflood forms small braided channels; if unobstructed, it may cover the entire surface.

sheeting A form of rupture, similar to jointing, that occurs in massive rock bodies. It is characterized by tabular surfaces that are somewhat curved and essentially parallel to the topographic surface, except in regions of recent rapid erosion. Sheeting is best exposed in openings such as quarries. At the surface of the earth, the fractures are close together; the interval increases with depth and, at a point, the visible sheeting disappears. At greater depths, the planes of weakness parallel to the sheeting are utilized in working the quarry. Sheeting is a result of the release of pressure as overlying material is removed. The rock mass, which was formed under conditions of hydrostatic stress, will now tend to expand as its distance from the surface decreases.

sheet sand *Blanket sand.*

shelf ice Syn: *barrier ice.*

shield A large area of the earth's crust consisting mostly of Precambrian rocks and named for its form, which is generally similar to a shield; i.e., broad and gently sloping from the center. Shields are the exposed areas of the larger structural units known as *cratons.* A Precambrian shield is a mass of rocks that are part of a Precambrian orogenic belt whose active life has been completed within the Precambrian, and has therefore been relatively stable over a long period of time. Shields were once the sites of Precambrian mountain belts, and consist predominantly of igneous rock such as granite, or of highly metamorphosed rock such as gneiss. Each continent shows at least one shield, one of the best known being the Canadian Shield.

shield basalt See *shield volcano.*

shield volcano A gently sloping volcano, resembling a flattened dome. It is built by flows of very fluid basaltic lava erupted from numerous, closely spaced vents and fissures, coalescing to form a single unit. It is generally of smaller area than a *flood basalt.* Cf: *volcanic cone.*

shoal An accumulation of sediments in a river channel or on a continental shelf; it is potentially dangerous to ships. Sediment collects along the upcurrent face of a breakwater, forming a shoal behind it, which is then protected from wave erosion by the breakwater.

shock wave A compressional wave that is formed whenever the speed of an object, relative to the medium in which it is traveling, exceeds the speed of sound transmission for that medium. Its amplitude is greater than the elastic limit of the medium through which it is propagating, and it is characterized by a region within which there occur abrupt and drastic changes in the temperature, pressure, and density of the medium. If the medium is a rock, a shock wave is capable of melting, vaporizing, mineralogically altering, or profoundly deforming the rock materials. See also: *hypervelocity impact.*

shoestring sand An elongate, buried sand body that may be an offshore bar, a buried coastal beach or bar, a stream channel filling, or some other geomorphic feature. Particular shoestring sands may hold oil, but the specific origin of the formation must first be recognized. For example, an offshore bar differs considerably from a stream channel filling in cross-section, pattern, and trend with respect to the ancient shoreline. Offshore bars are likely to be discontinuous or segmented; bar segments may be oil-productive, whereas the intervening gaps will not. See also: *channel sand.*

shore 1. The narrow strip of land along any body of water. 2. The zone ranging from the low-tide to the landward limit of wave action. 3. *Shoreline.* See also: *backshore; foreshore.*

shoreline The line along which water of a sea or lake meets the shore or beach. It marks the position of the water level at any given time, and varies between high-tide and low-tide shoreline positions. A shoreline whose features result from the dominant relative submergence of a land mass is a *shoreline of submergence*. Such a shoreline is usually very irregular, and is either of *ria* or *fjord* type. There is no implication in the term *submergence*, as used here, as to whether it is land or sea that has moved. (Syn: *positive shoreline*.) If an absolute subsidence is implied in a submerged shoreline, it is called a *shoreline of depression*.

A *shoreline of emergence* is one whose features are the result of the dominant relative emergence of a lake or ocean floor. Such shorelines are generally less irregular than submergent shorelines. The most pronounced change along an emergent shoreline involves the migration of its *offshore bar*. Whether it is the land or the sea that has moved cannot be inferred from the term. (Syn: *negative shoreline*.) A shoreline of emergence that implies an absolute subsidence of land is a *shoreline of depression*. The succession of changes through which a shore profile passes during its development is called the *shoreline cycle*.

shoreline cycle See *shoreline*.

shoreline of depression See *shoreline*.

shoreline of elevation See *shoreline*.

shoreline of emergence See *shoreline*.

shoreline of submergence See *shoreline*.

shot break In seismic exploration, a record of the moment of seismic wave inception, as by explosion.

shot depth In seismic work, the vertical distance from the surface to an underground explosive charge.

shot point That point at which a charge of dynamite is exploded in order to generate seismic energy. In field work the point includes the hole in which the charge is planted and its immediately surrounding area.

shrinkage crack A crack formed in fine-grained sediment by moisture loss during drying; e.g., a *mud crack*.

sial A petrologic name suggested by Edouard Suess for the upper layer of earth's crust, composed of rocks rich in silica and alumina. It is an acronym for silica and alumina. Syn: *granitic layer*. adj. *Sialic*. Cf: *sima*.

sialma The layer of the earth's crust intermediate in both composition and depth between the *sial* and the *sima*. It is an acronym for *si*lica + *al*umina + *ma*gnesia.

siderites 1. Another name for *iron meteorites*. 2. mineral, $FeCO_3$, belonging to the calcite group of carbonates.

siderolites Another name for *stony iron meteorites*.

sideromelane *Tachylyte*.

siderophile See *affinity of elements*.

siderosphere The central iron core of the earth.

sienna Any of various brownish-yellow, earthy limonitic pigments for paints and oil stains. When burnt, it becomes dark orange-red to reddish brown, and is referred to as *burnt sienna*.

sierra Etymol: Span., sierra, "saw." A chain of mountains or hills, the peaks of which resemble the teeth of a saw; e.g., the Sierra Nevada in California.

Sigillaria See *scale trees*.

silcrete 1. A conglomerate composed of sand and gravel cemented by opal, chert, and quartz; formed by water evaporation in a semi-arid climate. 2. Siliceous *duricrust*. Cf: *calcrete; ferricrete*.

silica Silicon dioxide, SiO_2, is dominant in sand, chert, and diatomite, and occurs as crystalline quartz in a great variety of forms. It also occurs in cryptocrystalline chalcedony and amorphous opal.

silica concentration Silica, SiO_2, with few exceptions, is the principal constituent of igneous rocks; thus, it is often used as a basis for classification. A commonly employed technique uses the concentration of SiO_2. The categories below have no direct correlation with the modal quantity of quartz in a rock, although, generally speaking, acidic rocks contain quartz and ultrabasic ones do not. Two rocks having equal concentrations of silica may have very different quantities of quartz, and two rocks of similar quartz content may have different silica concentrations.

Silica Concentration (wt.%)	Description
60% or more	Acid (acidic)
52 to 60%	Intermediate
45 to 52%	Basic
45% or less	Ultrabasic

silica sand An industrial term for a sand containing a high percentage of quartz. It is a raw material for glass and other products.

silica saturation The degree of silica saturation is a function of the concentration of silica relative to the concentrations of other chemical components in a rock that combine with it to form silicate minerals. By using the normative calculation, two classes of minerals and three classes of igneous rocks are recognized: 1. *Silica-saturated* minerals—can exist at equilibrium with quartz. These minerals include feldspar, amphiboles, micas, and many others. 2. *Silica-unsaturated* minerals—never found in quartz. These include all feldspathoids, Mg-olivine, and corundum. 3. *Silica-oversaturated* rocks—contain quartz or its polymorphs; ex: granite. 4. *Silica-saturated* rocks—contain neither unsaturated minerals nor quartz; ex: diorite. 5. *Silica-unsaturated* rocks—contain unsaturated minerals; ex: nepheline syenite. Cf: *silica concentration*.

silicate A compound whose basic structural unit is the silicon-oxygen tetrahedron, SiO_4, which is either isolated or joined by one or more oxygen atoms to form configurations of independent or double groups, rings, chains, sheets, or three-dimensional frameworks. Silicates are classified according to the manner in which the basic tetrahedral groups are joined. See *nesosilicate; sorosilicate; cyclosilicate; inosilicate; phyllosilicate; tectosilicate*.

silication The process of replacing by or converting to silicates; e.g., the formation of skarn minerals in carbonate rocks. adj. *Silicated*. Cf: *silicification*.

siliceous Referring to a rock or other substance containing abundant silica, particularly as free silica.

siliceous ooze See *pelagic deposits*.

silicic Describes an igneous rock or magma that is rich in silica. The amount of silica generally recognized as minimum is at least 65% of the rock. In addition to combined silica, silicic rocks usually contain free silica in the form of quartz. Granite and rhyolite are typical examples. Syn: *acidic; silica-oversaturated*. Cf: *basic; intermediate; ultrabasic*.

silicification 1. The introduction of, or replacement by, silica, particularly in the form of fine-grained quartz, chalcedony or opal, which may fill cavities and pores and replace existing minerals. (Cf: *silication*.) Silicification may result from the filling of cavities with silica-saturated groundwater. The silica is often deposited in concentric rings around a central nucleus. 2. Process of fossilization wherein the original hard parts of an organism are replaced by quartz, chalcedony or opal. adj. *Silicified.* See *fossil; replacement.*

silicified wood A material formed by the silica permineralization of wood in such a manner that the original shape and structural detail (grain, rings, etc.) are preserved. The silica is generally in the form of chalcedony or opal. Syn: *petrified wood; opalized wood.*

sill 1. A tabular igneous intrusion with boundaries parallel with the planar structure of the surrounding rock. (Cf: *dike.*) Sills having a small area may be restricted to a single plane; larger ones may be *transgressive*; i.e., cut the bedding or foliation of the surrounding rock. Sills may be multiple or composite and are generally medium-grained, although larger ones may be coarse-grained. 2. A submarine ridge, located at a shallow depth, separating one basin from another or from the open sea; e.g., at the Straits of Gibraltar. 3. A ridge at shallow depth near the mouth of a fjord. It functions as a barrier between the deep water of the fjord and that of the ocean. Syn: *threshold.* See *intrusion.*

silled basin An area on the ocean floor set off by a rise that prevents maximum water circulation over it, a condition often resulting in oxygen depletion. Syn: *barred basin.*

sillimanite An orthorhombic nesosilicate, Al_2SiO_5, trimorphic with andalusite and kyanite. Its colors are gray, brown, or pale green, and it often occurs in silky, fibrous aggregates. Sillimanite is widespread in high-temperature metamorphic rocks, and also occurs in contact metamorphic types; it is a sensitive indicator of the temperature and pressure at which the host rock formed. One of its main uses is in the manufacture of refractories and high-temperature minerals.

silt 1. A detrital particle, finer than very fine sand and coarser than clay, in the range of 0.004 to 0.062 mm. 2. A loose clastic sediment of silt-sized rock or mineral particles. 3. Fine earth material in suspension in water.

silt flat *Alkali flat.*

siltstone A sedimentary rock composed of indurated silt, having the composition and texture of shale, but lacking its fissility or fine lamination. It occurs in layers, rarely thick enough to be classified as formations. Siltstones are intermediate between sandstones and shales but less common than either. They contain less alumina, potash, and water than shales, but more mica. Although many shales contain more than 50% silt, not all are siltstones; siltstones differ from these shales in that they are often chemically cemented and show evidence of cross-bedding, flowage within a stratum, and structures resulting from current flow.

Silurian A division of the Paleozoic Era extending from 430 to 395 million y.b.p., named in 1835 by Murchison after the Silures, an ancient Celtic tribe of the Welsh borderland. Here the rock system was studied and subdivided according to its graptolitic (*Monograptus*) and shelly fauna. Silurian System rocks are identified by comparison of their contained fossils with those of the marine sequence type in Wales and Shropshire, or in correlation with similarly dated rocks. The Taconic orogeny of central North America and Caledonian orogeny of northwest Europe are indicated in the Silurian System. Evidence of volcanism is found in the Acadian region of North America and the eastern Appalachian Trough. The Silurian, however, was a period of relative orogenic and geosynclinal quiesence in North America. Rocks of this system have yielded iron, salt, gold and tin ores, and petroleum. The stratified rocks are usually divided into: 1. *platform carbonates*— characterized by relatively thin layers (300 m) of either limestone and calcareous shale or dolomite; 2. *platform* *mudstone*—thin layers of mudstone, shale, and siltstone; 3. *terrigenous* and *volcanic*—restricted to geosynclinal regions. These are characterized by a thickness of 3100 to 4600 m; the rock types are predominantly sandstone containing silt and mud, and graywacke with bedded chert; 4. *nonmarine deposits*—which may occur as thin sheets of mudstone, siltstone, and sandstone.

Silurian invertebrate life is an expansion of Ordovician lines. Echinoderms, brachiopods, and *graptolites* show distinctive features. Crinoids are the most successful Silurian echinoderms; tabulate and horn corals are abundant. Bryozoa and the pelecypods continue to expand. Silurian plants and fishes are the oldest preserved, except for fragmental findings in the Ordovician of the Rocky Mountains. Trilobites continue to decline from the Cambrian maximum. Ostracods, an index fossil for the Silurian in the Appalachian region, expand slightly, and eurypterids become much more common. Ostracoderms (a jawless freshwater fish) of a much larger size are found; the first fish with jaws appears in fresh water at this time. Upper Silurian fossils from Australia indicate that psilopsids and lycopsids grew on land.

The existence of a relatively thin sheet of platform mudstone in certain areas implies that this region formed one continental block during the Early Paleozoic (Cambro-Ordovician relationships are similar). Widespread reef formation by Silurian corals suggests warm shallow seas and little seasonal temperature variation. See also: *continental drift.*

silver A mineral, the native element Ag, distributed in small amounts as a secondary mineral in the *oxidized zone* of ore deposits. It is found in larger amounts deposited from hydrothermal solutions. Native silver is found in deposits associated with zeolites, calcite, quartz, arsenides, and sulfides of cobalt, nickel, and silver.

sima The petrologic name suggested by Eduard Suess for the lower layer of the earth's crust. It consists of silica- and magnesia-rich rocks, and is equivalent to the oceanic crust and the lower part of the continental crust beneath the sial. It is an acronym for *silica + magnesia.* adj. Simatic. Syn: *basaltic layer.* Cf: *sialma; sial.*

similar folding See *fold.*

simple shear A deformation consisting of a displacement in one direction of all straight lines that are initially parallel to that direction; it is closely similar to shearing a deck of cards. Cf: *pure shear.*

singing sand See *sounding sand.*

sinistral fault *Left-lateral fault.*

sink 1. *Sinkhole.* 2. A depression on the flank of a volcano, resulting from collapse. 3. A generally depressed area where a desert stream ends or disappears by evaporation.

sinkhole See *karst.*

sinter A term applied to a white, porous opaline variety of silica, deposited as an incrustation by precipitation from geyser and hot-spring waters (siliceous sinter). The term is also used for *tufa* or *travertine,* which are calcareous spring deposits.

skarn A metamorphic rock composed of calcium, magnesium, and iron silicates that has been derived from nearly pure limestone or dolomite into which abundant amounts of silicon, iron, aluminum, and magnesium were metasomatically introduced. Skarns are often host rocks for deposits of magnetite and copper sulfide.

slag 1. A cindery or scoriaceous pyroclastic rock. 2. Formerly, a waste material from an iron blast furnace; it is now utilized in construction work.

slaking 1. The crumbling and disintegration of earth materials on exposure to moisture or air. 2. The addition of water to lime to yield hydrated (slaked) lime.

slate 1. A fine-grained metamorphic rock derived mostly from shale. It is characterized by *slaty cleavage*; i.e., the ability to be split into large, thin, flat sheets. Such cleavage is due to the alignment of platy minerals into parallel planes, which occurs during metamorphism. 2. A coal miners' term for shale that is associated with coal in a bed or vein.

slaty cleavage A parallel foliation of fine-grained, platelike minerals, developed in slate or other homogeneous rock by deformation and low-grade metamorphism. Most slaty cleavage is also *axial-plane cleavage*. Syn: *flow cleavage*.

slickenside A rock surface that has become polished and striated from the grinding or sliding motion of an adjacent rock mass, commonly found along fault planes or zones. The striations or flutings may indicate the direction of relative motion, and tend to end abruptly in the form of small steps in the direction of movement. They sometimes bear a superficial resemblance to ancient coal-forming plants, and can be mistaken for fossils.

slip The relative displacement of formerly contiguous points on opposite sides of a fault, as measued on the fault surface. Syn: *total displacement*. 2. *Crystal gliding*.

slip fold *Shear fold.*

slope stability The resistance of a slope, natural or artificial, to landsliding.

slope wash Earth material moved down a slope principally by the action of gravity, aided by non-channeled running water; also, the process itself by which such material is moved. Cf: *colluvium*.

sluicing The concentration of heavy minerals by flushing unconsolidated material through boxes (sluices) equipped with grooves or slats of wood (raffles) that trap the heavier minerals on the bottom of the box.

slump 1. The downward sliding of rock debris as a single mass, usually with a backward rotation relative to the slope along which movement takes place. 2. The slipping or sliding down of a mass of sediment relatively soon after its deposition on a subaqueous slope. Syn: *subaqueous gliding*.

slump fault *Normal fault.*

smectite A group of clay minerals characterized by a three-layer crystal lattice, by deficiencies of charge in the tetrahedral and octahedral positions, and by swelling upon wetting. The smectite minerals are the primary constituents of *fuller's earth* and *bentonite*. *Montmorillonite*, formerly used as a group name in the above sense, is now classified as a mineral of the smectite group.

smithsonite An ore of zinc, $ZnCO_3$, variously colored white, pale blue or green, yellow or pink, depending on the impurities present. It is found in mammillary, botryoidal, reniform, or stalactitic aggregates, and is a typical sedimentary precipitate produced by the reaction of waters rich in zinc sulfate on carbonate rocks. Smithsonite is characteristically found in the oxidation zone of sulfide deposits.

smoky quartz See *quartz.*

soapstone 1. A massive metamorphic rock composed mostly of talc, with varying quantities of mica, chlorite, and other minerals. It can be cut, sawed, and carved. 2. A miners' term for any soft, yielding rock such as micaceous shale. 3. *Steatite.*

soda feldspar A misnomer for "sodium feldspar" (*albite*).

soda lake An *alkali lake* containing large quantities of dissolved sodium salts, especially sodium carbonate, together with sodium chloride and sulfate. Some examples are found in Mexico and Nevada.

sodalite A mineral of the feldspathoid group, $Na_4Al_3Si_3O_{12}Cl$; it occurs as compact masses in bright blue, white, or gray, with green tints in undersaturated plutonic igneous rocks. Sodalite is widely used in jewelry and for carved ornaments.

soda niter Chile saltpeter.

sodium bentonite See *bentonite*.

soft coal *Bituminous coal.*

soil horizon A layer of a soil that can be distinguished from adjacent layers by physical properties (structure, texture, color) or chemical composition. Syn: *horizon; soil zone.* See *soil profile.*

soil mechanics The application of the concepts of mechanics and hydraulics to engineering problems associated with the nature and behavior of soils, sediments, and other unconsolidated accumulations.

soil profile A vertical arrangement of the various discrete horizontal layers, called horizons, that make up any soil from the surface downward to the unaltered parent material or bedrock. As commonly defined, the ideal soil profile is divided into A, B, and C horizons and their subdivisions. 1. *A horizon*—the layer from which soluble substances have been removed by leaching or eluviation, and carried downwards to the B horizon. 2. *B horizon*—a zone of accumulation, enriched in the clay minerals leached from the A horizon. Approx. syn: *subsoil.* 3. *C horizon*—the layer of weathered bedrock at the profile base, which has undergone little alteration by organisms.

soil stripe See *patterned ground.*

soil zone *Soil horizon.*

sol That component of a colloidal system that is more fluid than a *gel.*

sole 1. The undersurface of a stratum or vein. 2. The fault plane underlying a thrust sheet. 3. The lower parts of the shear surface of a landslide. 4. The basal ice of a glacier.

sole mark Any structure found on the bottom (sole) of sandstone or siltstone; e.g., *flute casts.* The presence of such features in a bed would thus indicate which side of a bed was the bottom.

solfatara A sulfurous fumarole. See *fumarole.*

solid earth geophysics See *geophysics, solid earth.*

solid flow Flow in a solid by means of rearrangement among or within the component particles. Cf: *liquid flow; viscous flow.*

solid solution A single crystalline phase in which composition, temperature, and pressure may be varied within limits without the appearance of an additional phase. In most rocks, only quartz is a pure phase.

solidus See *liquidus-solidus.*

solifluction The slow downslope movement of water-logged earth materials. Solifluction can occur in the tropics, but is particularly characteristic of cold regions where permafrost is found. In areas of frozen terrain, the term *gelifluction* is used. Because permafrost is impermeable to water, the overlying soil may become oversaturated and will slide down under the force of gravity. Soil that has been weakened by frost action is most susceptible to this process. Original soil profiles become greatly altered or destroyed. *Solifluction deposits* are usually poorly sorted, but can contain a preferred orientation of rock fragments.

solution breccia A collapse breccia resulting from the removal of soluble material by solution, thus permitting fragmentation of overlying rock.

solution load *Dissolved load.*

solution mining 1. The mining of soluble rock material, esp. salt, from underground deposits by pumping water downward into contact with the deposit, removing the brine, and then processing it. 2. The in-place dissolution of mineral constituents of an ore deposit by trickling down a leaching solution through the broken ore to collection chambers further down.

solution valley *Karst valley.*

solvus In a phase diagram, the curve of surface that separates a domain of stable solid solution from a domain of two or more phases that may form from the stable one by unmixing.

sonic log See *well logging.*

sorosilicate Those silicates each of whose tetrahedral silicate groups shares one of its oxygen atoms with a neighboring silicate group. The ratio of silicon to oxygen is two to seven, S_2O_7.

sorted 1. Said of a sediment composed of particles of uniform size. Syn: graded. 2. Said of patterned ground features that show a border of stones surrounding finer material such as sand, silt, or clay.

sorted circle See *patterned ground.*

sorted net See *patterned ground.*

sorted polygon See *patterned ground.*

sorted step See *patterned ground.*

sorted stripe See *patterned ground.*

sorting The process by which sedimentary particles are naturally separated according to size, shape, and specific gravity from associated but dissimilar particles.

soufrière 1. A common name, meaning "sulfur mine," used in French-speaking regions for volcanoes giving off sulfurous gases, or for a *solfatara.* See *fumarole.* 2. A volcanic crater on St. Lucia Island in the East Caribbean Sea. It derives its name from its volcanic sulfur springs. 3. La Soufriere—a volcanic peak on southern Basse-Terre Island, Guadeloupe, West Indies.

sound A relatively long arm of the ocean or sea that forms a channel between a mainland and an island, or that connects two larger bodies of water.

sounding 1. The vertical distance from the surface of a body of water to its floor or some specific point beneath the water surface. 2. Procedure by which the depth of a body of water is determined. This is accomplished either by (a) direct physical measurement using a lead line or (b) indirectly, by measuring the time it takes for a sound pulse, generated at the surface, to reflect from the bottom and return to the surface as an echo (*echo sounding*). Sounding is performed for navigational safety, scientific investigation, and economic or engineering purposes. See *seismic reflection; echogram.*

sounding sand A sand which, when disturbed in some fashion, emits sounds audible to the human ear. The nature of the sounds, described as booming, singing, roaring, and musical, evidently varies according to the volume of the sand involved and its velocity following disturbance. When sand is of such a type, sound can be initiated by stepping on the slip face of a dune, or treading on a beach. Although the ultimate cause of the phenomenon still remains unexplained, it is known that the sound is caused by the impact of sand grains upon other sand grains. The production of sound is favored by sand that is clean, dry, and extremely well-sorted.

spall, spalling See *exfoliation.*

spar A term imprecisely applied to any transparent or translucent light-colored mineral that is somewhat lustrous and usually shows easy cleavage.

sparite 1. A term used for the coarsely crystalline interstitial cement of limestone. It may be transparent or translucent, and consists of calcite or aragonite that precipitated during deposition, or was later introduced as a cement. Its crystalline structure is much coarser than that of *micrite*. 2. A limestone having more sparite cement than micrite matrix.

sparker A device for generating underwater seismic energy by means of a high-voltage electrical discharge.

spathic Resembling spar, particularly in showing good cleavage.

spatter Agglutinized masses of primary magmatic ejecta larger than *lapilli*, erupted in a plastic or fluid state. It often accumulates around vents as mounds or cones, or along fissures, and is produced chiefly by the frothing of erupting magma in lava fountains. Essentially synonymous with *driblet*. See *lava; spatter cone*.

spatter cone A low, steep-sided mound or hill of spatter formed by lava fountains along a fissure or about a central vent. It consists mainly of *spatter*, the glassy skin of which adheres to form agglutinate. True *bombs* and *cinder* may also be present, and there is usually some *Pele's hair* and *scoria*. Spatter cones are usually less than 15 m high and are generally elongate. They are characteristic of basaltic eruptions. Cf: *hornito*. See *lava*.

specific gravity The ratio of the weight of a given volume of a substance to the weight of an equal volume of water.

spectrographic analysis Analysis of a substance by matching the lines of its spectrum with lines in the spectra of elements whose wavelengths are known.

specularite A gray or black variety of hematite that occurs in aggregates of tabular crystals or foliated masses; it has a brilliant metallic luster. Syn: *specular hematite*.

speleologist A scientist engaged in speleology. Cf: *caver*.

speleology The exploration and scientific study of caves and caverns. As a science, it involves exploration, surveying, biology, geology, and mineralogy.

speleothem Greek, "cave deposit." Any one of variously shaped mineral deposits formed in a cave by the action of water. Almost all speleothems are made of calcium carbonate crystals in the form of calcite. See *cave onyx; dripstone*.

spelunker *Caver*.

spermatophyte A vascular, seed-bearing plant; e.g., angiosperm or gymnosperm. Spermatophytes range from the Carboniferous to the present. Cf: *pteridophyte*.

spessartine The manganese-aluminum end-member of the garnet series, $Mn_3Al_2(SiO_4)_3$. It ranges in color from pale orange-yellow to deep red. Spessartine is relatively rare, and is found combined with almandine in granite, acidic lavas, and thermal metamorphic rocks. Syn: *spessartite*. See also: *garnet*.

spessartite 1. *Spessartine*. 2. A lamprophyre composed of phenocrysts of clinopyroxene or green hornblende in a matrix of sodic plagioclase with accessory biotite, olivine, and apatite.

sp. gr. *Specific gravity*.

sphalerite An isometric sulfide mineral, $(Zn,Fe)S$; its color varies from yellow or reddish brown to blackish. It occurs in aggregates of distorted crystals and banded concretionary masses. Sphalerite is found mainly in pegmatitic pneumatolytic veins, and in hydrothermal veins associated with galena, argentite, and chalcopyrite. It is the main ore for zinc, and generally also provides cadmium and other minerals as byproducts. Syn: *blende; zincblende; blackjack*.

sphene A yellow, green, or brown monoclinic mineral, $CaTiSiO_5$, that occurs as an accessory mineral in igneous rocks, esp. nepheline syenite. Syn: *titanite.*

sphenopteris A fossil seed fern with small, symmetrical, lobed leaflets that show radiating veins. Range, Devonian to Pennsylvanian. See *fossil plants.*

sphericity The degree to which the form of a sedimentary particle approaches that of a sphere; it is not a syn. for *roundness.*

spheroid In geodesy, a mathematical figure closely approximating the *geoid* in size and form, and used as a reference surface for geodetic surveys.

spheroidal weathering See *exfoliation.*

spherulite, spherulitic Generally spherical aggregates, commonly consisting of alkali feldspar fibers radially arranged about some nucleus, such as a *phenocryst.* The usual size is between 1 mm and 2 to 3 cm, but some may be as great as 3 m in diameter. They occur mainly in acid glasses, such as volcanic rhyolites, but are present in some partly or wholly crystalline rocks that include shallow intrusive types. They can form as the result of rapid crystallization in a quickly cooled magma, or by devitrification. A rock body containing spherulites is said to have *spherulitic texture.* See also: *devitrification; lithophysae; variolitic.*

spilite A greenstone of basaltic textural aspect, comprised of low-metamorphic-grade minerals, including albite, chlorite, actinolite, sphene, and calcite. It is thought that spilites form from sea-floor basalts near oceanic rifts, where a metasomatic interchange with heated sea water occurs.

spine 1. Volcanic spine. 2. A pointed projection found on the shell surface of certain invertebrates such as the echinoids.

spinel 1. An isometric mineral, $MgAl_2O_4$, that occurs as crystals, aggregates, or rounded grains in white, pink, light blue, black, and colorless. It is found in contact metamorphic rocks, esp. in many Mg-rich dolomites. Some varieties of spinel are used as semiprecious gemstones. 2. A member of the spinel group.

S-P interval In seismology, the time interval between the first arrivals of transverse and longitudinal waves, which is a measure of the distance from the earthquake source.

spit A narrow ridge or embankment of sediment forming a fingerlike projection from the shore into open water. Spits are formed by the action of waves and currents, and are frequently found on irregular coastlines; in America they are common along the New England, New York, and New Jersey coasts.

splendent luster Mineral luster of the highest intensity.

spodumene A mineral of the pyroxene group, $LiAlSi_2O_6$, occurring as prismatic crystals, sometimes as large as 16 m, or rodlike aggregates of compact cryptocrystalline masses; its colors include whitish, yellow, gray, pink (kunzite), and emerald green. Spodumene is found in lithium-bearing pegmatites associated with quartz, feldspars, beryl, and tourmaline. It is a source of lithium and its salts.

sponge The most primitive multicellular aquatic invertebrate, having an internal skeleton consisting of interlocking spicules of calcite or silica and chitinous fibers. Of the thousands of species known, all but about 20 are marine. The division of fossil sponges is based on (1) the nature of the spicules, and (2) the way they interlock. The complexity of the water current system of a particular sponge connotes the degree of its advancement. Although sponges are sometimes used as horizon markers, they are not used as zone fossils. Range, Precambrian to present. Syn: *poriferan*.

spontaneous potential method See *well logging*.

spring tide A tidal pattern in which the difference in level between high and low tides is a maximum. Spring tides occur twice each month, at full or new moon, when the earth, moon and sun are aligned. Cf: *neap tide*.

stabile Resistant to chemical change or to decomposition. Ant: *labile*.

stabilized dune An *anchored dune*.

stable 1. Said of a constituent of a sedimentary rock that is able to resist mineralogic change, and which represents a terminal product of sedimentation; e.g., quartz. 2. Said of a part of the earth's crust that manifests neither subsidence nor uplift. 3. Said of a substance that is not spontaneously radioactive.

stack Erosional remnant in the form of a small bedrock island a short distance off shore. Stacks are pillar-like formations detached from a headland by marine erosion. During the erosion of a coast, the more resistant parts will usually form headlands; progressive erosion forms caves, then arches or bridges. If the seaward part of the headland remains after collapse, a stack is formed. Further erosion reduces the stack to a *wave-cut platform*. Stacks are found off coasts around the world. American examples are found near San Simeon, Calif., and Seal Rocks, Ore. Fine examples are the Flowerpot Islands of Tobermory, Ont. See *marine abrasion; natural bridges and arches*.

stadia A surveying technique used to determine distances and elevation differences. The direction of a point is established by sighting the point through a telescope, after which a direction line is drawn along a straightedge, and its distance scaled. Distances are read by noting the interval on a graduated rod (*stadia rod*) intercepted by two parallel hairs (*stadia hairs*) mounted in the telescope of a surveying instrument. The rod is placed at one end of the distance to be measured, and the surveying instrument at the other.

stadial moraine *Recessional moraine*.

stadia tables Mathematical tables from which may be determined, without calculation, the vertical and horizontal components of a reading made with a *stadia rod* and an *alidade*.

stage 1. A *chronostratigraphic unit* ranking above *substage* and below *series*; it is based on biostratigraphic zones considered to approximate deposits that are equivalent in time, and includes the rocks formed during an age of geologic time. 2. A major subdivision of a glacial epoch, which includes glacial and interglacial stages. 3. The height of a water surface above a datum plane.

stalactite and stalagmite *Stalactites* are conical or cylindrical mineral deposits, usually calcite, that hang from ceilings of limestone caves and range in length from a fraction of a centimeter to several meters. They are chemical precipitates deposited from water that is supersaturated with calcium bicarbonate, and which enters the caves through cracks and joints in the ceiling. Depending on the cave temperature when deposition occurred, the calcium carbonate precipitate may be calcite or aragonite. As the water drips through the ceiling, successive rings of crystals form a minute tube ("soda straw"); if the tube is blocked, the water flows down the outside, and the deposits thicken the tube into a stalactite.

Water dripping onto the cave floor builds *stalagmites*, which are usually blunter than stalactites. A stalagmite and stalactite may eventually join to form a column. The growth of stalactites is limited by the tensile strength of calcite, whereas the height of stalagmites is limited by the massiveness of the calcite. The limiting effect of the second factor is greater, and so stalactites are able to grow to a greater length than stalagmites. Both types of deposit may be white, translucent or, if impurities are contained, shades of brown, yellow, or gray.

Growth rates for both forms are variable, since they are affected by many factors; thus, the age of either type cannot be accurately determined by size, current growth rate, or weight. Stalactites may also form in subways, tunnels, under bridges, and in mines. Conical formations of lava that rise from the floor of a cavity in a lava flow are also referred to as stalagmites.

stalagmite See *stalactite* and *stalagmite*.

standard mineral *Normative mineral.*

standard section A *reference section* that shows as fully as possible a sequence of all strata within a certain area, in proper order, thus providing a correlation standard; it supplements the *type section* and sometimes replaces it.

standing wave A wave, the form of which oscillates vertically between two fixed points or nodes, without progressive lateral movement. Syn: *stationary wave.*

stanniferous Containing or yielding tin.

stationary wave *Standing wave.*

staurolite A red-brown to black orthorhombic silicate mineral, $(Fe,Mg)_2Al_9Si_4O_{23}(OH)$. It occurs as stubby prismatic crystals, or frequently in cruciform twins; very rarely massive. Staurolite is a metamorphic mineral typical of medium-temperature conditions associated with garnet and pegmatite; it is used as a means of defining the metamorphic type and grade of the host rock.

steatite 1. A compact, massive rock composed mainly of talc, but also con-

taining many other materials. See also: *soapstone.* 2. A term originally used to mean *talc,* esp. the gray-green or brown massive variety that can be easily carved. 3. *Steatite talc.*

steatite talc A high-grade variety of talc, the purest commercial form. Syn: *steatite.*

S-tectonite See *tectonite.*

steinkern (stone-kernel) See *mold.*

steppe A broad, treeless, generally flat area of an arid region on which scattered bushes and short-lived grasses grow and furnish scant pasturage. Although by some definitions most deserts of North America would be steppes, the term is usually restricted to the semi-arid mid-latitudes of southeastern Europe and Asia.

stereographic projection See *map projection.*

stibnite A steel-gray orthorhombic mineral, Sb_2S_3, the main ore of antimony. It is opaque with a bright metallic luster, and occurs as prismatic or acicular crystals. Antimony is found in low-temperature hydrothermal veins with silver, lead, and mercury minerals.

stigmaria Root stocks of lycopod trees, e.g., *Sigillaria* and *Lepidodendron.* The root system of these trees consisted of horizontally spreading main trunk roots without a tap root. The real rootlets sprang directly from the sides of these trunk roots and extended radially to distances of several meters. Stigmaria are often found in the fire clays under coal beds, and sometimes in the coal itself (Pennsylvanian Period). See also: *scale trees.*

stilbite A white, gray, or reddish-brown zeolite mineral, occurring usually in sheaflike aggregates. Stilbite is a late hydrothermal mineral found in cavities of basaltic rocks, and is associated with calcite and other zeolites.

stishovite A high-pressure, extremely dense polymorph of quartz, SiO_2, that is produced under static conditions at pressures greater than 100 kb. Stishovite is found only in shock-metamorphosed, quartz-bearing rocks; its presence provides a criterion for meteorite impact.

stock An intrusive body of deep-seated igneous rock, usually discordant and resembling a *batholith*, except for its size. It covers less than 100 square km in surface exposure. Contacts with the country rock generally dip outward and may vary from a steep to a shallow angle. A stock is more or less elliptical or circular in cross-section. Many have a porphyritic texture intermediate between granitic and volcanic rocks. Syn: *boss. See intrusion.*

Stokes' law An expression for the rate of settling or rising of spherical particles in a fluid; e.g., crystal-liquid fractionation or gravitational segregation in a magma.

$$v = \frac{2r^2 g \Delta\rho}{9\eta}$$

where v is the velocity of the crystals, g the acceleration due to gravity, $\Delta\rho$ the difference of densities between crystal and melt, and η the viscosity of the melt.

stomach stone *Gastrolith.*

Stone Age See *Neolithic; Paleothic.*

stone circle *Sorted circle.*

stone field Syn: *block field.*

stone net Sorted *polygon.*

stone polygon *Sorted polygon.*

stone ring Syn: *sorted circle; sorted polygon.*

stone stream Br. *block stream.*

stone stripe See *patterned ground.*

stony iron meteorites One of three major divisions of meteorite types. Stony irons contain substantial amounts of both stone (silicate) and metal (Fe and Ni) intermixed, and represent an intermediate type between *stony* and *iron* meteorites. In some stony irons, the Ni-Fe is a coherent mass containing discrete stony parts. Others are similar to stony meteorites but contain large quantities (up to 40%) of interspersed metal. See also: *pallasite; mesosiderite; meteorite.*

stony meteorite (stones) One of three major divisions of meteorite types. Stony meteorites consist largely of silicate minerals, and are the most abundant of meteorites seen to fall (>80%). This group is divided into two subgroups: *chondrites* (containing chondrules) and *achondrites* (lacking chondrules).

stoping 1. A process by means of which magma displaces country rock fragments, which then sink or are assimilated. *Piecemeal stoping* usually occurs on a minor scale, and is inefficient as an emplacement mechanism for large igneous rock masses. It serves rather to modify contacts of intrusions implaced in other ways. Cf: *cauldron subsidence. See country rock; magma.* 2. In ore mining, any excavation made to remove the ore made accessible by *shafts* or *adits* (approximately horizontal passageways used in underground mining).

stoss adj. The side of a hill facing the direction from whence a glacier moves. The stoss side of hills (also *stoss-seite*) is most exposed to the glacier's abrasive action, and so will have rounded edges and gentle slopes. Ant: *lee (lee-seite).*

stoss-and-lee topography A grouping of hills or prominences in a glaciated area, having gradual slopes on the stoss side and sharper slopes with a rougher profile on the lee side. Cf: *crag and tail.*

strain The change in the shape or volume of a body in response to *stress,* or a change in relative position of the particles of a substance. Syn: *deformation*.

strain ellipsoid Deformation, or strain, in a rock can be conceptualized by visualizing the change in shape of an imaginary sphere in the rocks. Such a sphere, in a body of granite, would become deformed into an oblate spheroid if the granite were compressed from top to bottom. The most general solid resulting from the deformation is an ellipsoid called the *strain ellipsoid*. The sphere is considered to have unit radius, and the ellipsoid has principal semi-axes of length equal to the principal strains. Syn: *deformation ellipsoid*.

strait A narrow body of water joining two larger bodies of water.

strand The land at the edge of any large body of water. The term is often used as a syn. of *shore* and *beach*. A *strandline* is the ephemeral mark or level at which a body of water, such as a sea or lake, meets the land.

stratification A bedded or layered arrangement of materials. The term is used esp. in reference to sedimentary rock. Layering can also be seen in lava flows, metamorphic rocks, masses of snow, firn, or ice. Waters of a lake may be arranged in layers of differing temperature or density.

stratiform Having the configuration of a layer or bed, but not necessarily being *bedded*.

stratigraphic correlation The procedure by which the mutual correspondence of stratigraphic units in two or more removed locations is shown. It is based on fossil content, geological age, lithographic features, or some other property. The term usually implies the identification of rocks of equivalent age in different places.

stratigraphic geology *Stratigraphy.*

stratigraphic range The duration and distribution of any taxonomic group of organisms through geologic time, as indicated by their presence in strata whose geologic age is known. Syn: range; *geologic range*.

stratigraphic trap See *oil trap*.

stratigraphic unit A stratum or body of strata recognizable as a unit that may be used for mapping, description, or correlation; it does not constitute a time-rock unit. A *lithostratigraphic unit* is a body of rock having certain unifying lithologic features. Although it may be sedimentary, igneous or metamorphic, it must meet the critical requirement of an appreciable degree of overall homogeneity. A lithostratigraphic unit has a two-part designation consisting of a locality name with a rock name; e.g., Ohio Shale, Beckmantown Limestone. Syn: *rock-stratigraphic unit*. A body of strata that is identified by particular fossil content is a *biostratigraphic unit*. The basic unit is the *biozone*.

A *chronostratigraphic unit* is one that was formed during a defined interval of geologic time; it represents all, and only, those rocks formed during a designated time span of geological history. Chronostratigraphic units are, in descending order: *erathem, system, series, stage, chronozone*. Syn: *chronolithologic unit; time-stratigraphic unit; time-rock unit; chronolith*.

stratigraphy 1. The branch of geology concerned with all characteristics and attributes of rocks as they are in strata, and the interpretation of strata in terms of derivation and geological background. Syn: *stratigraphic geology*. 2. That aspect of the geology of an area that pertains to the character of its stratified rock.

stratotype The original or later designated type representative of a named stratigraphic unit or of a stratigraphic boundary, established as a point in a distinctive sequence of rock strata. The stratotype is the standard for the definition and recognition of a particular stratigraphic unit or boundary. See also: *type section; boundary stratotype.*

strato-volcano A *composite cone.* See *volcanic cone.*

stratum A defined layer of sedimentary rock that is usually separable from other layers above and below; a bed. Pl: *strata.* Cf: *lamina.*

streak The color a mineral shows when rubbed on a piece of unglazed porcelain (streak plate). Streak is one of the important characteristics in mineral identification.

stream beheading See *stream capture.*

stream capture The diversion of one stream into another as a result of the erosional encroachment of one upon the drainage of the second; sometimes referred to as *stream beheading.* The stream whose waters are intercepted is the *beheaded stream.* There are three

A is more powerful than B, hence its valley is therefore deeper than that of B.

A's tributary is extending its valley toward B.

River B

River A

A's tributary eventually captures the headwaters of B.

This part of B's valley is now dry and forms a dry or wind gap.

B is now a beheaded river. The beheaded part is called a misfit.

River B

River A

Process of stream capture

main ways of stream capture: abstraction, headward erosion, and subterranean diversion. *Abstraction* occurs when two streams join as a result of the erosion of the divide between them. It takes place in ravines and gullies at the higher end of drainage lines. Stream capture by *headward erosion* takes place where two streams oppose each other on a ridge; one lowers its valley in the upper areas at a faster rate than the other, because there is a steeper gradient at its head, or because it cuts through softer bedrock. *Subterranean capture* occurs where water from a higher stream percolates down to a lower stream through soluble rock, and subsequently forms a diversion tunnel. Syn: *capture; piracy.*

streamline flow *Laminar flow.*

stream load All the material that is transported by a stream. The *dissolved load* consists of material in ionic solution; the ions are part of the fluid and move with it. They are derived from groundwater that has filtered through weathering soil and rocks. The *suspended load* is comprised of clay, silt, and sand particles so fine that they remain in suspension almost indefinitely. (Syn: *suspension load; wash load.*) Material that is too coarse to be lifted by the stream water is pushed along the bottom; this is the *bed load.*

stream piracy See *stream capture.*

stress In a solid, the force per unit area that acts on or within a body. Given a vertical column of material, along any imaginary horizontal plane within the column, the material above the plane, because of its weight, exerts a force on the material below the plane. Similarly, the part of the column below the plane exerts an equal upward force; the mutual action-reaction along a surface then constitutes a stress.

stress difference The algebraic difference between the greatest and least principal stresses.

stress ellipsoid A geometric representation of a state of stress as defined by three mutually perpendicular principal stresses and their magnitudes.

striation One of many thin lines or scratches, generally parallel, incised on a rock by some geological agent such as a glacier or a stream. adj. *Striated; striate.*

strike 1. The direction taken by a structural surface such as a fault or bedding plane as it intersects the horizontal; it is the compass direction of the horizontal line in an inclined plane. Cf: *trend; trace.* 2. y. To be aligned in a direction at right angles to the direction of *dip.*

strike fault A fault that strikes essentially parallel to the strike of the adjacent rocks. Cf: *dip fault; dip joint.*

strike joint A joint that strikes nearly parallel to the strike of the bedding of a sedimentary rock, schistosity of a schist, or gneissic structure of a gneiss. Cf: *dip joint.*

strike separation The distance between two previously adjacent beds on either side of a fault surface, measured parallel to the strike of the fault. Cf: *dip separation.*

strike shift Relative displacement of rock units parallel to the strike of a fault, but beyond the fault zone itself; partial syn. of *strike slip.*

strike-shift fault *Strike-slip fault.*

strike slip The component of movement parallel with the strike of a fault. Cf: *dip slip; strike separation; strike shift.*

strike-slip fault A fault along which the movement has been predominantly horizontal, parallel to the strike. The movement of a strike-slip fault is described by looking straight across the fault and observing in which direction the block on the opposite side has moved. If the opposite block has moved to the left, it is called a *left-lateral*, or sinistral, fault. If the opposite block has moved to the right, it is a *right-lateral*, or dextral, fault. No vertical displacement of horizontal strata occurs during this type of faulting. Syn: *strike-shift fault.* Cf: *dip-shift fault.* See *fault.* See also: *transform fault; transcurrent fault.*

stripe See *patterned ground.*

striped ground See *patterned ground.*

strip mining *Open-cut mining.*

stromatiporoid Any of a group of extinct, sessile, benthic marine organisms of undetermined biological affinity. They were spongelike colonial structures of calcium carbonate, tabular, domelike, or bulbous in form. The external view shows a "pored" texture, while the external structure is that of a dense, laminated mass. Stromatiporoids were especially abundant in reefs of the Ordovician-Devonian. Range, Cambrian to Cretaceous.

stromatolite Laminated calcareous sedimentary formations produced by lime-secreting, blue-green algae. Living stromatolites (Hamelin Pool, Shark Bay, Australia) are in the shape of stony cushions or massive columns. Fossilized (silicified) stromatolites dating well back into the Precambrian are found in the Gunflint chert of Lake Superior, in cherts of Africa and Australia, and in calcareous sediments. Until the presence of organic material in these cherts was ascertained, the forms embodied therein were called *cryptozoa.*

Strombolian-type eruption See *volcanic eruption, types of.*

structural Of or relating to rock deformation or to features that result from it.

structural feature A feature produced by the displacement or deformation of rocks, such as a fault or fold.

structural geology The branch of geology concerned with the description, spatial representation, and analysis of structural features ranging from microscopic to moderate. As such, it includes studies of the forces that produce rock deformation, and of the origin and distribution of these forces. Structural features may be primary, i.e., those acquired in the genesis of a rock mass (e.g., horizontal layering), or secondary, i.e., resulting from later deformation of primary structures (e.g., folding or fracturing). Cf: *tectonics.*

structural relief The difference in elevation between the lowest and highest points of a bed or stratigraphic horizon in a specified region.

structural trap See *oil trap.*

structure 1. The altitude and positions of rock masses of an area; the sum total of structural features resulting from processes such as folding and faulting. 2. In geomorphology, a general term for underlying rocks of a landscape. 3. A feature of a rock that embodies mutual relations of aggregates of grains such as bedding or foliation. Cf: *texture.* 4. In petroleum geology, any physical arrangement of rocks that may hold oil or gas accumulations. 5. The form assumed by a mineral; e.g., cruciform.

stylolite An irregular surface within a bed, usually of carbonate rocks, characterized by pits and teethlike projections on one side that fit into counterpart negatives on the other. Stylolite seams are formed of material less soluble than limestone, such as clay or organic matter. A diagenetic formation of stylolites is indicated by their cross-cutting of allochems such as fossils, in which the lower or upper part has apparently dissolved away. The generally parallel disposition of the stylolite seams implies that dissolution was caused by an interaction between overburden pressure and pore waters of the same type that forms the pressure-solution surfaces in some quartzarenites.

sub- 1. A prefix meaning "under" or "beneath," such as subaqueous—beneath the water. 2. Indicating a condition that is slightly outside the lower limit of a given condition; e.g., *subtropical; subangular.* 3. A division of a larger category: *subgenus; subgroup.*

subalkalic 1. Used as a group term for rocks of the tholeiitic and calc-alkaline series or suites. 2. Said of an igneous rock containing no alkali minerals other than feldspars.

subaqueous gliding *Solifluction* or *slump* under water.

subarkose Although precise definitions vary, the term generally refers to a sandstone that does not contain enough feldspar to be called an *arkose*; it also refers to a sandstone that is intermediate in composition between arkose and pure quartz sandstone. Approx. syn: *feldspathic sandstone; arkosic sandstone.*

subautomorphic 1. adj. Referring to the texture of an igneous or metamorphic rock that is characterized by crystals only partly bounded by their typical (rational) faces. Syn: *hypidiomorphic.* 2. Syn. of *subhedral*, in European usage. Cf: *automorphic; xenomorphic.*

subbituminous Official U.S. Geologic Survey term that covers the rank between brown woody lignite and bituminous coal. It is characterized by a higher carbon content than lignite and a lower moisture content. The calorific value ranges between 8000 and 10,000 BTU.

subduction Movement of one crustal plate (lithospheric plate) under another. The descending plate is frequently said to be "consumed." Subduction refers to the process, not the site. See *subduction zone.*

subduction trench See *plate tectonics.*

subduction zone An extended region along which one crustal block descends relative to another, and along which deep oceanic trenches occur. Subduction zones are one of three kinds of boundaries (see *mid-oceanic ridges; transform faults*) of crustal plates. During the tectonic process of two plates moving toward each other, one dips downward and is forced under the edge of the other, penetrating the earth's mantle at a rather sharp angle. The bend in the submerging plate creates an *ocean trench.* The downward action of this plate is responsible for deep-focus earthquake activity, and the resultant friction produces volcanoes and intrusions on the far side of the trench. See also: *plate tectonics.*

subgraywacke The most common type of sandstone, intermediate between *graywacke* and *orthoquartzite.* It differs from graywacke in having less feldspar and more and better rounded quartz grains; it also has less matrix and is lighter-colored and better-sorted than graywacke.

subgroup See *group.*

subhedral Said of a mineral grain bounded partly by its typical (rational) faces and partly by surfaces formed against pre-existing grains as a result of crystallization or recrystallization. Also said of the shape of such a crystal. Cf: *euhedral; anhedral.*

submarine canyon A V-shaped submarine valley resembling land canyons that have been cut by streams. It is characterized by high, steep walls and an irregular floor that slopes continually outward; such gorges have been found throughout the world. All submarine canyons head on the continental shelf, and most debouch at the base of the continental slope. Alluvial fans are often found at the mouths. These valleys vary considerably in wall height, length and slope of the valley floor; their average length is about 55.5 km and average wall height about 915 m, but the Great Bahama Canyon has a wall height of 4285 m (Grand Canyon—1700 m). Several mechanisms have been proposed for the origin and maintenance of submarine canyons. There is also some question about whether these formations all share a common origin. At present, most geologists seem to subscribe to the following concepts: 1. Most canyons have been cut by the action (either subaerial or submarine) of downslope sediment movement. 2. Many of those canyons that were originally formed by subaerial erosion have since been drowned, and are now actually kept free by submarine processes such as turbidity currents, slumping, or gravity flow. See *turbidity current; submarine fans.*

submarine eruption An eruption from a volcano located on the oceanic crust. Because of their location, the magmas of such volcanoes are basaltic, and the type of eruption is a modification of the *Hawaiian* or *Strombolian* type. However, the submarine eruption is much more explosive than its continental counterpart, as a result of the contact between magma and seawater. See also: *volcanic eruption.*

submarine fans Fan- or cone-shaped submarine features that are accumulations of terrigenous sediment. They are also called abyssal cones, deep-sea fans, submarine deltas, etc. These fans are found offshore from most of the world's great rivers, and extend downward to abyssal depths. Periodic *turbidity currents* that flush sediments through submarine canyons are responsible for their buildup.

submergence A rise of the water level relative to the land, so that areas that were formerly dry land become inundated; either sinking of the land or a net rise of sea level is responsible for submergence. Ant: *emergence.*

subsequent Said of geological or topographic features whose development is governed by differences in erodibility of the underlying rocks. For example, a *subsequent valley* or a *subsequent stream* is one that has shifted from its original *consequent* course to belts of more readily erodible rock.

subsidence The downward settling of material with little horizontal movement. The most common cause is the slow removal, either by geological processes or human activity, of material from beneath the subsiding mass. Processes such as compaction, solution, withdrawal of fluid lava from beneath a solid crust, and subsurface mining can all be responsible for subsidence. *Kettles* and sinkholes of limestone regions are subsidence landforms. Subsidence may also refer to the downwarping of a large part of the earth's crust relative to the surroundings; e.g., the formation of a *rift valley.* See also: *cauldron subsidence.*

subsolidus A chemical system that is below its melting point, and in which reactions may take place in the solid state.

substage 1. A subdivision of a stage that includes the rocks formed during a subage of geologic time. 2. The subdivision of a glacial stage during which there occurred a secondary fluctuation in glacial advance and retreat.

subterranean diversion See *stream capture.*

subterranean water See *groundwater,* def. 2.

sucrose *Saccharoidal.*

sulfate A chemical compound containing the sulfate radical, $(SO_4)^{-2}$.

sulfide One of a group of compounds in which sulfur is combined with one or more metals; a mineral example is cinnabar, HgS.

sulfide enrichment See *secondary enrichment.*

sulfide zone See *secondary enrichment.*

sulfur An orthorhombic material, the native element sulfur (S). It occurs in bright yellow crystals and granular aggregates, often at or near hot springs and fumaroles. Sulfur is associated with sedimentary deposits of the evaporite type, and with oil-bearing deposits. See *sulfur bacteria.*

sulfur bacteria Bacteria by means of whose life activities sulfur is formed. One type, which is chemosynthetic, oxidizes hydrogen sulfide to elemental sulfur, utilizing some of the released energy in manufacture of their foods. Another type, a photosynthetic organism, combines carbon dioxide (in the presence of light energy) to form carbohydrates and elemental sulfur. Cf: *iron bacteria.* See *bacteriogenic.*

summit concordance The equal or nearly equal elevation of mountain summits or hilltops over a certain region. Such coincidence of elevation is generally thought to suggest that the summits are the preserved high features of a former erosion plan. See also: *accordant summit level; peneplain.*

summit eruption See *volcanic eruption, sites of.*

sump 1. An excavation in which quarry waters may be collected and later pumped out. 2. A surface pit in which mined material is combined with water to form a slurry in order to facilitate removal.

sun opal *Fire opal.*

sunstone An *aventurine* feldspar, usually a translucent variety of oligoclase that emits a reddish glow. Its appearance is due to minute, platelike inclusions of hematite, oriented parallel to one another. Cf: *moonstone.*

super- A prefix meaning "above" or "over," either in position or condition; e.g., *supergene; supersaturated.*

supergene A term applied to ores or ore minerals formed by processes that almost always involve water, with or without dissolved material, descending from the surface. The word connotes "origin from above." Typical supergene processes are hydration, solution, oxidation, and deposition from solution. Cf: *hypogene.*

supergene enrichment See *secondary enrichment.*

supergroup See *group.*

superposition See *law of superposition.*

superstructure The upper structural layer of an orogenic belt, which is subjected to relatively shallow or near-surface deformation processes, as distinct from an underlying and more profoundly deformed *infrastructure.*

surf 1. Refers to wave activity in the surf zone, which is the area bounded by the landward limit of wave uprush (*swash*) and the farthest seaward breaker. 2. Collective term for *breakers.*

surface of no strain A surface along which the original configuration remains unchanged after deformation of the body in which the surface is contained. For example, if a sheet is bent, the convex side is subject to tension, whereas the concave side is subject to compression; between the two loci of points there is a surface of no strain. Syn: *neutral surface.*

surface wave See *seismic wave.*

surf zone See *surf.*

surging glacier A glacier that exhibits a period (usually brief) of very rapid flow, perhaps 100 or more times its normal velocity. Usually, surging glaciers have been stable or even retreating beforehand. During this time, ice that would normally move downglacier by normal flow mechanisms accumulates in an ice-reservoir area, and surges downward when a critical threshold is reached. It appears that most glaciers that have surged have done so in cycles ranging from 15 to 100 years.

susceptibility *Magnetic susceptibility.*

suspended load See *stream load.*

suspended water *Vadose water.*

suspension A type of sediment transport in which sediment particles are indefinitely held in the surrounding water by the upward component of eddy currents associated with turbulent flow.

suspension current *Turbidity current.*

swale 1. A low place in a land area, usually marshier than the adjacent higher land. 2. A depression in an undulating ground moraine, due to uneven glacial deposition. See *swell-and-swale topography.* 3. A long, narrow depression between *beach ridges.*

swallow hole A depression, closed or open, into which all or part of a river or stream disappears underground. The term is often used to mean *sinkhole*, but is not synonymous with the latter.

swamp A type of wetland characterized by mineral soils with poor drainage, and by plant life dominated by trees such as gums, willows, and maples in temperate swamps, and mangroves and palms in tropical regions. Swamps can develop in any area with poor drainage and a water supply sufficient to keep the ground waterlogged. Flood plains, abandoned river channels, and oxbows all have conditions that support swamps and marshes. Salt swamps are formed when flat intertidal land is subjected to the flooding and draining of seawater. In tropical and subtropical regions, regularly flooded protected areas develop mangrove swamps. The Great Dismal Swamp of North Carolina and Virginia is a mixture of waterways, swamps, and marshes. Okefenokee Swamp of Georgia consists of swamps and marshes. The Florida Everglades is a marsh-swamp combination growing on a limestone base in a region near sea level. Its waters, unlike those of marsh or swamp, are clear and alkaline. The mineral supply in swamp water is sufficient to stimulate decay of organisms and prevent accumulation of organic materials. However, if the rate of oxygen supply becomes too low, decay will be incomplete, and organic matter will accumulate to form peat. A slow, continuous rise of the water table, and protection of the swamp area against major sea inundations, help to promote thick peat deposits and the consequent formation of coal seams. Cf: *bog; marsh.*

swamp theory *In-situ theory.*

swash See *breaker.*

S wave See *seismic wave.*

swell-and-swale topography A gently undulating surface that may be formed by silty or clay till, and which is characterized by rounded, usually symmetrical ridges and depressions that form a sinusoidal curve. Its patterns vary from subparallel to parallel, and ridges may be forked or crooked. This topography is found in basins of extinct glacial lakes, e.g., southern Ontario, and on till plains such as found in Iowa.

syenite A group of plutonic rocks usually containing microcline, orthoclase, and small amounts of plagioclase, hornblende, and/or biotite, and little or no quartz. Syenite is the intrusive equivalent of *trachyte*; with an increase in quartz, it grades into *granite.*

syenodiorite *Monzonite.*

symmetrical fold See *fold.*

symmetry axis *Axis of symmetry.*

synchronous Formed or occurring at the same time; contemporary, such as a rock surface on which every point is of the same geological age.

syncline A generally U-shaped fold or structure in stratified rock, containing stratigraphically younger rocks toward the center of curvature, unless it has been overturned. A *synclinorium* is a large compound syncline composed of several minor folds. (Cf: *anticlinorium*. See also: *geosyncline*.) The *synclinal axis* is that which, if displaced parallel to itself, generates the form of a syncline. A mountain whose underlying structure is that of a syncline is called a synclinal mountain. Ant: *anticline*. See also: *synform; fold*.

synclinorium See *syncline*.

synecology The study of relationships between communities of organisms and their environment. Cf: *autoecology*.

synegabbro An intrusive rock differing from gabbro by the presence of alkali feldspar.

synform A U-shaped fold in strata of unknown stratigraphic sequence. Cf: *syncline*. Ant: *antiform*.

syngenesis See *diagenesis*.

syngenetic 1. adj. Describing mineral deposits formed at the same time as the enclosing rock, and by the same process. (Cf: *epigenetic*.) 2. Describing small markings in sediments, such as ridges, formed at the same time as the deposition of the sediment.

syntectite A rock formed by *syntexis*. See also: anatexite.

syntexis Generation of magma by the melting of two or more rock types, and the assimilation of host rocks. adj. *Syntectic*. Cf: *anatexis*.

synthetic fault A minor normal fault that is oriented in the same direction as the major fault with which it is associated. Cf: *antithetic fault*.

synthetic group A rock stratigraphic unit composed of two or more formations that are associated because of certain similarities between or among their fossils or lithologic features. Cf: *analytic group*.

system 1. The fundamental unit of *chronostratigraphic* classification that is taken from a type area and correlated chiefly by its fossil content; it includes the rocks formed during a *period* of geologic time. In rank, system is above *series* and below *erathem*. 2. A group of related natural features; e.g., a *fault system* or a *mountain system*. 3. *Crystal system*. 4. *Magmatic system*.

T

tablemount *Guyot*.

tabular 1. Said of a form having two dimensions much greater than the third. For example, any igneous dike is a tabular pluton. Cf: *massive*. 2. A reference to the shape of a sedimentary body whose ratio of width to thickness is greater than 50 to 1, but less than 1000 to 1. Cf: *blanket deposit*. 3. A crystal shape having one dimension markedly smaller than the other two is said to be *tabular*. Cf: *prismatic*. 4. Said of a metamorphic texture characterized by a large proportion of tabular grains disposed in approximately parallel orientation.

tachygenesis See *acceleration (Biol.)*.

tachylyte A highly unstable *volcanic glass* of basaltic composition. Syn: *sideromelane*. See also: *pseudotachylyte*.

Taconian (Taconic) orogeny A Mid-Ordovician orogeny named for the Taconic Range of eastern New York State. This disturbance culminated in a chain of fold mountains extending from Newfoundland through the Canadian Maritime Provinces and New England, reaching as far south as Alabama. Evidence of volcanic activity during the Taconian Disturbance is found from Alabama to New York, and as far west as Wisconsin and Iowa. In Canada, Quebec and Newfoundland show evidence of the greatest volcanic activity of this time. In northwest Europe, the *Caledonian orogeny* corresponds to the Taconian plus *Acadian orogenies* of North America. The Taconian may be considered an early phase of the Caledonian. It is now generally thought to have consisted of several pulses, extending from the Mid-Ordovician to Early Silurian.

tactite An approx. syn. of *skarn*.

talc 1. A very soft white, greenish-white, gray, or brownish mineral, $Mg_3Si_4O_{10}(OH)_2$, with a hardness of 1 on the Mohs scale. Talc is found in foliated, fibrous, or granular masses as an alteration product of magnesium silicates or ultramafic rocks; it is also formed by metasomatism in impure dolomitic marbles. See also: *steatite*. 2. In commercial usage, a rock containing talc, chlorite, tremolite, and related minerals; it is used as a filler, dusting agent, and coating in ceramics and lubricants.

talus A heap of coarse debris, a result of weathering (frost action), at the foot of a cliff. (Cf: *scree*) The slow downslope movement of talus or scree produces *talus-creep*. If the debris assumes the flowing motion and form of a glacier, it is referred to as a *talus glacier*.

talus cone A steep-sided pile of rock fragments lying at the base of a cliff from which they have been derived. Talus cones are formed primarily by the movement of materials aided by gravity.

tantalite A black mineral, (Fe,Mn)(Ta,Nb)$_2$O$_6$, the principal ore of tantalum. It is isomorphous with *columbite* and occurs in pegmatites.

taphrogeosyncline A deeply depressed crustal block filled with sediment and bounded by faults.

tarn A small lake formed in a cirque; also called a *cirque lake*.

tar pit An accumulation of natural bitumen that is exposed at the land surface, and which becomes a trap into which animals sink; the hard parts of the carcasses are preserved. Ex: La Brea tar pits, Los Angeles, Calif.

tar sand A sand area that is large enough to contain a commerical store of asphalt; it may be an *oil sand* from which more volatile materials have escaped.

T.D. 1. *Total depth.* 2. *Time-distance.*

T.D. curve *Time-distance curve.*

T Δ T analysis A method of detecting such lateral variations of velocity as might distort seismic interpretation when calculating the average velocity to a reflecting horizon. In the T Δ T method, records are picked from shots distributed over such a wide area that the net dip of the reflecting surface can be reasonably assumed to be zero. Velocities calculated by the T Δ T method have been shown to be of considerably lower magnitude than well-shooting velocities at all depths.

tectogene A narrow, elongate unit of downfolding of *sialic* crust believed to be related to mountain-building processes. Cf: *downbuckle*.

tectogenesis *Orogeny.*

tectonic Adj. Relating to structures of or forces associated with *tectonics*.

tectonic axis *Fabric axis.*

tectonic breccia An aggregation of angular rock fragments that have formed as a result of tectonic movement; a *crush breccia.*

tectonic conglomerate *Crush conglomerate.*

tectonic fabric *Deformation fabric.*

tectonic facies A term for rocks that owe their present features chiefly to tectonic activity.

tectonics A branch of geology that is closely related to but not synonymous with structural *geology.* Whereas structural geology is concerned primarily with the geometry of rocks, tectonics deals not only with larger features of the earth, but also with the forces and movements that produced them. See also: *plate tectonics.*

tectonism *Diastrophism.*

tectonite A deformed rock, the fabric of which reflects the coordination of the componential movements responsible for the deformation. Conversely, a *nontectonite* is a rock resulting from many separate components, each of which functions independently; all undeformed sedimentary and igneous rocks belong to this category. Two principal types of tectonites are referred to in the literature: *B-tectonite* and *S-tectonite.* A B-tectonite, frequently called *L-tectonite,* is one in which lineation is prominent. In an *S-tectonite,* the fabric is characterized by planar elements caused by deformation; e.g., foliation.

tectosilicates Any of the three-dimensional silicates, each tetrahedral group of which shares all its oxygen atoms with neighboring groups, the ratio of silicon to oxygen being one to two, (SiO_2). Ex: quartz. See also: *silicates.*

tectosphere A gravitationally stable subcontinental zone above the mantle, in which crustal or tectonic movements

originate. The tectosphere concept answers, at least indirectly, the geophysical question, based on heat-flow studies, of how a necessarily cooler subcontinental mantle can be in a condition of hydrostatic equilibrium with a hotter suboceanic mantle.

tektite A small, rounded mass consisting almost entirely of glass. Prior to the Apollo missions, tektites were believed to be of lunar origin, but this was not supported by evidence from lunar studies. It is thought that tektites result from large meteoritic impacts on earth; under the force of impact, solid rock is instantaneously changed into drops of molten liquid, which then solidify as a glass.

telluric current *Earth current.*

telluride A chemical compound that is a combination of tellurium with a metal. A mineral example is calaverite.

temperate glacier See *glacier.*

temperate gradient See *geothermal gradient.*

temporary base level See *base level.*

tennantite A dark, lead-gray mineral, $(Cu,Fe)_{12}As_4S_{13}$, the arsenic end-member of the tetrahedrite-tennantite series.

tenor The *grade* of an ore body.

tension Stress that tends to pull a solid body apart. adj. *Tensile.*

tephra All fragmental material ejected from a volcanic vent and transported through the air; e.g., ash, pumice, bombs, etc. A synonym for *pyroclastic material.* Cf: *tuff.*

tephrochronology A method of age determination by means of layers of volcanic ash (*tephra*). By identifying and correlating several tephra units over a wide area, a chronology can be established. Tephrochronological studies in Iceland have provided an effective technique for the correlation of geological and archaeological events there.

terminal moraine See *glacial moraine*.

terminus The extremity or outer margin of a glacier.

ternary system See *phase diagram*.

terra An upland region on the lunar surface characterized by a rough texture and lighter color than that of a *mare*. It is thought that the terra is either a result of lunar volcanic activity, or an ancient surface subjected to the impact of meteorites.

terra cotta Etymol: Ital. "baked earth." A fired clay of a particular yellowish-red or brownish-red color, used for ornamental objects and decorative work on the outside of buildings.

terrain A region of the earth's surface that is treated as a physical feature or as a type of environment.

terrain correction In geophysical surveys, a correction that accounts for the attraction of all material higher than the gravity station, and also eliminates the effect of that material required to fill in below-station hollows in order to "level up" the infinite slab hypothesized when applying the *Bouguer correction*. Syn: *topographic correction*. See *gravity anomaly*.

terrane A term that is variously dying and being revived, depending upon the particular geological literature. As it has been used, the term applies to a rock or group of rocks together with the area of an outcrop.

terra rossa Etymol: Ital., "red earth." Residual red clayey soil, mantling limestone bedrock and extending down into open joints. It is a result of surface solution by descending groundwater.

terrestrial deposit A sedimentary deposit formed on land without the agency of water; e.g., sand dunes.

terrigenous deposits Land-derived marine sediments that are carried down the continental shelf by gravity and by geostrophic currents; their composition is influenced by the region of derivation. *Red mud* is reddish-brown silt and clay that extends to ocean depths of about 2000 m. Its calcium oxide content, from 6 to 60%, varies with locality. The red color, due to the presence of ferric oxide, seems to be largely confined to the surface. (Cf: *red clay*.) *Blue* and *gray muds* range in depth from just below sea level to slightly greater than 5000 m. At the surface, their colors are reddish to brownish, but beneath the surface the colors range from gray to blue. These muds are composed of about 15% organic matter, a calcium oxide content ranging from 0 to 35%, and varying amounts of hydrogen sulfide; glauconite may or may not be present. *Green mud* is found at depths from less than 100 m to over 2500 m, near the edge of the continental shelf. Apart from the presence of glauconite, from which it derives its color, green mud is similar in composition to gray and blue muds. Its calcium carbonate content reaches about 50%. Cf: *terrestrial deposit*. See also: *ocean-bottom deposits; black mud*.

Tertiary A geological period that is part of the Cenozoic Era. See *Cenozoic Era*.

Tethys A sea that lay between the northern and southern continents of the Eastern Hemisphere from the Permian through the early Tertiary periods. It occupied the general region along the Alpine-Himalayan orogenic belt, and separated *Laurasia* and *Gondwana*. The Alpine-Himalayan continental collision obliterated all but the last vestige of Tethys—the Mediterranean Sea. At present, the continued movement of the African and Eurasian Plates toward each other is shrinking the Mediterranean as well. See also: *plate tectonics*.

tetracoralla Extinct coral with a four fold asymmetry; i.e., septa arranged in quadrants.

tetragonal system See *crystal system*.

tetrahedrite An isometric mineral, $(Cu,Fe)_{12}Sb_4S_{13}$, isomorphous with tennatite. It is usually found in veins associated with copper, silver, lead, and zinc minerals.

textural maturity An end condition reached by detrital sediment, defined in terms of uniformity of particle size and degree of rounding. The textural maturity of a sediment is a result of certain modifying processes acting on it; it is a sequence of clay removal, sorting, and rounding, the rate of which depends on such factors as the current strength at the site of deposition, and the mechanical resistance of the grains. Cf: *immaturity*.

texture 1. The general character or appearance of a rock as indicated by relationships between its component particles; specifically, grain size and shape, degree of crystallinity, and arrangement. The term is applied to smaller features of a rock; *structure* is generally applied to larger features. 2. The physical nature of a soil expressed in terms of relative proportions of sand, silt, and clay. 3. *Topographic texture*.

thallophyte A nonvascular plant without roots, stems, or leaves, and non-seed-bearing. Ex: fungi, algae, lichens. Cf: *bryophyte; pteridophyte*.

thanatocoenosis An assemblage of fossils brought together after death by sedimentary processes rather than because they had all shared the same habitat during life. Syn: *death assemblage*. Cf: *biocoenosis*.

theodolite A precision surveying instrument equipped with telescopic sight for determining horizontal and sometimes vertical angles.

thermal center *Hot spot*.

thermal conductivity See *conductivity, thermal*.

thermal gradient See *geothermal gradient*.

thermal resistivity The reciprocal of thermal conductivity.

thermal spring A spring that discharges water heated by natural processes. If the water has a temperature higher than that of the human body (27° C), the spring is called a *hot spring*. The heat of thermal springs is thought to be derived from hot igneous rocks at depth, the deep circulation of meteoric waters, heat generated by friction caused by crustal movements, and radioactive decay. When such a spring ejects steam and boiling water, it is called a *geyser*. Thermal springs are usually characterized by the presence of travertine, silica, sulfur, and many other substances that are precipitated.

thermal stratification Stratification of lake waters due to temperature changes at different depths; this results in the formation of horizontal layers of different densities. See also: *density stratification*.

thermocline A layer of water with a sharper vertical gradient in temperature than that of the layer below or above it; originally applied to a lacustrine environment. In the oceans, the boundaries of this sharp gradient layer usually form the division between the surface water and the deep and intermediate depth water below. These layers vary in depth, thickness, area, and permanence. Thermoclines can be affected by almost all physical processes occurring in oceans and lakes, and by above-surface meteorological processes.

thermohaline A term applied to a slow vertical movement of seawater, generated by density differences. Such movements are due to temperature and salinity variations which, in turn, induce convection and mixing.

thermoluminescence The property, exhibited by many minerals, of emitting light when heated.

thermoremanent magnetism (TRM) See *natural remanent magnetism.*

thin section A slice of rock or mineral that has been mechanically ground to a thickness of about 0.003 mm and mounted as a microscopic slide. In such thin sections most minerals are rendered translucent or transparent, thus allowing their optical properties to be studied.

thixotropy A property of fluids and plastic solids, characterized by high viscosity at low stress but decreased viscosity when the applied stress is increased. Certain colloidal substances, e.g., bentonitic clay, weaken when disturbed, but increase in strength upon standing.

tholeiite *Tholeiitic basalt.*

tholeiitic basalt A term derived from Tholey, Saarland, Germany, for a variety of *basalt* that is poor in alkalis. It is characterized by the presence of orthopyroxene in addition to calcic plagioclase and clinopyroxene. Tholeiitic basalt is the commonest lava on the ocean floor, but is not abundant on the continents.

threshold pressure *Yield point.*

threshold velocity The minimum velocity at which wind or water in a specified location, and under given conditions, will begin to move particles of sand, soil, or other material.

throw On a fault, the vertical component of the dip. Cf: *heave.*

thrust *Thrust fault.*

thrust fault A fault with dip of 45° or less, on which the *hanging wall* is displaced upward and laterally relative to the footwall. Its characteristic feature is horizontal compression rather than vertical displacement. Partial syn: *reverse fault.* Syn: *thrust; overthrust.*

thunder egg A common term for a small, spherical or nearly spherical geode that has weathered out of the welded tuffs of central Oregon. See *geode.*

tidal bore See *bore.*

tidal flat A broad, flat, marshy or barren tract of land that is alternately uncovered and covered by the tide. It consists of unconsolidated sediment and may form the top surface of a deltaic deposit. See also *tidal marsh; mud flat.*

tidal friction A tidal effect, particularly evident in shallow waters, that lengthens the tidal epoch and tends to retard the rotation of the earth, thus very slowly increasing the length of the day.

tidal marsh See *marsh.*

tidal prism The total volume of tidally-moved water, excluding any freshwater, that flows into a harbor or out again.

tidal wave See *tsunami.*

"tidal wave" theory A dualistic theory of the origin of the planetary system. The first such theory (1745) came from Le Clerc, who proposed that the impact of a massive "comet" tore material from the sun, which later dispersed and condensed into planets. More than a century later, the idea was modified into a close encounter of the sun with a second star. The tidal effect of such an encounter would draw a gaseous filament from the sun, which would then break up into planetary fragments. Cf: *Nebular Hypothesis.*

tiger's eye A yellowish-brown gem variety of quartz, pseudomorphous after crocidolite. The chatoyancy of the mineral is due to penetration of the quartz by asbestiform fibers of crocidolite.

tight fold *Closed fold.*

till Nonstratified material deposited directly by glacial ice. Sometimes called *boulder clay*, till is made up of clay and intermediate-size boulders, which are usually angular because they have undergone little or no water transport. Till is of two types: *basal*—that which was carried in the glacier base and usually laid down under it, and *ablation*—that which was carried on or near the glacier surface and finally melted out. See also: *glacial drift; moraine.*

tillite A sedimentary rock formed by the compaction and cementation of *till*, a glacial deposit.

time break *Shot break.*

time-depth chart A chart, used in seismic work, by means of which time increments can be connected to corresponding depths. It is based on the relationship between the velocity and arrival time of vertically traveling seismic reflections that have been initiated by a series of shots. Syn: *time-depth curve.*

time-depth curve Time-depth chart.

time-distance curve In seismology, a relation between the travel times of various seismic phases and their epicentral distances. Data for establishing such relations can come from records of the same earthquake at a number of stations with different epicentral distances, or from records noted at one station of several earthquakes at various distances. The paths of different phases, distribution of intra-earth seismic velocities, and locations of discontinuities are deduced from these curves.

time-rock unit *Chronostratigraphic unit.*

time-stratigraphic Said of rock units the boundaries of which are based on geologic time.

time tie The principle involved in continuous profiling, wherein seismic events on different records are identified by the coincidence of their travel times.

time-transgressive *Diachronous.*

tin 1. The native metallic element Sn. 2. An informal term to designate cassiterite and concentrates containing cassiterite and minor amounts of other minerals.

tinstone *Cassiterite.*

titaniferous Containing titanium.

titanite *Sphene.*

tombolo Etymol: Ital. "sand dune," from Latin *tumulus*, "mound." A bar or barrier joining an island with a mainland or another island. Tombolos occur along shorelines of submergence, where islands are common.

tonalite *Quartz diorite.*

tongue 1. Any projection or offshoot of a larger body, such as a glacier tongue or a lava flow extending from a larger flow. 2. A minor (informal) lithostratigraphic unit. See *lentil.*

topaz An orthorhombic silicate mineral, $Al_2SiO_4(F,OH)_2$. Found as sometimes enormous crystals in colors of yellow, blue, green, violet, or colorless. Topaz is typical of pegmatitic-pneumatolithic conditions found in gneisses, granites, and microlitic cavities.

topaz quartz *Citrine*, the yellow variety of quartz used as a gem.

topographic correction *Terrain correction.*

topographic maturity *Maturity.*

topographic texture The disposition or general size of the topographic elements composing a particular topography. It usually refers to the spacing of drainage lines in stream-dissected areas.

topography The general configuration of a land surface, including size, relief, and elevation.

topset bed See *delta*.

topsoil The fertile, dark-colored surface soil, or a horizon. See *soil profile*.

torbanite A mineral substance intermediate between oil shale and coal. Coals produce aliphatic hydrocarbons, whereas torbanite, on destructive distillation, produces paraffinic and olefinic hydrocarbons. As its petroleum content increases, torbanite grades into cannel coal. It is named from its type locality, Torbane Hill, in Scotland. Torbanite is often referred to as a type of *boghead coal*, or as a synonym for boghead coal.

torque The movement of a force or system of forces tending to cause rotation.

torsion The twisting of a body by two equal and opposite torques.

torsion balance See *Eötvös torsion balance*.

torsion fault *Wrench fault*.

total displacement *Slip*.

total reflection *Reflection* in which all of the incident wave is returned.

tourmaline A mineral group of complex borosilicates with variable chemical composition, occurring as prismatic crystals or aggregates of parallel or radiating individuals. Tourmaline is a common accessory mineral in igneous and metamorphic rocks, and very common in pegmatites, where it occurs in crystals of sometimes enormous size.

T phase An occasionally recorded seismic phase for which the corresponding ray has most of its length in the ocean, where the wave velocity is about 1.5 km/sec. The T phase has been recorded in California, from Hawaiian earthquakes.

trace element An element that is found in a mineral in small quantities, significantly less than 1.0% of the mineral, and which is not essential to the composition of the mineral.

trace fossil A fossilized track, trail, boring, or burrow formed by the movement of an animal in soft sediments. Syn: *ichnofossil*.

trachyandesite An extrusive rock with a composition between trachyte and andesite. It contains sodic plagioclase, alkali feldspar, and one or more mafic minerals (amphibole, biotite, pyroxene). Cf: *latite*.

trachybasalt An extrusive rock with a composition intermediate between trachyte and basalt. Both calcic plagioclase and alkali feldspar are present, as well as augite and olivine. Leucite or analcime may be minor constituents. Cf: *latite*.

trachyte A fine-grained, extrusive alkaline rock, sometimes porphyritic, approximately silica-saturated, and with a wide compositional range. Its main components are alkali feldspar, minor mafic minerals, and a small amount of quartz. Trachytes range from slightly oversaturated types with less than 10% quartz to undersaturated types containing feldspathoids. With greater or less silica, they grade into rhyolites and phonolites. Trachytes are the extrusive equivalent of syenite. See *trachytic*.

trachytic Textural term applied to the groundmasses of volcanic rock in which feldspar microlites are arranged in parallel or near-parallel manner, bending around phenocrysts (when these are present); the pattern corresponds to the flow lines of the lava from which the rock was formed. Such texture is common in trachytes, but "trachytic" is sometimes applied to non-trachytes with similar texture. See also: *trachytoid; pilotaxitic.*

trachytoid A term describing the texture of phaneritic igneous rocks in which the feldspars are disposed in parallel or subparallel fashion, thus resembling the *trachytic* texture of some volcanic rock; e.g., varieties of nepheline syenite.

traction load *Bed load.*

train 1. A narrow glacial deposit extending for a long distance, such as *boulder train.* 2. A series of oscillations on a seismograph record.

transcurrent fault A *strike-slip fault* of large scale in which the fault surface is steeply inclined.

transform fault 1. A particular type of *strike-slip fault* along which displacement stops abruptly or changes direction. It derives its name from its association with the offsetting of mid-oceanic ridges. Transform faults mark spots where the motion of the sea floor is transformed from a shearing movement between the offset ridge segments into a spreading movement away from the ridge. 2. A *plate boundary* characterized by pure strike-slip displacement in the ideal sense.

transgression The incursion of the sea over land areas, or a change that converts initially shallow-water conditions to deep-water conditions. Syn: *invasion; marine transgression.* Ant: *regression.*

transit A *theodolite* in which the telescope can be reversed in direction by rotating it 180° about its horizontal transverse axis.

transition zone Any of several zones within the earth in which there is a sharp increase in seismic velocity corresponding to phase or chemical changes.

translational movement Apparent fault-block displacement in which there has been no rotation of the blocks relative to each other; features that were parallel before movement are still parallel afterwards. Cf: *rotational movement.*

translation gliding See *gliding.*

transmutation The transformation of one element into another, such as during radioactive decay.

transverse dune See *dune.*

transverse section *Cross-section.*

transverse valley A valley whose course cuts across geological structure, or one whose direction is at right angles to the general strike of the underlying strata. Cf: *longitudinal valley.*

transverse wave *S wave.*

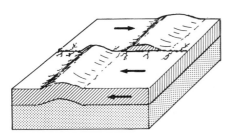

Transcurrent fault

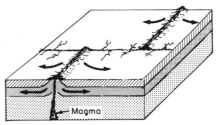

Transform fault

trap 1. A general term for a dark, fine-grained igneous rock, such as basalt or diabase. Syn: *trap rock*. 2. *Oil trap*.

trap rock *Trap*.

travertine Etymol: Ital., "tivertino," the old Roman name for Tivoli, where travertine deposits are extensive. A finely crystalline calcium carbonate deposit of white, cream, or tan color, formed by chemical precipitation from solution by inorganic processes. It occurs in caves as banded deposits of flowstone or dripstone, and also forms in springs as *tufa*, a coarsely crystalline, spongy limestone. See also: *onyx marble*.

tree ring See *dating methods*.

trellis drainage pattern See *drainage patterns*.

tremolite A white to pale grayish-green monoclinic mineral of the amphibole group, $Ca_2Mg_5Si_8O_{22}(OH)_2$. The opaque fibrous variety is commonly known as asbestiform amphibole. It is found in dolomitic marbles, serpentinites, and talc schists associated with magnesite and calcite.

trench 1. A long, relatively straight depression between two mountain ranges. 2. A long narrow excavation, natural or artificial, in the earth's surface. 3. *Ocean trench*.

triangular diagram See *AFM* diagram; *ACF diagram; A'KF diagram*.

triangulation A method of surveying that is generally used where the area to be surveyed is large, and requires the use of geodetic techniques. It establishes the distance between any two points by using such points as vertices of a triangle or series of triangles, such that each triangle has a side of known length (the *base line*). This permits the angles of the triangle and length of its other two sides to be determined by observations taken from the two ends of the base line.

Triassic A period of geological time that began 225 million y.b.p. and lasted 35 m.y.; it is the earliest period of the Mesozoic Era. The term *Trias*, later modified to Triassic, was proposed in 1834 for a sequence of strata in Germany. It refers to a threefold domain of the strata into a lower unit of non-marine red sediments, a middle unit of marine limestone, and an upper unit of non-marine rock. Zonal schemes for this period vary with location and are not well correlated. In some areas, e.g., Great Britain, the Triassic is represented by continental deposits exclusively. In North America, Triassic continental sediments and red-bed facies make up much of the scenery of Utah, Wyoming, and Colorado. The Triassic deposits have little economic importance.

Large-scale extinctions took place in the Permian. *Ammonites* are the dominant invertebrates of the Triassic. Bivalves and brachiopods are much less abundant, and no corals are found. Nautiloids show no change in evolutionary pattern across the Permian-Triassic boundary. Tetrapod faunas of the Early Triassic are dominantly those surviving the Permian. New tetrapods that appear later are the proanurans (ancestors of frogs), chelonians (turtles), early dinosaurs of various kinds, and the first mammals. Some late Permian sharks, as well as one family of coelacanth and one type of lungfish, lived into the Triassic. Gymnosperms typify Triassic flora; cycads and primitive conifers flourished in upland areas. Fossilized remains of their logs are found in the Petrified Forest of Arizona. Ferns and *scouring rushes* thrived in lower, moist areas, whereas seed ferns had vanished. *Cordaites* are not conspicuous, and *Lepidodendron* is not represented.

The Triassic was a time of extensive continental emergence. During the late period, the Appalachian Mountain regions from Nova Scotia to the Carolinas underwent faulting. In Nevada, thrust faulting continued along the Antler orogenic belt through central Nevada. Supporters of the *continental drift hypothesis* favor the Triassic as the time when the universal supercontinent of *Pangaea* began to break up. The absence of Triassic marine deposits on either side of the South or North Atlantic, south of Spitzbergen, lends strong support to this hypothesis. Fossil and rock records suggest highly equable climatic conditions. There are no continental glacial deposits of this age. Triassic *red beds* of sandstones and shales, found in places associated with *evaporite* deposits, were very likely formed in warm, seasonally wet areas.

tributary Any stream that contributes water to another stream. Syn: *feeder.* Ant: *distributary.*

trichroism See *pleochroism.*

triclinic system See *crystal system.*

tridymite A monoclinic mineral, SiO_2, a high-temperature polymorph of quartz. It occurs in small, white, bladelike crystals, spherulitic masses, or rosettes in cavities in volcanic rocks. Tridymite is stable between $870°$ and $1470°$ C. Cf: *cristobalite.* See also: *alpha quartz; beta quartz.*

trilobite An extinct marine arthropod of the class Trilobita, characterized by a body divided longitudinally into three lobes and transversely into three regions from head to tail; the cephalon, thorax, and pygidum (tail region). Each region bore a pair of jointed appendages. Trilobites had a chitinous exoskeleton, which was shed as it grew. Some forms were eyeless, others had a pair of compound eyes. Little is known of the ventral surface. Trilobites showed considerable diversity in shape and size. Their adult size ranged from 6 mm to 45 cm; some attained sizes of 70 to 75 cm. However, the intrinsic construction plan seems to have remained the same. For this reason, no single character (e.g., facial sutures) has proved satisfactory for a major classification of these animals, since none demonstrates a progressive change within the group (other than adaptational change). Trilobites range from the Lower Cambrian to Permian, and many kinds are used for stratigraphic correlation. Because they appear fully developed in the Cambrian, it seems likely that ancestral forms originated in the Precambrian.

triple junction A point at which three tectonic (lithospheric) plates meet. Stable triple junctions retain their geometry as plates drift, while unstable junctions change with drift. At present, six different triple junctions of spreading centers (ridges), subduction zones (trenches), and transforms are found, although sixteen types are theoretically possible. Magnetic patterns in sea-floor rocks provide clear evidence that old triple junctions have disappeared in subduction zones.

tripoli A light-colored powdery or siliceous sedimentary rock produced by the weathering of siliceous limestone, or chert. It is sometimes erroneously called *diatomaceous earth.*

tritium A radioactive isotope of hydrogen, having one proton and two neutrons in the nucleus.

trona A monoclinic mineral, $Na_2CO_3 \cdot NaHCO_3 \cdot 2H_2O$. It occurs in white or yellowish-white tabular crystals, earthy crusts, or efflorescences as evaporite deposits. Trona is an important source of sodium compounds.

troposphere The lowest layer of the earth's atmosphere, characterized by a decrease in temperature with altitude and by the formation of clouds and convection currents; it is in this layer that earth's weather originates.

true dip Synonym of *dip,* used in comparison with *apparent dip.*

true north That direction from any point on the earth's surface toward the geographic north pole; that is, the northerly direction of any geographic meridian, or of that meridian passing through the point of observation. It is the universal zero (or 360)-degree point for mapping reference. True north differs from magnetic north by the degree of magnetic declination at the selected point.

truncated spur A spur that projected into a valley and was completely or partially cut off by a moving glacier.

truncation The shortening or removal of a part of a landform or geological structure, as by erosion; e.g., a truncated soil profile is one in which one or both upper horizons have been removed by erosion. Cf: *leveling*.

tsunami Often erroneously called "tidal wave," a tsunami is the Japanese name for the gravity-wave system that follows any short-duration, large-scale disturbance of the free sea surface. When the wave breaks over the coastline, water piles up, causing great destruction. Tsunamis seem to occur principally after earthquakes of magnitude greater than 6.5 (Richter scale) and focal depth of less than 50 km. However, not all such earthquakes produce them. Once the tsunami is formed, the wave system closely resembles that which is produced by throwing a stone into a pond. The wave configuration is axi-symmetric at the early stage, and consists of concentric rings of crests and troughs. It is bounded at the outside by a kind of front. This front expands everywhere at the limiting velocity, $C = gh$, for free waves in water of depth h (g = gravitational acceleration). The oscillation period of a tsunami is of the order of 1 hr, and the wavelength may measure several hundred kilometers. Wave velocities reach speeds of 900 km/hr over deep ocean. It is not clearly understood how these waves are caused by an earthquake. It is possible that a tsunami originating in the vicinity of the earthquake epicenter is produced by vertical faulting on the sea floor, which draws in the water surface and produces a seismic sea wave. Another possible

reason is a submarine landslide caused by the earthquake, but not necessarily at the epicenter. See also: *seiche*.

tufa A chemical sedimentary rock of calcium carbonate, precipitated by evaporation; it commonly occurs as an incrustation around the mouth of a spring or along a stream. The compact, dense variety is *travertine*. Not to be confused with *tuff*. Cf: *sinter*.

tuff General term for the pyroclastic rock composed of volcanic ash (fragments <2 mm in diameter) cemented or consolidated by the pressure of overlying material. The consolidated mixtures of ash and lapilli, or of ash and volcanic blocks are called *lapilli tuff* and *tuff breccia*, respectively. Not to be confused with tufa. adj. *Tuffaceous*. See also: *palagonite tuff; welded tuff; pyroclastic material; pyroclastic rocks*.

tuffite A consolidated mixture of pyroclastic and sedimentary detritus; usually well-bedded and sorted according to grain size.

tufflava Extrusive rock showing both lava-flow and pyroclastic characteristics; e.g., lenses of obsidian lying in a tuffaceous matrix. Considered to be intermediate between a lava-flow and a welded-tuff type of *ignimbrite*, it is often used as a synonym for *welded tuff*.

tundra A generally level or undulating, treeless region found mainly in the Arctic lowlands of Europe, Asia, and North America. Vegetation is restricted to mosses, grasses, lichens, and sedges; the subsoil is permanently frozen.

tungstate A chemical compound characterized by the WO_4 radical. An example of a tungstate mineral is wolframite, $(Fe,Mn)WO_4$. Tungsten and molybdenum may substitute for each other. Cf: *molybdate*.

turbidites, turbidity currents Horizontal differences in density within gaseous or fluid bodies can cause currents. Turbidity currents are caused by an excess density, the result of a suspended load of sediment, and are thus also referred to as *density currents*. These currents flow downslope at very high speeds and spread along the horizontal. The distance covered is determined by the turbulence and velocity of the current. The sedimentary load is gradually dropped as the current slackens and the water comes to rest. Such deposits are called *turbidites*, and are characterized by moderate sorting, graded bedding, and well-developed primary structures. Examples of turbidity currents in the atmosphere are *nuées ardentes* and dry avalanches. The flow of a muddy river, however, is independent of the load and is, therefore, not a turbidity current. It is generally believed that submarine canyons are at least partially eroded by turbidity currents.

turbulent flow The motion of a liquid or gas is turbulent when the velocity at any point changes direction and magnitude in a random manner. Its diffusive action distinguishes turbulence from an irregular wave motion, i.e., turbulence tends to increase rates of transfer of momentum, heat, water vapor, or salt. When the velocity of a fixed-size flow is increased, a smooth or *laminar* motion will become unstable and turbulent. (Pipe flow is turbulent for *Reynolds numbers* over about 2000.)

turnover 1. An interval, esp. fall or spring, of uniform vertical temperature, when convective circulation occurs in a lake; the period of an *overturn*. See also: *circulation*. 2. A process by which some species in a region become extinct and are replaced by other species.

turquoise A triclinic mineral, $CuAl_6(PO_4)_4(OH)_8 \cdot 5H_2O$, usually occurring in light-blue or green microcrystalline masses, nodules, and veins. Turquoise is a secondary mineral produced by the alteration (in arid regions) of aluminum-bearing rocks such as apatite and

chalcopyrite. It is a valuable ornamental stone.

turtlestone See *septarium*.

turrilite A Cretaceous cephalopod, about 13 cm in length, with a sharp-spired shell on which the whorls barely touch. It resembles a gastropod (cf: *Turritella*), but the presence of septa and its complex suture pattern differentiate it.

Turritella A genus of gastropods ranging from Cretaceous to Recent time. Members of the genus have high-pointed shells, about 10 cm long, that are frequently incised with grooves, ridges, or spiral lines. Cf: *turrilite*.

twin gliding See *gliding*.

twin law See *twinning*.

twinning Twinned crystals are a special case of two or more individuals of the same mineral species in accordance with the *twin law* (law of twinning), a statement defining the relationshiip of two crystallographic elements. These can be twins along an axis or a plane, but the orientation cannot coincide with the symmetry elements of the two twinned parts. A *contact twin*, as occurs in gypsum, is made up of two crystals united along a cystal plane. In a *penetration twin*, two crystals interpenetrate one another; e.g., the "Greek cross" of staurolite. A *Carlsbad twin*, common in feldspar, is a penetration twin in which the twinning axis is the *c* crystallographic axis and the composition surface is irregular.

type concept A principle for regulating the application of scientific nomenclature by recognizing and describing some unit which can then be used as a point of reference.

type locality 1. The location in which a stratotype is situated and from which it usually takes its name. It is contained within the type area and contains the *type section*. Cf: *reference locality*. 2. The place where some geological feature, e.g., a particular kind of metamorphic rock or the type specimen of a fossil species, was originally recognized and described.

type section 1. The sequence of strata originally described as constituting a *stratigraphic unit*, and which serves as a standard of comparison when identifying geographically separated parts of a stratigraphic unit. A type section should be selected in an area where at least the top and bottom of the formation are exposed. Cf: *reference section.* 2. Syn. of *stratotype.*

U

uintahite *Gilsonite* that occurs mainly in veins in Utah's Uinta Basin.

ulexite A triclinic mineral, $NaCaB_5O_9 \cdot 8H_2O$, occurring as light, spongy, rounded masses composed of aggregates of white, silky, hairlike fibers. Ulexite, an ore of boron, is produced by evaporite precipitation in lake basins of arid regions, and is generally associated with borax and saltpeter.

Ulsterian Lower Devonian of North America.

ultimate base level See *base level.*

ultimate landform The theoretical or "ideal" landform that coincides with the end of a cycle of erosion.

ultrabasic See *silica concentration.*

ultramafic See *silica saturation.*

ultraviolet That part of the electro-

magnetic spectrum corresponding to wavelengths in the range of 40 to 4000 angstroms.

umber A brown earth consisting of manganese oxides, hydrated ferric oxide, alumina, silica, and lime. It is widely used as a pigment, either in the greenish-brown natural state, "raw umber," or as "burnt umber," the dark-brown calcined state. It is darker than either ocher or sienna.

unaka A large residual form rising above a *peneplain,* and occasionally showing in its summit or surface the remnants of an even older peneplain. It is greater in height and size than a *monadnock.* The name derives from the Unaka Mountains of Tennessee and North Carolina. See also: *catoctin.*

unakite A granite rich in epidote; its type locality is the Unaka Range of the Great Smoky Mountains.

unconformable Referring to those strata that do not follow the underlying rocks in immediate age sequence, or whose position is not parallel with the rocks beneath. In particular, it applies to younger strata that do not have the same dip and strike as the underlying rocks. Syn: *discordant.* Cf: *conformable.* See also: *unconformity.*

unconformity A buried erosion surface separating younger rock strata from older rocks that remained exposed to erosion long before the younger layers were deposited. An unconformity may be the result of uplift and erosion, an interruption in sedimentation, or non-deposition of sedimentary material. The absence of rocks normally present in a sequence indicates a break in the geologic record. There are four basic types of unconformity. In a *disconformity,* the buried erosion surface lies between two parallel series of strata. If the older rocks have been folded, the dip discordance is a maximum over the limbs of the fold, and disappears over crests and troughs, producing an appearance of conformity at these points. Where the beds beneath an unconformity are not parallel with those above, an *angular unconformity* exists. This indicates folding or faulting of the lower layers before being leveled by erosion and covered by younger layers. An unconformity between overlying stratified sediments and underlying unbedded metamorphic or plutonic rocks is a *nonconformity*. A *paraconformity* is similar to a disconformity in that the strata are parallel; however, there is little discernible evidence of erosion or prolonged non-deposition. Cf: *conformity*.

underground water See *groundwater, def. 2.*

undersaturated 1. Said of an igneous rock composed of silica-unsaturated minerals such as olivine and feldspathoids. 2. Said of a rock whose norm contains olivine and feldspathoids. Syn: *silica-undersaturated*. Cf: *oversaturated; saturated*. See *silica saturation*.

uniaxial Said of crystals having a single optic axis, e.g., crystals of the tetragonal or hexagonal systems. Cf: *biaxial*.

uniformitarianism The principle of the physician-turned-geologist, James Hutton, that states, "The present is the key to the past." It postulates that the laws of nature now prevailing have always prevailed and that, accordingly, the results of processes now active resemble the results of like processes of the past. The principle does not imply the invariance through time of conditions now prevailing; what is meant is that changes or modifications of the earth's crust are produced only in accordance with invariant physical laws. Cf: *catastrophism*.

univalve n. A mollusk having a single valve; e.g., cephalopod, gastropod. adj. Having only one valve. Cf: *bivalve*.

universal stage A device used in a polarizing microscope that allows the thin

section under study to be positioned about two horizontal axes at right angles to one another. It is used in the optical study of low-symmetry minerals.

unmixing *Exsolution*.

unsaturated Said of a mineral that does not form in the presence of free silica; e.g., nepheline, leucite, feldspathoids. Syn: *silica-unsaturated*. Cf: *undersaturated; saturated; oversaturated*. See *silica saturation*.

unstable 1. Referring to a portion of the earth's crust that has experienced subsidence or uplift. 2. A constituent of a sedimentary rock that is a result of rapid erosion and deposition is said to be unstable. 3. Said of sedimentary rock that consists of poorly sorted particles of rock fragments and feldspar; i.e., an immature rock. 4. Said of a radioactive substance. Cf: *stable*.

uphole shooting In seismic exploration, the shooting of successive shots at several hole depths. The time required for the seismic impulse to travel from a given depth to the surface is the *uphole time*. This is one of the commonest methods for finding *weathering* and subweathering velocity.

upright fold See *fold*.

uprush Syn: *swash*.

upthrown Referring to that side of a fault that appears to have moved upward relative to the other side. The amount of upward vertical displacement is called the upthrow. Rocks on the upthrown side of a fault are referred to as the *updip block* and are described as *upfaulted*. Cf: *downthrown; heave*.

upwelling A process of vertical water motion in the sea, whereby subsurface water is transported toward the surface. The converse downward displacement is called *downwelling* or sinking. The *upwelling area* is the geographic location of the vertical motion, but upwelled water and its influence on oceanographic conditions may extend for hundreds of kilometers. It may occur anywhere, but is most common along the western coasts of continents.

uraninite An isometric uranium mineral, UO_2, that is highly radioactive and is the primary ore of uranium. It usually occurs in granular masses or aggregates. Uraninite is found in pegmatites and high-temperature hydrothermal veins in sedimentary clastic deposits with minerals of lead, tin, and copper. See also: *pitchblende.*

uranium-thorium-lead-age-method See *dating methods.*

U.S.G.S. United States Geological Survey.

uvala Etymol: Serbo-Croatian. Sometimes used as a syn. of *karst valley,* but most commonly applied to larger depressions resulting from the extensive collapse of roof sections above underground water flows.

uvarovite The calcium-chromium end-member of the garnet series, $Ca_3Cr_2(SiO_4)_3$. It occurs as small, brilliant green crystals and is the rarest of all the garnets. See also: *garnet.*

V

vadose water Water that occurs between the ground surface and the *water table*; i.e., in the *zone of aeration.*

valley A linear, low-lying tract of land bordered on both sides by higher land and frequently traversed by a stream or river. All valleys have been cut by running water over time; their cross-profiles change because of various factors. A *V-shaped* valley may be widened by sheetwash, weathering, and mass wasting. Glacial erosion produces a valley that is *U-shaped* in cross-profile; e.g., a glacial trough.

valley glacier *Alpine glacier.*

valley sink *Karst valley.*

valley train A long, relatively narrow, trainlike deposit of outwash sand and gravel deposited by meltwater streams, usually in front of a terminal moraine.

vanadinite A mineral, $Pb_5(VO_4)_3Cl$, of the apatite group, an ore of vanadium. It occurs in reddish-orange to yellow or brown hexagonal prisms or fibrous, radiate masses or crusts. Vanadinite is found as a secondary mineral in the oxidation zone of lead deposits.

variation diagram A method used in the interpretation of petrochemical data, in which compositional variations are plotted in terms of various chemical parameters. For example, a group of basalt lavas may exhibit compositional variations that can be related to accumulations of different proportions of olivines by crystal-liquid fractionation. Two basic variation diagrams are used: A Cartesian graph of two variables, or a triangular diagram of three variables. The *Harker diagram* plots the weight percentage of SiO_2 along the X-axis of a Cartesian graph, and concentrations of other oxides along the ordinate axis.

varietal mineral A mineral that distinguishes one variety of rock from another; it may be present in small or considerable amounts.

variolitic A term used to describe a texture or structure that occurs only in basaltic rocks; equivalent to spherulitic texture in rhyolite. Variolitic structure is a common feature of the margins of certain pillow lavas. The varioles, which sometimes stand out as small knobs, are commonly composed of fanlike sprays of feldspar fibers.

Variscan (Hercynian) orogeny Upper Paleozoic orogenic period of central Europe during the Carboniferous and Permian periods. It corresponds to the Appalachian orogeny of North America. Folded Hercynian mountains extended from England to Ireland through France and Germany. The more westerly folded areas are commonly called Amorican, and the remainder, Variscan. Amorican folds trend from northeast to southwest in central France, but Variscan folds run southwest to northeast. It is thought, therefore, that two separate deformational systems were involved in their formation. The folding and faulting that occurred during this disturbance were accompanied by large-scale igneous activity in England and on the continent.

varve See *rhythmites*.

vascular plant A plant characterized by a fully developed circulatory system and structural differentiation into roots, stems, and leaves.

vein A tabular or sheetlike body of one or more minerals deposited in openings of fissures, joints, or faults, frequently with associated replacement of the host rock. Cf: *lode*.

veld See *savanna*.

velocity log *Sonic log*.

velocity profile A technique used to determine seismic velocity from the time-distance relationships of seismic reflections that are generated over a wide range of distances between a shot and a geophone.

vent See *volcanic vent*.

ventifact Any stone shaped by the ac-

tion of windblown sand. Ventifacts are generally pitted, polished, and grooved. Abrasive wind action sometimes facets the windward side of large fragments. Two or more facets are often formed on smaller stones, and a ridge or edge is formed where facets intersect. If a ventifact has one edge it is sometimes called an *einkanter*. (Ger.: "one edge"); if it has three edges, a *dreikanter* ("three edges").

Venus hair Needlelike crystals of yellow or reddish-brown rutile that form tangled masses of inclusions in quartz. See also: *sagenite*.

verd antique A dark-green massive variety of serpentine, usually veined. It is commercially considered a marble. Syn: *green marble; serpentine marble*.

vermiculite A group of micaceous clay minerals related to chlorite and montmorillonite, with the same general formula: $(Mg,Fe,Al)_3,(Al,Si)_4O_{10}(OH)_2 \cdot 4H_2O$. It is frequently pseudomorphic after biotite. Vermiculite is a hydrothermal alteration product of phlogopite and biotite. When heated to about 300° C, it loses water quickly and expands to about 18 to 25 times its original volume.

vertical accretion The vertical accumulation of a sedimentary deposit, such as the settling of sediment suspended in a stream that is prone to overflow. Cf: *lateral accretion*.

vertical separation In a fault, the vertical distance between two sections of a displaced marker. Cf: *horizontal separation*.

vertical slip The vertical component of the net slip in faulting; it is equal to the vertical component of the dip slip. Syn: *vertical dip slip*. Cf: *horizontal slip*.

vesicle; vesicular A small cavity in a glassy or aphanitic igneous rock, formed by the expansion of a gas bubble during solidification of the rock. As magma moves upward, decompression during extrusion causes the magmatic gases to come out of solution and form bubbles. Vesicles are generally less than 2 cm in diameter, but may be larger where they coalesce. They tend to be spherical if formed in stationary lava, due to surface tension of the glass. If formed in moving lava they are elongated in the direction of flow, due to stretching of the glass. *Pipe vesicles* are hollow, upward-rising tubes that form in some lava flows. The general opinion is that they are formed by a stream penetrating the flow from the wet rocks at the site of lava extrusion. Vesicles are generally larger in basaltic lava flows than in silicic to intermediate lavas. Their size seems to be a function of the lower-viscosity type of melt and higher temperature of basaltic extrusion, both of which permit the gas bubbles to expand before solidification. Rocks containing vesicles, such as pumice, are said to have *vesicular texture* or *fabric*. See *lava*.

vesuvianite A silicate mineral occurring as brown or yellow crystals, or compact granular masses in brown or olive green. Syn: *idocrase*.

Virgilian Uppermost Pennsylvanian of North America.

viscosity The property of a substance that offers internal resistance to the force tending to cause the substance to flow.

viscous flow In structural geology, flow in which the rate of shear strain varies directly with the shear stress. Cf: *liquid flow; solid flow*.

vitrain See *lithotype*.

vitreous Having the appearance of glass.

vitric Of or pertaining to glass or containing glass; e.g., *vitric-tuff*.

vitrification Formation of a glass.

vitroclastic Referring to pyroclastic rock structure characterized by fragmented glass.

vitrophyre A porphyritic igneous rock having a glassy matrix. adj. *Vitrophyric*.

void *Interstice*.

volatile constituent 1. A substance, originally present in a magma, but which, because of its loss during crystallization, now resides chiefly in a gaseous or vaprous state, and does not ordinarily appear as a rock constituent. 2. *Volcanic gases*.

volatile matter Those substances, other than moisture, that are driven off as gases and vapors during the combustion of coal. Syn: *volatiles*.

volatiles 1. *Volatile matter*. 2. Volatile *constituents*.

volcanic 1. Pertaining to the activities (e.g., volcanic eruption), structures (e.g., *volcanic cones*) or rock types (e.g., *volcanic breccia*) of a volcano. 2. Synonym of *extrusive*.

volcanic ash Fine pyroclastic material under 2 mm in diameter. The term usually refers to the unconsolidated material, but is sometimes also confusingly used for its consolidated form, *tuff*. See *pyroclastic material; pyroclastic rocks*.

volcanic belt *Volcanic chain*.

volcanic bomb A clot of lava, usually of mafic to intermediate composition, ejected from a volcanic vent while viscous, and which assumes a rounded shape as it solidifies in flight. (Cf: *block*.) Bombs range in size from 64 mm to several meters in diameter, and show great variation in shape; they may be vesicular to hollow. See also: *pyroclastic material; pyroclastic rocks*.

volcanic breccia 1. Pyroclastic rock consisting of abundant volcanic fragments greater than 64 mm in diameter (*blocks*). It is comprised of older volcanic material brecciated by later explosive volcanic action and then consolidated; it may or may not have a matrix. Cf: *agglomerate*. See also: *pyroclastic material; pyroclastic rocks*. 2. Rock composed of fragments (volcanic or non-volcanic) in a volcanic matrix.

volcanic chain A linear arrangement of several volcanoes, apparently associated with an oceanic ridge, subduction zone, or other major geological feature.

volcanic cloud *Eruption cloud*.

volcanic conduit The channel through which volcanic material rises from depth. Cf: *volcanic vent*.

volcanic cone A conical hill formed of lava and/or pyroclastic materials deposited around a volcanic vent; it may be simple or composite. Simple volcanic cones are perhaps best typified by *cinder cones*. These are *monogenetic*, generally symmetrical landforms composed of cinders and other pyroclastics, rarely exceeding a few hundred meters in height. The exterior inclination is about 33°, the *angle of repose* for loose cinder. *Composite cones* (strato-volcanoes) are much

larger than the simple type, being built up by many eruptions, usually of andesitic lava, over a long period (*polygenetic*). A greater proportion of pyroclastic material is erupted along with the lavas, resulting in a cone that is composed of layers of lava flows alternating with pyroclastics. Cf: *shield volcano*.

volcanic conglomerate A water-deposited conglomerate, generally derived from coarse pyroclastic rocks and debris. Its consolidation is due to cementation.

volcanic crater A broad, shallow-rimmed depression, sometimes steep sided, at the top or on the flanks of a *volcanic cone*; formed by eruption, explosion, or subsidence. *Eruption craters* are funnel-shaped depressions formed by the accumulation of material around the vent, and by the explosive ejection of that material. *Explosion craters* usually occur in areas of past or present volcanic activity, but volcanic material is not always directly involved. Such craters, which are almost always filled with water, are produced by *phreatic explosions* resulting

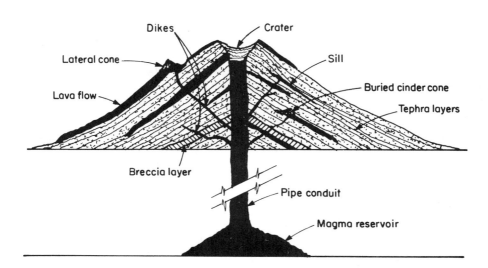

Diagram of volcanic cones

from contact between downward-percolating surface water and magma. (See also: *maar.*) *Subsidence craters,* which are commonly vertical sided and without raised rims, are formed when the roof of a magma chamber collapses following the subterranean draining off of a large volume of underlying magma. See also: *caldera.*

volcanic dome An extrusion of highly silicic magma, commonly forming a thick, bulbous dome above and around a volcanic vent. Some domes grow by internal expansion, forcing masses of semisolid lava out of and away from the vent in fan like fashion. *Volcanic spines* sometimes develop during the growth of such domes. Blockage of the vent by a dome may result in a lateral blast of volcanic material, sometimes generating a *nuée ardente.* Some effusions from the vent are almost solid, and are thrust gradually upward out of the vent with little lateral expansion; these are sometimes called *plug domes.* Slightly less viscous magma may extrude as a bulbous mass, growing with flow layering that is more concentric than radial.

volcanic earthquake A volcanic disturbance whose origin is found beneath or near a volcano, whether it be active, dormant, or extinct. Although earthquakes and volcanoes both occur along plate margins, most of these earthquakes are considered to be related to grinding of the plate edges. In areas of *hot-spot volcanoes* (e.g., Hawaii), however, earthquakes accompany volcanism, even though they may be thousands of miles distant from plate margins.

volcanic ejecta A synonym for *pyroclastic material.*

volcanic eruption The nature of a volcanic eruption is largely determined by the viscosity and gas content of its magma. A low-viscosity magma will be easily extruded as lava. For a more viscous magma, the deciding factor in the degree of explosiveness is the gas content, because a viscous magma by itself cannot be explosive. Viscosity is a function of temperature and composition, namely silica content. A basaltic lava, which contains about 50% silica, is much less viscous than a rhyolitic lava, which contains more than 65% silica. The viscosity of andesitic lava, whose silica content is about 60%, is intermediate. Basalts are produced at *ocean ridges,* whereas rhyolites and andesites are associated with *subduction zones.* Volcanoes that occur along ocean ridges are thus generally less explosive than subduction-zone volcanoes. See also: *volcanic eruption, types of; lava; volcano.*

volcanic eruption, sites of Volcanic eruptions may occur at various sites relative to a pre-existing volcano. In a *summit eruption,* activity is confined to the summit crater. *Flank eruption* is a condition in which some explosive activity takes place from within the summit crater, but most of the lava is erupted quietly from a fissure vent along the flank of a volcano. During a *lateral eruption,* there may be very little summit activity, but the feeder channel may branch off and most of the eruption may take place along the flanks, with a small cone being formed. In the extreme case, there may be no activity at the summit crater, and the new eruption may take place independently of the primary volcano, possibly resulting in the formation of a *parasitic cone.* See also: *central vent eruption; fissure eruption.*

volcanic eruption, types of Volcanic eruptions have traditionally been arranged and classified into a sequence based on progressive explosiveness. The labels provided by this are accurate, however, only if they are applied to recognizable phases of an eruption, as the complete process may exhibit several eruptive patterns. (See *volcanic eruption, sites of.*) Although the grouping described here is still useful, it has been generally replaced by a more objective method based on the *pyroclastic material* associated with a particular eruption.

A *Hawaiian-type eruption*, the mildest member of the sequence, is characterized by large quantities of very fluid basaltic magma from which gases readily escape, along with subordinate pyroclastic material. *Lava fountains* are commonly associated with this type of eruption. Although typical of Hawaiian volcanoes, similar activity has been noted elsewhere, usually on oceanic-island volcanoes.

The *Strombolian-type eruption* is slightly more explosive. Less fluid basaltic lava permits spasmodic gas escape in the form of minor explosions that eject pyroclastic material.

The *Vulcanian-type eruption* is violent and characterized by a much more viscous lava, large quantities of pyroclastic material (ash, blocks), and ash clouds. These eruptions may occur infrequently during the life cycle of a volcano, but once they do occur, may continue intermittently for several months.

The increased violence of the *Plinian-type eruption* is due to extremely viscous, gas-filled magma that is blasted out of the vent at nearly twice the velocity of sound, ejecting tremendous volumes of pyroclastic material. It is common for large parts of the volcano to blow away or collapse after such blasts. The eruption of Mt. Vesuvius in 79 A.D. was of Plinian type.

The 1902 eruption of Mt. Pelée was the first recognized example of a *Peléean-type eruption*. This type shares many features with the Plinian, esp. vast amounts of pyroclastic material. Its uniqueness is defined by the presence of a *nuée ardente* (glowing cloud), which is associated with the growth of a *volcanic dome*. See also: *volcanic eruption; volcano; lava.*

volcanic focus The apparent or assumed center of activity beneath a volcano, or in a volcanic region.

volcanic gases Compounds and elements that were previously dissolved in the magma while under great pressure, and which are released as volatiles during a volcanic eruption. The gases are released when pressure is decreased as the magma reaches the surface. Most common constituents of the gases are water vapor and carbon dioxide; other constituents include sulfur dioxide, hydrogen sulfide, hydrogen chloride, and nitrogen as a free element. In addition to its dependence on viscosity and composition of the lava, the explosiveness of a volcanic eruption is to a large degree determined by the proportion of gaseous material in the magma. See *volcanic eruption.*

volcanic glass Any glassy rock produced from molten lava or some liquid fraction of it. The cooling of such molten material is accompanied by an increase in viscosity. Because high viscosity inhibits crystallization, solidification at this stage by sudden cooling, such as after expulsion from a volcanic vent, tends to chill the material to a glass rather than crystallize it. Volcanic glass is unstable and devitrifies in geologically short periods of time. A streaked or swirly structure is characteristic of many natural glasses. See also: *obsidian; tachylyte.*

volcanicity *Volcanism.*

volcaniclastic 1. A term used to describe the texture of rocks whose clastic fabric is a result of any volcanic process. Cf: *pyroclastic texture.* 2. Pertaining to fragmental rocks containing volcanic material in any proportion whatever, without regard to origin. Also a synonym for *fragmental volcanic.* Cf: *pyroclastic material.*

volcanic mud Mud formed by the mixture of volcanic ash with water. If the mixture is in a hot and fluid state, it will flow down the flanks of a volcanic edifice as a *mudflow* (lahar).

volcanic neck See *volcanic plug.*

volcanic pisolites *Accretionary lapilli.*

volcanic plug Typically, a roughly

cylindrical mass of congealed magma or tephra that fills the conduit of an inactive volcano. It may overlie a still-fluid magma column, and fragments of the solid plug are usually ejected when volcanic activity is resumed. Volcanic plugs are generally more erosion-resistant than the enclosing rock. The remnants that persist after much or all of the volcano has been destroyed are called *volcanic necks*. It should be noted, however, that the term "neck" is used synonymously with "plug," regardless of whether or not the formation stands up in relief. A large neck is often surrounded by smaller necks from *parasitic cones*. Volcanic necks are usually less than 1 km in diameter, but may be as high as 450 km. Long, low dikes may radiate from them; these are the erosion-resistant, exposed feeding cracks of the eroded volcano. Volcanic necks are found in the U.S. in Arizona, New Mexico (Shiprock), Utah, and Wyoming.

volcanic products Generally separated into three categories: *volatiles* (gaseous products, including water vapor), *lava* (liquid products other than water), and *pyroclastic material* (solid products). The second two may overlap, since some pyroclastic matter has solidified from molten lava.

volcanic rent A very large volcanic depression bordered by fissures that are commonly concentric in form. It is a result of magmatic activity or the overburdening of cone material on a weaker substratum.

volcanic rocks 1. All extrusive rocks and associated high-level intrusive ones. The group is entirely magmatic and dominantly basic. 2. Very fine-grained or glassy igneous rock produced by volcanic action at or near the earth's surface, either extruded as lava (e.g., basalt) or expelled explosively. Cf: *pyroclastic rock*.

volcanic sand Sand-sized volcanic fragments of either pyroclastic or detrital origin.

volcanic spine An obelisk-like, monolithic protrusion of extremely viscous lava squeezed up through the solidified surface crust of a volcanic dome or thick lava flow. Spines range from a few centimeters to hundreds of meters in height. A famous example is the spine that formed on Mt. Pelée in Martinique (300 m). Megaspines, which are even larger, include the pitons of the West Indies. See also: *volcanic domes*.

volcanic vent An opening at the earth's surface through which volcanic materials are extruded. Cf: *volcanic neck*.

volcanic water Water within or derived from magma at the earth's surface or at a comparatively shallow depth. Also, *juvenile* water of volcanic origin.

volcanism Also spelled *vulcanism*. Any of various processes associated with the surficial discharge of magma or hot water and steam; included are volcanoes, geysers, and fumaroles.

volcano Etymol: Vulcan, the Roman god of fire and forging. A vent in the earth's surface through which magma and associated ash and volatiles erupt. The hill or mountain formed by the eruptive debris is also often called a volcano. (See *volcanic cone.*) Almost all recently *active volcanoes* of the world are found along several distinct narrow chains, some of which extend along the border mountains of continental land masses, some along island arcs, and some through mid-ocean zones. Very few are found in the center of continents.

As related to plate-tectonic processes, there are three types of volcano: 1. Rift volcanoes, typified by the submarine volcanoes along the mid-Atlantic Ridge. Along the rifts, convection within the mantle pulls the plates apart, and upwelling magma extends their edges. 2. Volcanoes formed along *subduction zones*. Ex: the Andes Mountains and the *island arc* of Japan. In such areas, two plates converge and one plate plunges beneath the other, thus melting parts of both plates and sending magma to the surface. 3. Hot-spot volcanoes, such as Hawaii and the Hawaiian Ridge, occur in the middle of a plate as it slides over a hot plume of magma that originates deep beneath the plate and maintains the same location for millions of years. Such thermal plumes are possibly generated by the decay of radioactive isotopes.

Although the first two types of volcano are associated with mountain-building regions, not all young mountain ranges exhibit volcanic activity; for example, volcanoes are absent from the Himalayas and the Alps, both of which are being formed by collisions of continental plates. (See *plate tectonics.*)

Subduction volcanoes form island arcs and high mountain chains rather than the submarine ridges formed by rift volcanoes; they are more explosive than rift volcanoes, and produce large volumes of ash as well as lava flows. (See *volcanic eruption; pyroclastic material.*) In addition, the products of subduction volcanoes show greater compositional variation, and their landforms differ from those of rift volcanoes. Because of such diversity, volcanoes may be variously classified according to type of eruption, pyroclastic material, and lava composition and flow pattern, as well as according to the nature, structure, and appearance of their external form. See also: *volcanic eruption, types of; volcanic cones.*

volcanogenic Formed by processes directly related to volcanism. In particular, sand or mineral deposits that can be shown to be associated with volcanic phenomena.

volcanology Study of the structure, origin, and petrology of volcanoes, and their role as contributors to earth's atmosphere and hydrosphere, and to the rock structure of earth's crust.

volume elasticity *Bulk modulus.*

vug (also vugh, vugg) Etymol: Cornish, *vooga,* "cave." 1. A cavity in a rock or lode, commonly not joined to other cavities; often lined with euhedral,

vapor-phase crystals, the mineral composition of which differs from that of the surrounding rock. (Cf: *druse.*) 2. Crystalline-encrusted cavity in a rock. (See *geode.*) In both 1 and 2, only cavities that remain unfilled are called vugs. 3. As used by petroleum geologists, any opening from the size of a pea to a boulder. See also: *amygdule; druse; geode; vuggy.*

vuggy 1. Referring to porosity due to vugs in calcareous rocks; e.g., the cavities formed by the dissolution of oöliths in limestone. 2. A particular type of structure in volcanic rocks. This vuggy structure results from the formation of irregularly shaped pockets of exsolved gas in an already crystalline matrix. A similarly appearing structure in phaneritic plutonic rocks is called *miarolitic.* See also: *vug; druse; miarolitic.*

Vulcanian-type eruption See *volcanic eruption, types of.*

vulcanism *Volcanism.*

W

wacke A texturally and usually mineralogically immature sandstone, consisting of poorly sorted mineral and rock fragments in a matrix of fine silt and clay. In particular, it refers to an impure sandstone containing more than 10% argillaceous matrix. The term is used to designate a category of sandstone, as distinguished from *arenite.* Its use as a shortened form of *graywacke* is a subject of considerable controversy. Cf: *graywacke.*

wad A black or dark brown impure mixture of manganese and other oxides. Syn: *bog manganese.*

warm glacier *Temperate glacier.*

wastage 1. General term for denudation of the earth's surface. 2. *Ablation.*

water cycle *Hydrologic cycle.*

waterfall A steep to vertical descent of a stream course where it drops over the edge of a plateau or where erosion has worn away overhanging softer rock. Because of the force of the water, all waterfalls must necessarily be relatively ephemeral features. See also: *cascade.*

water gap A deep pass through a mountain ridge, occupied by a river or stream. Water gaps, in general, may result from two sets of conditions; in one case, the river forms its channel prior to uplift of the ridge, and maintains its course as the ridge rises across the path. In the second case, the river forms its channel on rock lying above the structure, but maintains its course, while cutting down into the ridge below. A famous example in the U.S. is the Delaware Water Gap, Penna. Cf: *wind gap.*

water of imbibition 1. The quantity of water above the water table that a rock can contain. 2. The quantity of water that a water-bearing substance can absorb without an increase in volume.

water of retention That portion of interstitial water in a sedimentary rock that remains in its pores under capillary pressure.

watershed *Drainage basin.*

water table The position in the earth's crust below which cracks and other openings are filled with water. Above the water table (*zone of aeration*), the interstices in earth materials are partly filled with air. In the *zone of saturation*, the interstices are completely filled with water. This zone extends from the water table downward. The position of the water table below the ground surface is a function of the topography of an area and of local climatic conditions.

water-table well A well that taps into unconfined groundwater. Its water level does not necessarily lie at the water-table level. Cf: *artesian well.*

water witch Dowser. See *dowsing.*

Waucoban Lowest of three Cambrian subdivisions in North America. A very long section of early Cambrian rocks, the Waucoban Series, is located in Waucoban Springs, Calif.

wave-cut cliff A *sea cliff.*

wave-cut platform Also called *abrasion platform.* A gently sloping rock ledge that extends from high-tide level at a steep cliff base to below the low-tide level. It develops as a result of wave abrasion at the sea-cliff base, which causes overhanging rocks to fall. Wave-cut platforms depend on rock structure and type.

wavefront 1. A surface that is the locus of points representing the position of a progressing seismic disturbance at a specified time. 2. The locus of all points reached by light that is emitted outward in all directions from a point.

wavelength The distance between two successive crests or troughs in a series of harmonic waves. In symmetrical periodic fold systems, the term is applied to the distance between synformal or antiformal hinges.

wavemark *Swash mark.*

wave platform *Wave-cut platform.*

wave ripple mark *Oscillation ripple mark.*

weathered layer A term used in seismology for a zone of irregular thickness immediately below the surface. It is characterized by low seismic-wave velocities, which are referred to as *weathering velocity.* Unless some correction is made, variations in the weathered zone near the surface might indicate fictitious "formations" along the refraction horizon that is being mapped. The simplest method "removes" the weathered layer by determining its thickness and velocity, and then subtracting the delay time related to it from the observed intercept time; this is called a *weathering correction.*

weathering Destructive natural processes by which rocks are altered with little or no transport of the fragmented or altered material. Mechanical weathering occurs with the freezing of confined water and the alternate expansion and contraction due to temperature changes. Chemical weathering produces new minerals. The main chemical reactions are oxidation, hydration, and solution.

weathering correction See *weathered layer.*

weathering velocity See *weathered layer.*

welded tuff Pyroclastic rock, also called *welded ash-flow tuff,* formed by the welding together of pumice, lapilli and glass shards present in an ash flow. Welding is effected by the activity of contained hot gases and the weight of overlying material, which creates collapsed bubbles and flattened, elongated fragments. The flattening produces laminations in the tuff, which cause it to appear banded. The most intense compaction and welding occur in the lower portion of the ash flow. The banded appearance of welded tuffs sometimes resembles flow-banding, and thus welded tuffs are often mistaken for rhyolites. Syn: *tufflava.* See also: *ignimbrite.*

well log See *well logging.*

well logging Well logging denotes an operation wherein a continuous recording, called a *log,* is made of the depth of some typical datum of a formation. Many types of well logs are recorded by *sondes,* which are lowered into the borehole. Signals from the sonde are transmitted to the surface, where a continuous log is made.

Electrical logging is basically the recording (made in uncased sections of a borehole) of the *resistivities* or *conductivities* of subsurface formations, and of the *spontaneous potentials (S.P.)* generated in the borehole. In any means of electrical logging it is useful to classify reservoir rocks according to porosity, composition, and anisotropy. *The S.P. log* is a record of the naturally occurring potentials in the mud at different depths in the borehole. Uses of the S.P. log include the detection of permeable beds by obtaining values for the formation-water resistivity. *Induction logging* is a method wherein the conductivity or resistivity of formations is measured by means of induced currents, without the aid of electrodes.

Radioactivity logging involves the use of gamma rays or neutrons. Since radiations from the rocks must penetrate fluids, and sometimes cement and casing, only gamma rays will do; the weak penetration capacity of alpha and beta particles renders them useless for this purpose. In *gamma-ray logging,* differences in the radioactive content of the various rock layers surrounding a well is shown by corresponding differences in gamma radiation within the well bore. *Neutron logging* is an important technique for obtaining information relating to the fluid content of the rocks; the method, however, does not distinguish between water and oil. A neutron-emitting source bombards the rocks in a borehole with a constant flux of neutrons, the varying intensity of which is detected as gamma radiation. This is therefore properly described as a *neutron-gamma log.*

Unlike electrical logs, which are greatly affected by saltwater or by oil in the fluid column within the well, the radioactivity log is equally reliable with any ordinary well fluids. However, other logging methods must be used to locate petroleum, since its natural radioactivity is low, and accumulations of it cannot be observed on the gamma-ray curve. With rather few exceptions, similar rocks throughout the world provide similar responses on a radioactivity log. On this basis, a particular formation can usually be identified on the log with a reasonable degree of certainty.

The purpose of a *directional log* is to ensure that a straight hole is being drilled. It is a record of the hole drift, i.e., derivation of the wellbore from the vertical, and direction of that drift. It is usually obtained with a *dipmeter* log. (See *dipmeter*). An acoustic-velocity log (sonic log) measures the time required for a sound wave to pass through a specified length of formation. The method is used extensively as a porosity and correlation log. it may be run in uncased boreholes containing oil, water, or any type of mud; it cannot be recorded in air- or gas-filled boreholes. The velocity of sound through a formation depends mainly on the velocity of sound through both the rock matrix and interpore fluid, and the amount of porosity. Sound sources include magnetostrictive metal alloys, piezoelectric quartz crystals, and the electromechanical hammer and anvil. Dectectors include geophones that use piezoelectric quartz crystals, and magnetostrictive metal alloys.

well shooting In seismic prospecting, a procedure to determine velocity as a function of depth. Charges are exploded in a shallow drill hole alongside a deep exploratory borehole; an in-hole detector, placed at a number of depths from top to bottom, receives the waves, whose arrival times are then recorded.

Wentworth grade scale A scale to measure the grain size of sedimentary rocks. The scale ranges from less than one two-hundred-fifty-sixth mm (clay particles) to greater than 256 mm (boulders).

whistling sand See *sounding sand*.

Widmanstätten pattern A system of interesecting bands seen on a cut, polished, and acid-etched surface of an *iron meteorite*. Iron meteorites are composed chiefly of nickel and iron minerals, and the acid etches the iron minerals more rapidly than the ones containing nickel. The Widmanstätten structure cannot be developed in artificial alloys in any feasible cooling time; it is therefore used as conclusive proof that a piece of iron is meteoritic in origin.

wildcat well An exploratory well drilled for gas or oil in a geological tract not yet shown to be productive.

wind gap A fairly shallow cut or notch in the upper part of a mountain ridge. Some wind gaps are the sites of former water gaps that have been abandoned by the rivers or streams that eroded them.

window See *nappe*.

wind polish *Desert polish*.

wind shadow The area in the lee of some obstacle, where air movement is impeded to the degree that it cannot move material such as sand. As a result, accumulations of the material may reach gigantic size; e.g., the great sand dunes of the southwestern U.S.

Wisconsin Pertaining to the recognized classical fourth glacial stage of the Pleistocene in North America.

witherite A white or gray orthorhombic mineral, $BaBO_3$, used as a barium ore when its quantities are large enough.

wold Syn: *cuesta*.

wolframite An iron-manganese-tungstate mineral, $(Fe,Mn)Wo_4$, the principal ore of tungsten. Wolframite occurs as crystals or granular masses in pegmatites and high-temperature quartz veins.

wollastonite A triclinic silicate mineral,

$CaSiO_3$, usually occurring as white or grayish radiating masses. It is typical of contact-metamorphic calcareous rocks, and also occurs in low-pressure regional-metamorphic rocks.

wood opal See *silicified wood*.

wrench fault A near-vertical fault along which strike separation has occurred. Syn: *torsion fault*.

wulfenite Lead molybdate, $PbMoO_4$, a secondary ore of molybdenum. Found as massive granular or earthy aggregates, yellowish or reddish orange in color. It is a secondary mineral occurring in the oxidation zone of lead deposits.

Wyoming bentonite *Sodium bentonite*.

X

xeno- A prefix meaning "guest" or "stranger."

xenoblast See *crystalloblast*.

xenocryst A crystal resembling a phenocryst in an igneous rock, but unrelated to the rock body in which it occurs.

xenolith A foreign inclusion in an igneous rock. In some cases the assimilation of xenoliths is only physical, whereas in other instances these foreign rock bodies are melted or dissolved, and produce chemical contamination. Syn: *accidental inclusion*. Cf: *autolith*.

xenomorphic Descriptive term for the texture of an igneous or metamorphic rock characterized by minerals whose crystals do not have their own faces. This is a result of interference of the individual minerals with each other during crystallization. Xenomorphic crystals form when growth conditions are favorable for a continued increase in crystal size, but growth is ultimately limited by intercrystal contact. The texture is commonly shown by granitic rock. Cf: *subautomorphic*.

X-ray Electromagnetic radiation of very short wavelength, between 0.1 and 100 angstroms.

Y

yardang An elongated ridge carved mostly from relatively weak earth materials by wind abrasion in a desert region. Yardangs appear as sharp-edged topographic forms up to 15 m in height, but some in Iran stand over 200 m. The ridges are separated by steep-sided troughs, both the ridges and troughs lying parallel to the prevailing winds. Yardangs are common in the deserts of Central Asia and on the Peruvian coast.

Yarmouth The name given to the classical second interglacial stage of the Pleistocene Epoch in North America; it follows the Kansan glacial stage and precedes the Illinoisan.

yazoo stream A tributary stream that flows for a long distance parallel to a main stream before joining it. The type example, the Yazoo River in western Mississippi, is prevented from joining the Mississippi by the latter's elevated bed, making it higher than the backswamp area where the Yazoo spills onto a floodplain. There the Yazoo must continue alongside the main stream to the place where the Mississippi crosses over to the valley side (Vicksburg, Miss.).

yellow ocher A limonite mixture used as a pigment; it usually contains clay and silica.

yield point *Elastic limit.*

yoked basin Syn: *zeugogeosyncline.*

Young's modulus A *modulus of elasticity* expressing the relation between tension or compression and deformation in terms of change in length. In the equation, $E = \sigma/\epsilon$, E is Young's nodulus, σ is the tensile or compressive stress, and ϵ is the tensile or compressive strain.

young valley A valley in its early stages of development; it is characterized by a relatively straight orientation, V-shape cross-profile, and high gradient.

Z

zenithal projection *Azimuthal projection.*

zeolite Etymol: Greek, "to boil." One of a large group of hydrous aluminosilicates with alkali metals (esp. Na,Ca,K), characterized by their reversible loss of water of hydration, and by their fusion with a bubbling action when strongly heated. Zeolites commonly occur as well-formed crystals in cavities in basalt, and also occur as poorly crystallized deposits in beds of tuff. The term *zeolite* covers distinct mineral species including natrolite, analcime, stilbite, and many others. As a result of their crystal configuration, an "open" structural pattern, zeolites manifest a strong capacity for *base exchange,* a property that makes them useful as water softeners.

zeugen Sing: *zeuge.* Tabular erosional remnants, often 30 m or more in height, formed in desert regions by the erosional effect of sand-laden winds. Zeugen are masses of resistant rock resting on supports of softer rock, having become separated by erosion from the once continuous horizontal stratum. See also: *dune.*

zeugogeosyncline See *geosyncline.*

zinc blende *Sphalerite.*

zinc bloom *Hydrozincite.*

zincite A hexagonal orange to dark-red mineral, (Zn,Mn)O, an ore of zinc; usually massive.

zircon A silicate mineral, $ZrSiO_4$, an important ore for zirconium, hafnium, and thorium; some varieties are used as gemstones. Zircon is a typical accessory mineral of acidic igneous rocks and their metamorphic derivatives.

Moisture and temperature changes together open the joints. Wind abrasion develops furrows in the soft rocks. Hard rock forms block-like ridges called zeugens, which can vary from 3 to 35 meters in height.

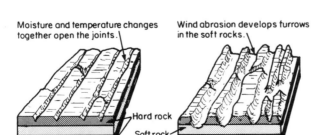
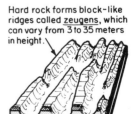

Hard rock
Soft rock

Formation of zeugen

zoisite An orthorhombic mineral of the epidote group, $Ca_2Al_3Si_3O_{12}(OH)$. Zoisite occurs in the form of prismatic crystals, rodlike aggregates, and granular masses. It is found in high-temperature and high-pressure metamorphic rocks (eclogites, granulites), and in hydrothermal veins.

zone of aeration See water table.

zone of capillarity Capillary fringe.

zone of mobility The asthenosphere.

zone of saturation See water table.

zone of weathering The layer of the earth's crust above the water table that is subject to destructive atmospheric precesses, and in which soils develop.

zoning 1. A core-to-margin variation in the composition of a crystal, due to a separation of crystal phases during growth in a continuous reaction series; higher-temperature phases form the core, and lower-temperature phases are near the margin. See also: normal zoning; reversed zoning. 2. In metamorphosed rocks, the development of regions in which particular minerals predominate, thus reflecting the metamorphic pattern of the rock. 3. The distribution pattern of minerals or elements in the region of ore deposits.

zooplankton All animals unable to swim effectively against horizontal ocean currents. The most abundant members are copepods, which are minute crustaceans. Some very large organisms, such as the Portuguese man-of-war, also belong to this group because they are poor swimmers. Zooplankton are part of the pelagic realm. See also: phytoplankton.

zweikanter See ventifact.